Second Edition

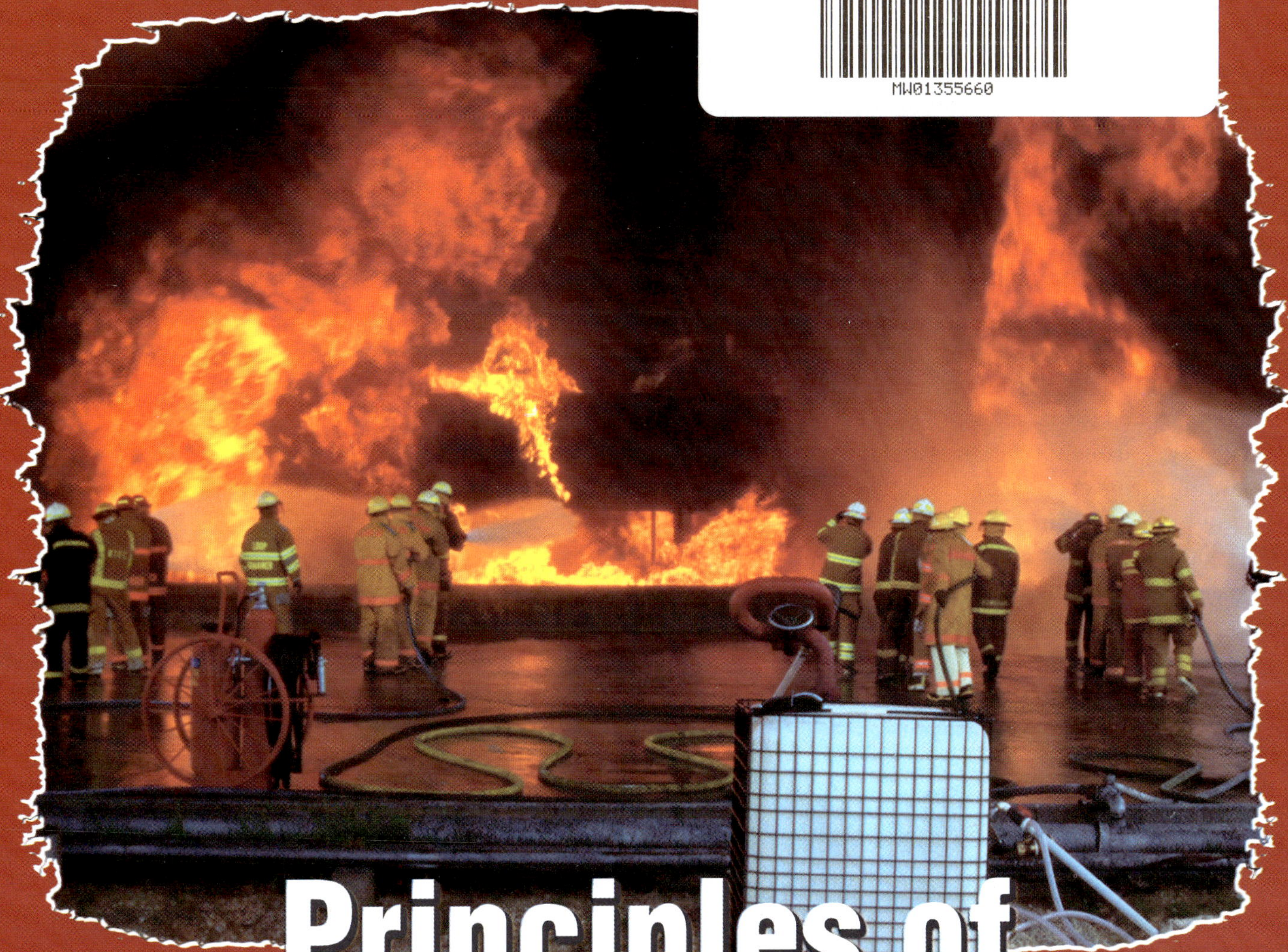

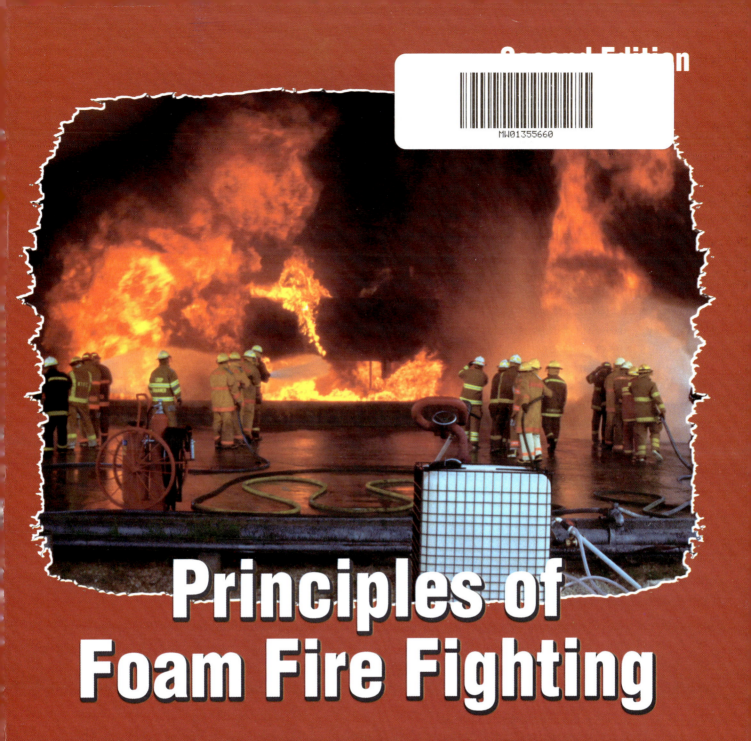

Principles of Foam Fire Fighting

Fred Stowell, Project Manager/Writer
Barbara Adams, Editor

Validated by the
International Fire Service Training Association

Cover photo courtesy of KK Productions, Inc.

RECYCLABLE

Published by
Fire Protection Publications
Oklahoma State University

The International Fire Service Training Association

The International Fire Service Training Association (IFSTA) was established in 1934 as a "nonprofit educational association of fire fighting personnel who are dedicated to upgrading fire fighting techniques and safety through training." To carry out the mission of IFSTA, Fire Protection Publications was established as an entity of Oklahoma State University. Fire Protection Publications' primary function is to publish and disseminate training texts as proposed and validated by IFSTA. As a secondary function, Fire Protection Publications researches, acquires, produces, and markets high-quality learning and teaching aids as consistent with IFSTA's mission.

The IFSTA Validation Conference is held the second full week in July. Committees of technical experts meet and work at the conference addressing the current standards of the National Fire Protection Association and other standard-making groups as applicable. The Validation Conference brings together individuals from several related and allied fields, such as:

- Key fire department executives and training officers
- Educators from colleges and universities
- Representatives from governmental agencies
- Delegates of firefighter associations and industrial organizations

Committee members are not paid nor are they reimbursed for their expenses by IFSTA or Fire Protection Publications. They participate because of commitment to the fire service and its future through training. Being on a committee is prestigious in the fire service community, and committee members are acknowledged leaders in their fields. This unique feature provides a close relationship between the International Fire Service Training Association and fire protection agencies which helps to correlate the efforts of all concerned.

IFSTA manuals are now the official teaching texts of most of the states and provinces of North America. Additionally, numerous U.S. and Canadian government agencies as well as other English-speaking countries have officially accepted the IFSTA manuals.

ISBN 0-87939-213-4 Library of Congress Control Number: 2002117654

Second Edition, First Printing, March 2003 *Printed in the United States of America*

10 9 8 7 6 5 4 3 2

If you need additional information concerning the International Fire Service Training Association (IFSTA) or Fire Protection Publications, contact:

Customer Service, Fire Protection Publications, Oklahoma State University
930 North Willis, Stillwater, OK 74078-8045
800-654-4055 Fax: 405-744-8204

For assistance with training materials, to recommend material for inclusion in an IFSTA manual, or to ask questions or comment on manual content, contact:

Editorial Department, Fire Protection Publications, Oklahoma State University
930 North Willis, Stillwater, OK 74078-8045
405-744-4111 Fax: 405-744-4112 E-mail: editors@osufpp.org

Table of Contents

Appendices

Preface

The second edition of IFSTA **Principles of Foam Fire Fighting** is written to assist fire and emergency services personnel in determining when foam-extinguishing agents are needed, what type of foam is required, and the proper methods of foam application. This manual is intended to help personnel meet the job performance objectives for foam fire fighting found in NFPA 1001, *Standard for Fire Fighter Professional Qualifications* (2002), NFPA 1002, *Standard for Fire Apparatus Driver/Operator Professional Qualifications* (1998), NFPA 1003, *Standard for Airport Fire Fighter Professional Qualifications* (2000), NFPA 1021, *Standard for Fire Officer Professional Qualifications* (1997), NFPA 1051, *Standard for Wildland Fire Fighter Professional Qualifications* (2002), NFPA 1081, *Standard for Industrial Brigade Member Professional Qualifications* (2001), and NFPA 472, *Standard for Professional Competence of Responders to Hazardous Materials Incidents* (2002).

Acknowledgement and special thanks are extended to the members of the validating committee. The following committee members contributed their time, wisdom, and knowledge to the development of this manual:

IFSTA Principles of Foam Fire Fighting, Second Edition, Validation Committee

Committee Chair
Jim Cottrell,
Cottrell Associates, Inc.
Fuquay-Varina, NC

Committee Secretary
Dave Coombs,
Lieutenant/Paramedic
St. Helens Rural Fire District
St. Helens, OR

Committee Members
Stephen Ashbrock,
Fire Chief
Maderia & Indian Hill Joint Fire District
Cincinnati, OH

V. Frank Bateman,
Training Manager
Kidde Fire Fighting
Martinez, CA

Brian Elliott,
Captain
Santa Rosa Fire Department
Santa Rosa, CA

Kenneth Gilliam,
Sr. Fire Safety Specialist
Federal Aviation Administration
Airport District Office
Orlando, FL

Ken Hanks,
Adjunct Instructor
Connecticut Fire Academy
Naugatuck, CT

Robert Lindstrom,
Fire Chief
WSI Will Rogers World Airport Fire Department
Oklahoma City, OK

Robert Moore,
Program Supervisor
TEEX-Emergency Services Training Institute
College Station, TX

Les Omans,
Fire Captain
San Jose Fire Department
San Jose, CA

Keith Phelps,
Senior Advisor
Planning and Training
Office of Homeland Security
Oklahoma City, OK

David "Skeeter" Rutledge,
Battlion Chief of Training
Memphis Fire Department
Memphis, TN

Mark Stuckey,
Deputy Chief
Owasso Fire Department
Tulsa Air National Guard Crash Rescue
Owasso, OK

Grant Wilson,
Platoon Chief/Fire Prevention Officer
Mt. Lebanon Fire Department
Mt. Lebanon, PA

The following individuals and organizations contributed information, photographs, and other assistance that made completion of this manual possible:

Air Line Pilots Association
American Airlines
Frank Bateman
Howard Chatterton
Christchurch International Airport, New Zealand
Conoco, Inc.
Michael T. Defina, Jr., Metro Washington Airports Authority Fire Department
Harry Eisner
Fire Wagons, Inc.
Richard W. Giles
Sam Goldwater
Martin Grube
Hawaii Department of Forestry and Wildlife
Mike Hildebrand
Ron Jeffers
Daniel J. Jordan
Kidde Fire Fighting
KK Products, Inc.
John F. Lewis
Michael R. Linville
Maryland Fire and Rescue Institute
Howard Meile III
Don Merkle, Maritime Institute of Technology and Graduate Studies
Lakeside Fire Department, Lakeside, AZ
Mesa (AZ) Fire Department
Metro Washington Airports Authority
Chris Mickal
Mount Shasta (CA) Fire Protection District
National Foam, Inc.
National Interagency Fire Center (NIFC)
Bob Norman
Edward J. Prendergast
Jim Rackl
Jim Reneau
Tom Ruane
Timothy Rugg
Stephen Sabo
Shell Oil Co., Deer Park, TX
William D. Stewart
Steven Stokely, Sr.
Mark Stuckey
Sunoco Refinery, Tulsa, OK
Texas A&M University System
Dan Thorpe
United States Coast Guard
United States Navy
Paul Valentine
Mike Wieder
Bill Wilcox
Joel Woods

Additionally, gratitude is extended to the following members of the Fire Protection Publications **Principles of Foam Fire Fighting** Project Team whose contributions made the final publication of this manual possible:

Project Manager/Writer
Fred Stowell, Senior Technical Editor

Editor
Barbara Adams, Senior Editor

Initial Staff Liaison
Richard Hall, IFSTA Projects Coordinator

Technical Reviewer
Richard Hall, IFSTA Projects Coordinator

Proofreader
Melissa Noakes, Curriculum Editor

Production Manager
Don Davis

Illustrators and Layout Designers
Ann Moffat, Graphic Design Analyst
Desa Porter, Senior Graphic Designer
Ben Brock, Senior Graphic Designer
Brien McDowell, Graphics Technician

Library Researchers
Susan F. Walker, Librarian
Shelly Magee, Assistant Librarian

Editorial Assistant
Tara Gladden

Introduction

The ancient Greeks believed that *fire* was one of the three basic elements that made up the world. The other two are *earth* and *water*. For fire and emergency responders of the 21st century, two of those elements, fire and water, are still very much a part of their world. While most people understand fire as a source of heat, power, light, and a method for cooking their food, fire and emergency responders see fire as a powerful source of destruction. Whereas water is understood to provide nourishment, energy, and a means of cleaning throughout the world, fire and emergency responders view it as a weapon to combat the effects of fire. Sometimes, however, water by itself is not an effective fire-extinguishing agent. Under certain circumstances, water supplemented by foam concentrates can speed fire extinguishment, reduce overhaul time, and increase personal safety.

The rapid explosion of technology that occurred in the late 19th and early 20th century resulted in industrialized societies that are irreversibly dependent on petroleum products and other chemicals. Since the introduction of the automobile in the earliest part of the 20th century, fire and emergency responders have been challenged with fires and spills involving these products on a regular basis. As the manufacture and use of these products expanded with technology, so did the frequency and severity of the incidents involving them. While water was, and still is, very suitable for fires involving ordinary combustibles, fire and emergency responders quickly learned that it was not particularly effective against petroleum products and other chemicals. First, water is heavier than petroleum products and when applied to a petroleum fire, it sinks to the bottom of the product and has little effect on extinguishing the fire. Second, water actually mixes with some chemicals, and even though water dilutes the chemical, it continues to burn. Foam solutions, which are created by combining foam concentrate with water, are now used to extinguish petroleum and chemical fires, wildland fires, transportation accident fires, and structure fires and to prevent fires at petroleum and chemical spills.

Before the introduction of foam, however, special fire-extinguishing agents were developed to combat the fires on which water was ineffective. Carbon dioxide, dry chemical, and loaded-stream extinguishing agents were the early answers to this problem. Loaded-stream agents were concentrations of water and salts (such as calcium acetate or potassium carbonate) that encapsulated flammable liquids within an emulsion (**Figure I.1**). However, all of these agents had their limitations as well. While carbon dioxide and dry chemical extinguishing agents were effective in rapidly extinguishing a fire, they were not able to keep the fuel from reigniting. In many cases, rapid reignition (burnback) of once extinguished fuels created a greater problem than the original fire. In addition, the reignition sometimes resulted in serious injury or death to fire and emergency responders operating in the area. In the case of loaded-stream agents, the formation of an emulsion to trap the flammable liquid took large volumes of water solution. Another limitation

Figure I.1 Before the introduction of foam extinguishing agents, loaded-stream portable extinguishers like these were used on fires where water was ineffective.

was that none of these agents were practical on large fires because of the difficulty in discharging large amounts of them over long distances. Air currents or thermal updrafts from large fires also easily disrupted these agent streams.

Eventually, researchers realized that there must be something that could be added to water to make it useful for combating flammable and combustible (burning) liquid or gas fires. This realization ultimately led to the development of *chemical foam:* the product of a reaction that occurs when two chemicals (commonly referred to as Chemical A and Chemical B) are mixed together. *Chemical* A, an acidic material, is a 13-percent aqueous (water-based) solution of aluminum sulphate. *Chemical B,* a basic or alkaline material, is an 8-percent aqueous solution of sodium bicarbonate with a licorice stabilizer. When these two solutions are mixed together, they produce carbon dioxide gas that acts as the expellant from a container. The licorice stabilizer causes the carbon dioxide to bubble and form foam. The first documented development of chemical foam occurred in 1877 with a patent application in England. Ten years later a chemical foam patent was issued in the United States. By the late 1880s, chemical foam was in use to combat coal mine fires.

Chemical foam, which became commonly known by its commercial name *Foamite*®, was commonly produced in 2½-gallon (10 L) portable fire extinguishers that closely resembled the old standard soda-acid extinguishers **(Figure I.2)**. Obviously, these extinguishers were only useful on very small fires. For large fires, a system was developed during the 1920s that allowed the foam to mix within and discharge through standard hoselines. In order to accomplish this process, a large device called a *hopper* was placed in the hoseline 100 feet (30 m) behind the nozzle. Dry-powdered versions of Chemicals A and B were dumped into the hopper **(Figure I.3)**. From there, the chemicals were inducted (admitted) into the hoseline at predetermined rates to provide a foam solution large enough to cover the burning liquid. A solid stream nozzle discharged the agent. Although chemical foam was successfully used for many years, it had one major limitation: It was affected by temperature change and performed poorly at low temperatures. Because of this limitation, chemical foams were replaced by mechanical foams and are now considered obsolete and rarely, if ever, found in use today.

Figure I.2 Portable chemical foam extinguishers, like this *Foamite®* brand unit, were used as early as the 1880s to combat fires in coal mines.

Figure I.3 During the 1920s, chemical foams were generated by adding two part dry-powder chemicals to a hopper that then mixed them into the water stream of a hoseline. *Courtesy of Mike Wieder.*

The development of mechanical foam liquid concentrates (foaming agents) occurred in the early 1930s. *Mechanical foam* is foam that is formed from mixing a liquid foaming agent and water with air (aeration) to create bubbles (finished foam). This mixing determines the foam's *expansion ratio:* the ratio of the final foam volume to the original foam solution volume. Applying mechanical foam involves adding air to the water and concentrate solution from either a compressed air source or by mixing air with the solution as it discharges from the nozzle (a process called *aspiration*). The military was the major user of mechanical foams during the 1940s (**Figure I.4**). Three limitations of mechanical foam were reignition or burnback, disruption by wind currents and updrafts, and submergence of the foam within a flammable liquid. Following World War II, mechanical foam research, production, and use decreased with the loss of military contracts. With the increase in the production, transportation, and use of petroleum products in the 1950s, however, new producers became involved in the development of mechanical foam concentrates. The results were mechanical foam concentrates based on protein, fluoroprotein, and synthetic detergent foaming agents. All are compatible with certain dry chemical extinguishing agents.

Figure I.4 During World War II, both the United States Merchant Marine and the United States Navy relied on foam extinguishing agents to fight shipboard fires, such as this one on the *SS Pennsylvania Sun* in 1942. *Courtesy of Maritime Institute of Technology and Graduate Studies.*

Protein foam concentrates consist of a protein hydrolysate plus additives to prevent the concentrate from freezing, prevent corrosion on equipment and containers, control viscosity (a liquid's resistance to flow), and prevent bacterial decomposition of the concentrate during storage. Fluoroprotein foam concentrates are similar to protein foams with the addition of a synthetic fluorinated surfactant. Synthetic foam concentrates such as aqueous film forming foam (AFFF), medium- and high-expansion foam concentrates, and wetting agents use foaming agents other than hydrolyzed proteins. The synthetic *medium-expansion* concentrates have a foam-to-solution ratio of approximately 20:1 to 200:1, while *high-expansion* concentrates have a foam-to-solution ratio of approximately 200:1 to 1,000:1 when generated with the appropriate equipment. High-expansion foam concentrates can totally flood a large compartment such as an aircraft hanger, warehouse, or ship's cargo hold. Aqueous film forming foam concentrate (initially called *Light Water*™) was developed through continuing research by the United States Navy Research Center in the early 1960s. AFFF was a new, *low-expansion* (expansion ratio less than 20:1) foam concentrate that actually allows an aqueous film to "float" on top of liquid fuels. AFFF is much superior to protein and fluoroprotein foam concentrates in extinguishing shallow liquid fuel spill fires because it forms a complete vapor-sealing blanket over the surface, which prevents reignition.

Foam technology exploded between the 1970s and 1990s. In 1972, 3-percent aqueous film forming foam concentrate was developed for use on Class B fires (flammable and combustible liquids and gases). A 3-percent concentrate results in a foam solution consisting of 3 percent foam concentrate and 97 percent water. By 1976 a new alcohol-resistant AFFF (AR-AFFF) concentrate was tested and certified (UL listed) by Underwriters Laboratories Inc. (UL). Class B foam concentrates designed to meet a variety of applications were developed, tested, certified, and manufactured. The 1980s saw the development of specialized hazardous materials foams (now replaced by better types of foam concentrates and no longer in use), stabilizers for AFFF, film forming fluoroprotein foam (FFFP) concentrate, and the certification (UL listing) of 3-percent AR-AFFF.

As research on foam continued, it became apparent that there could be benefits to adding foam concentrates to water when fighting fires involving ordinary Class A combustibles. Following the introduction of forestry foams by the Texas Forestry Service in 1978, Class A foam concentrates began to make their appearances in the industry, and the world of Class A foams was born. Apparent uses initially included wildland fires and deep-seated fires in refuge dumps and hay storage. Although initially developed for wildland fire applications, these foams have gained some acceptance for use in structure fires and those involving other ordinary combustibles. Compressed air foam systems (CAFSs) were developed to improve the application rates, delivery distances, and coverage of Class A foam (**Figure I.5**). Testing by various fire departments provided additional motivation to use Class A foam on all types of ordinary combustibles and in structures. Initially, much misinformation spread around the fire service about this foam concentrate and its capabilities. As with other foams, Class A foam really should be no mystery to fire and emergency responders. The benefits of reduced water damage, rapid fire extinguishment, and improved personal safety vastly offset the additional training required.

Foam is now the predominant extinguishing agent for fire departments and brigades that protect airports and petrochemical facilities. As well, most municipal fire departments have foam equipment and concentrate available for use on incidents involving transportation accidents and deep-seated fires in refuge dumps. Foam systems and equipment are found on vessels to protect decks and machinery spaces. Foam is the most commonly used extinguishing agent on Class B fires and spills on tank vessels. However, as widely as foam is used in today's fire service, its proper development and application remain a mystery to many fire and emergency responders. Many responders go through the motions of producing and applying foam to incidents without ever knowing if they are doing it properly. The basic knowledge of foam development and application is not a "voodoo" science that is difficult to understand. By mastering the basic knowledge of fire behavior, fire and emergency services personnel can determine the correct type of foam solution and application rate to effectively and efficiently control an incident.

Figure I.5 By the 1990s, compressed air foam systems (CAFSs) had been developed to improve the delivery of Class A foams. This photo illustrates the pump panel of a new CAFS unit.

Before the development of the first edition of this manual, little comprehensive written information on foam fire fighting was available to the fire service. Fire and emergency responders were limited to tidbits of information that they could find in other textbooks, manuals, trade journals, and foam concentrate and equipment manufacturers' product information literature. This second edition of the IFSTA **Principles of Foam Fire Fighting** has been rewritten to incorporate changes in the standards and industry. It provides fire and emergency responders and fire officers with the knowledge necessary to make informed decisions in the purchase, selection, and application of all types of fire-fighting foam concentrates and applicators.

Purpose and Scope

The purpose of this book is to educate municipal, rural, industrial, shipboard, and airport firefighters on the production and delivery of all types of foam. Its scope includes the following aspects of foam fire fighting:

- Class A and Class B foam concentrates
- High- and low-expansion foam systems
- Fixed, mobile, and portable foam proportioning (mixing) systems
- Tactics for fighting Class A and Class B fires with foam
- Tactics for preventing the ignition of Class B fuel spills with foam
- Foam use for aircraft, highway, pipeline, and maritime incidents
- Training for foam fire fighting
- Foam quality testing

This book is intended to help fire and emergency responders meet the foam requirements contained in the following National Fire Protection Association (NFPA) professional qualifications standards. See Appendix A, Standards Related to Foam, for lists of other standards.

- NFPA 472, *Standard for Professional Competence of Responders to Hazardous Materials Incidents* (2002)
- NFPA 1001, *Standard for Fire Fighter Professional Qualifications* (2002)
- NFPA 1002, *Standard for Fire Apparatus Driver/Operator Professional Qualifications* (1998)
- NFPA 1003, *Standard for Airport Fire Fighter Professional Qualifications* (2000)
- NFPA 1021, *Standard for Fire Officer Professional Qualifications* (1997)
- NFPA 1051, *Standard for Wildland Fire Fighter Professional Qualifications* (2002)
- NFPA 1081, *Standard for Industrial Fire Brigade Member Professional Qualifications* (2001)

Book Organization

This book follows a logical progression, leading the reader from the basics to the specific. Chapter 1, Fire Behavior and Extinguishment, is a review of the principles of fire behavior, fire-extinguishment theory, classifications of fires/fuels, and extinguishing properties of water and foam. Chapter 2 focuses on foam extinguishing concentrates: how they work, how they are stored, how they are proportioned, their characteristics, and the two basic categories of foam concentrates (Class A and Class B) and their applications. Chapter 3 describes the various proportioning and delivery equipment and the theories behind delivery devices. Chapter 4 provides an overview of foam delivery systems. Chapter 5 covers the use and application of Class A-type foams in structural and wildland situations. Chapter 6 deals with using foam to control unignited flammable and combustible liquid spills. Chapter 7 describes the use of Class B foam in controlling and extinguishing flammable and combustible liquid fires through fixed application systems. Chapter 8 discusses the use of foam concentrates in aircraft, maritime, and highway incidents through the use of handheld hoselines. Chapter 9 provides an overview of the training with foam applicators and concentrates.

The book's appendices include discussions on the standards and regulations that define and control the design, manufacture, and use of foam concentrates along with information on foam concentrate certification, regulatory organizations, quality control, and testing methods. Throughout the book, information of particular note that is more detailed, descriptive, or explanatory is located in a sidebar indicated by a shaded box.

NOTE: Some of the photographs used in this manual depict firefighters attacking exterior combustible liquid fires without wearing respiratory protection or self-contained breathing apparatus (SCBA). These photographs were made at a training facility that does not require the use of respiratory protection in these situations. The photographs are intended to depict the use of foam extinguishing agents. IFSTA recommends that respiratory protection be worn in accordance with NFPA standards when attacking interior- or exterior-type fires.

Chapter 1

Fire Behavior and Extinguishment

Job Performance Requirements

This chapter provides information that will assist the reader in meeting the following job performance requirements from NFPA 1001, *Standard for Fire Fighter Professional Qualifications*, 2002 edition; NFPA 1021, *Standard for Fire Officer Professional Qualifications,* 1997 edition; NFPA 1081, *Standard for Industrial Fire Brigade Member Professional Qualifications,* 2001 edition; NFPA 1003, *Standard for Airport Fire Fighter Professional Qualifications,* 2000 edition; and NFPA 1051, *Standard for Wildland Fire Fighter Professional Qualifications,* 2002 edition. Colored portions of the standard are specifically addressed in this chapter.

NFPA 1001

5.3.12 Perform vertical ventilation on a structure operating as part of a team, given an assignment, personal protective equipment, ground and roof ladders, and tools, so that ladders are properly positioned for ventilation, a specified opening is created, all ventilation barriers are removed, structural integrity is not compromised, products of combustion are released from the structure, and the team retreats from the area when ventilation is accomplished.

(A) *Requisite Knowledge:* **The methods of heat transfer; the principles of thermal layering within a structure on fire**; the techniques and safety precautions for venting flat roofs, pitched roofs, and basements; basic indicators of potential collapse or roof failure; the effects of construction type and elapsed time under fire conditions on structural integrity; and the advantages and disadvantages of vertical and trench/strip ventilation.

(B) *Requisite Skills:* The ability to transport and operate ventilation tools and equipment; hoist ventilation tools to a roof; cut roofing and flooring materials to vent flat roofs, pitched roofs, and basements; sound a roof for integrity; clear an opening with hand tools; select, carry, deploy, and secure ground ladders for ventilation activities; deploy roof ladders on pitched roofs while secured to a ground ladder; and carry ventilation-related tools and equipment while ascending and descending ladders.

5.3.16 Extinguish incipient Class A, Class B, and Class C fires, given a selection of portable fire extinguishers, so that the correct extinguisher is chosen, the fire is completely extinguished, and correct extinguisher-handling techniques are followed.

(A) *Requisite Knowledge:* **The classifications of fire; the types of, rating systems for, and risks associated with each class of fire**; and the operating methods of, and limitations of portable extinguishers.

(B) *Requisite Skills:* The ability to operate portable fire extinguishers, approach fire with portable fire extinguishers, select an appropriate extinguisher based on the size and type of fire, and safely carry portable fire extinguishers.

NFPA 1021

2-6.1 Develop a preincident plan, given an assigned facility and preplanning policies, procedures, and forms, so that all required elements are identified and the appropriate forms are completed and processed in accordance with policies and procedures.

(a) *Prerequisite Knowledge:* Elements of a preincident plan, basic building construction, basic fire protection systems and features, basic water supply, basic fuel loading, and **fire growth and development.**

(b) *Prerequisite Skills:* The ability to write reports, to communicate verbally, and to evaluate skills.

NFPA 1081

5.1.2 **Basic Incipient Fire Brigade Member Job Performance Requirements.** All industrial fire brigade members shall have a **general knowledge of basic fire behavior**, operation within an incident management system, operation within the emergency response operations plan for the site, the standard operating and safety procedures for the site, and site-specific hazards.

5.2 **Manual Fire Suppression.** This duty shall involve tasks related to the manual control of fires and property conservation activities by the incipient industrial fire brigade member.

5.2.1 Extinguish incipient fires, given an incipient fire and a selection of portable fire extinguishers, so that the correct extinguisher is chosen, the fire is completely extinguished, proper extinguisher-handling techniques are followed, and the area of origin and fire cause evidence are preserved.

(A) **Requisite Knowledge. The classifications of fire; risks associated with each class of fire;** and the types, rating systems, operating methods, and limitations of portable fire extinguishers.

(B) **Requisite Skills.** The ability to select, carry, and operate portable fire extinguishers, using the appropriate extinguisher based on the size and type of fire.

6.2.3 Operating as a member of a team, attack an exterior fire, given a water source, an attack line, personal protective equipment, tools, and an assignment, so that team integrity is maintained, the attack line is properly deployed for advancement, access is gained into the fire area, appropriate application practices are used, the fire is approached safely, attack techniques facilitate suppression given the level of the fire, hidden fires are located and controlled, the correct body posture is maintained, hazards are avoided or managed, and the fire is brought under control.

(A) **Requisite Knowledge.** Principles of fire streams; types, design, operation, nozzle pressure effects, and flow capabilities of nozzles; precautions to be followed when advancing hose lines to a fire; observable results that a fire stream has been correctly applied; dangerous conditions created by fire; principles of exposure protection; potential long-term consequences of exposure to products of combustion; **physical states of matter in which fuels are found**; and the application of each size and type of attack line, the role of the backup team in fire attack situations, attack and control techniques, and exposing hidden fires.

(B) Requisite Skills. The ability to prevent water hammers when shutting down nozzles; open, close, and adjust nozzle flow and patterns; apply water using direct, indirect, and combination attacks; advance charged and uncharged 38 mm (1½ in.) diameter or larger hose lines; extend hose lines; replace burst hose sections; operate charged hose lines of 38 mm (1½ in.) diameter or larger; couple and uncouple various handline connections; carry hose; attack fires; and locate and suppress hidden fires.

6.3.4 Operating as a member of a team, extinguish an ignitable liquid fire, given an assignment, an attack line, personal protective equipment, a foam proportioning device, a nozzle, foam concentrates, and a water supply, so that the correct type of foam concentrate is selected for the given fuel and conditions, a correctly proportioned foam stream is applied to the surface of the fuel to create and maintain a foam blanket, fire is extinguished, re-ignition is prevented, and team protection is maintained.

(A) Requisite Knowledge. Methods by which foam prevents or controls a hazard; principles by which foam is generated; causes for poor foam generation and corrective measures; **difference between hydrocarbon and polar solvent fuels and the concentrates that work on each**; the characteristics, uses, and limitations of fire-fighting foams; the advantages and disadvantages of using fog nozzles versus foam nozzles for foam application; foam stream application techniques; hazards associated with foam usage; and methods to reduce or avoid hazards.

(B) Requisite Skills. The ability to prepare a foam concentrate supply for use, assemble foam stream components, master various foam application techniques, and approach and retreat from fires/spills as part of a coordinated team.

8.2.2 Develop an initial action plan, given size-up information for an incident and assigned emergency response resources, so that resources are deployed to control the emergency.

(A) Requisite Knowledge. Elements of a size-up, SOPs for emergency operations, and **fire behavior**.

(B) Requisite Skills. The ability to analyze emergency scene conditions, to allocate resources, and to communicate verbally.

NFPA 1003

3-3.8 Ventilate an aircraft through available doors and hatches while operating as a member of a team, given PrPPE, an assignment, tools, and mechanical ventilation devices, so that a sufficient opening is created, all ventilation barriers are removed, the heat and other products of combustion are released.

(a) Requisite Knowledge: Aircraft access points; principles, advantages, limitations, and effects of mechanical ventilation; **the methods of heat transfer; the principles of thermal layering within an aircraft on fire;** the techniques and safety precautions for venting aircraft.

(b) Requisite Skills: Operate doors, hatches, and forcible entry tools; operate mechanical ventilation devices.

NFPA 1051

5.1.1 The Wildland Fire Fighter I shall meet the job performance requirements defined in Sections 5.1 through 5.5.

(A) Requisite Knowledge. Fireline safety, use, and limitations of personal protective equipment, agency policy on fire shelter use, **basic wildland fire behavior**, fire suppression techniques, basic wildland fire tactics, the fire fighter's role within the local incident management system, and first aid.

(B) Requisite Skills. Basic verbal communications and the use of required personal protective equipment.

5.5.3 Recognize hazards and unsafe situations given a wildland or wildland/urban interface fire and the standard safety policies and procedures of the agency, so that the hazard(s) and unsafe condition(s) are promptly communicated to the supervisor and appropriate action is taken.

(A) Requisite Knowledge. Basic wildland fire safety, **fire behavior**, and suppression methods.

(B) Requisite Skills. None specified.

5.5.6 Describe the methods to reduce the threat of fire exposure to improved properties given a wildland or urban/interface fire, suppression tools, and equipment so that improvements are protected.

(A) Requisite Knowledge. Wildland fire behavior, wildland fuel removal, structure protection methods, and equipment and personnel capabilities.

(B) Requisite Skills. The application of requisite knowledge to protect structures.

6.5.3 Effect the reduction of fire exposure to improved properties given a wildland or wildland/urban interface fire and available tools and equipment so that improvements are protected and the risk from fire is reduced.

(A) Requisite Knowledge. Knowledge of fire behavior in both wildland and improved properties, and the effects of fuel modification to reduce the hazard.

(B) Requisite Skills. The use of tools and equipment to protect the improved property.

6.5.7 Serve as a lookout given an assignment at a wildland fire as per agency procedures so that fire fighters are updated or warned when conditions change.

(A) Requisite Knowledge. **Basic fire behavior** and how to recognize hazardous situations, communications methods, equipment, and procedures.

(B) Requisite Skills. The ability to accurately describe fire behavior and changes in fire behavior through verbal communication, hand signals, or use of communication.

7.5.2 Size up an incident to formulate an incident action plan, given a wildland fire and available resources, so that incident objectives are set and strategies and tactics are applied according to agency policies and procedures.

(A) Requisite Knowledge. Size-up procedures, **fire behavior**, resource availability and capability, and suppression priorities.

(B) Requisite Skills. Identification of values at risk, objective setting, and selection of correct wildland fire-suppression strategies.

Chapter 1
Fire Behavior and Extinguishment

In order to control and extinguish fire, it is necessary to first understand the basic principles of fire behavior. Most emergency services personnel who engage in fire fighting have studied these principles in entry-level training courses or college fire protection courses. However, a brief review is appropriate in this manual before discussing the use of foam extinguishing agents. Fire is actually a by-product of a larger process called *combustion:* the self-sustaining process of rapid oxidation (chemical reaction) of a fuel, which produces heat and light. Fire is the result of this rapid oxidation reaction. The oxygen in air most commonly oxidizes fuels. The normal oxygen content of air is 21 percent, while 78 percent of air is nitrogen, and the remaining 1 percent is composed of trace amounts of other elements. A fire must be oxidized either by oxygen or some other chemical compound or mixture that contains oxygen. Some fuels do not need oxygen because they have their own oxidizers bound in their chemical formulas.

Combustion is defined as the process of rapid oxidation (resulting in fire), but oxidation is not always rapid. It may be very slow or it may be instantaneous. Neither of these extremes produces fire (combustion) as we know it, but they are common occurrences in themselves. Very slow oxidation is commonly known as *rusting* or *decomposition*. A light film of oil placed on metal prevents rusting by keeping air and its oxygen from the metal so that it cannot react and oxidize. Instantaneous oxidation is an *explosion* such as what occurs inside the casing of a bullet cartridge when the primer is ignited. The speed of the oxidation process determines the rate of released heat and the violence of the reaction. The presence of an oxidizer can accelerate a fuel that would normally burn slowly into an explosion.

This chapter provides a review of the chemical process of fire behavior, sources of heat energy, and heat transfer methods. It also discusses the classification, composition, and nature of fuels (materials that burn). The principles of fire development, extinguishment theories and methods, extinguishing properties of water and foam extinguishing agents are also included. A detailed explanation of these basic concepts may be found in IFSTA's **Essentials of Fire Fighting,** 4th edition.

Heat Energy Sources

Besides being a form of energy that raises temperature, *heat* may also be described as a condition of *matter in motion* caused by the movement of molecules. All matter contains some heat, regardless of how low the temperature, because molecules are constantly moving. When a body of matter is heated, the speed of the molecules increases; therefore, the temperature increases. Anything that sets the molecules of a material into faster motion produces heat in that material. It is important to understand how fuels become heated in order to reduce the possibilities of unwanted ignition and uncontrolled fire. The five general categories of heat energy are given in the following list and described in the sections that follow.

- Chemical
- Electrical
- Mechanical
- Nuclear
- Solar

Chemical Heat Energy

Chemical heat energy is generated as the result of some type of chemical reaction. Four types of chemical reactions result in heat production: (1) heat of combustion, (2) spontaneous heating, (3) heat of decomposition, and (4) heat of solution. The following sections describe these chemical reactions.

Heat of Combustion

Heat of combustion is the amount of heat generated by the combustion or oxidation reaction. The amount of heat generated by burning materials varies, depending on the material. This phenomenon explains why some materials (such as synthetic fabrics or paneling in houses) are said to burn "hotter" than others. For example, fiberglass burns hotter than wood (**Figure 1.1**).

Figure 1.1 Modern structures such as this residence contain a wide variety of combustible materials (including wood paneling, fiberglass insulation, plastic piping, and decorative fabrics) that burn at varying rates. *Courtesy of Chris Mickal.*

Spontaneous Heating

Spontaneous heating is the heating of an organic substance without the addition of external heat. Some hazardous materials (marked as *pyrophoric, spontaneous combustible,* or *dangerous when wet*) undergo spontaneous heating when they contact air or water. Spontaneous heating occurs most frequently where sufficient air is not present and heat produced cannot dissipate — heat that is produced by a low-grade chemical breakdown process. An example would be oil-soaked rags that are rolled into a ball and carelessly thrown into a corner. If ventilation (air circulation) is not adequate for the heat to escape, eventually it becomes sufficient to cause ignition of the rags. A further example is a fire that occurs in moist bailed hay stored in barns (**Figure 1.2**).

Figure 1.2 Hay bales stored in barns may ignite due to the spontaneous heating of the centers of the bales. *Courtesy of Mike Wieder.*

Heat of Decomposition

Heat of decomposition is the release of heat from decomposing (decaying) compounds usually due to bacterial action. In some cases, these compounds may be unstable and release their heat very quickly; they may even detonate. In other cases, the reaction and resulting release of heat is much slower. This reaction is commonly seen when viewing a compost pile (**Figure 1.3**). The decomposition of organic materials creates heated vapors and steam that can be seen on a cold day by jabbing holes into the pile. Wet brown (soft) coal loaded into a cargo hold of a ship undergoes a decomposition process that can result in a cargo-hold fire.

Figure 1.3 The decomposition of organic materials occurs in compost piles like this one as the material decays due to bacterial action.

Heat of Solution

Heat of solution is the heat released by the dissolving (solution) of matter in a liquid. Some acids can produce violent reactions, spewing hot water and acid with explosive force when they are dissolved in water (**Figure 1.4**).

Electrical Heat Energy

Electricity has the ability to generate temperatures capable of igniting combustible materials (those capable of burning) near the heated area, and it is not uncommon to see "electrical" listed as a cause of a fire. Electrical heating can occur in a variety of ways. The following sections highlight some of the more common methods: resistance heating, dielectric heating, leakage current heating, heat from arcing, and static electricity (**Figure 1.5**).

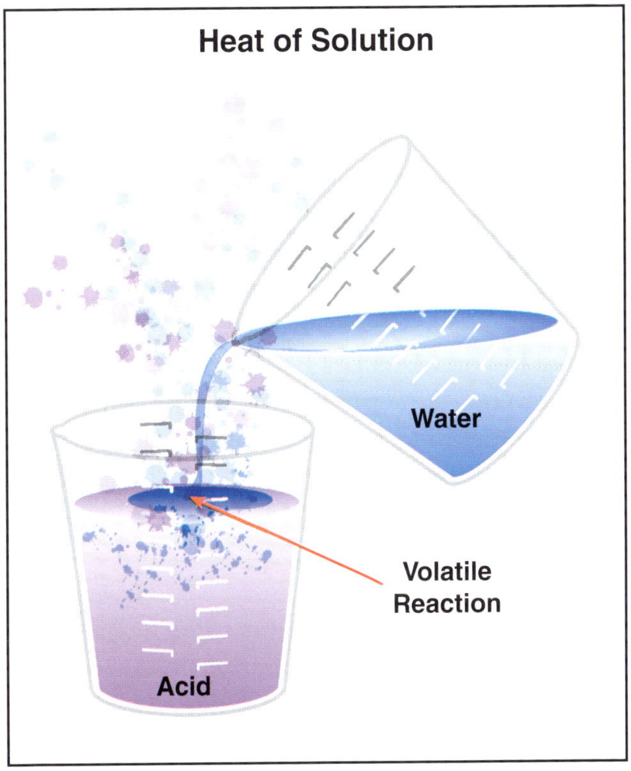

Figure 1.4 When water is added to acid, a volatile and explosive reaction can occur.

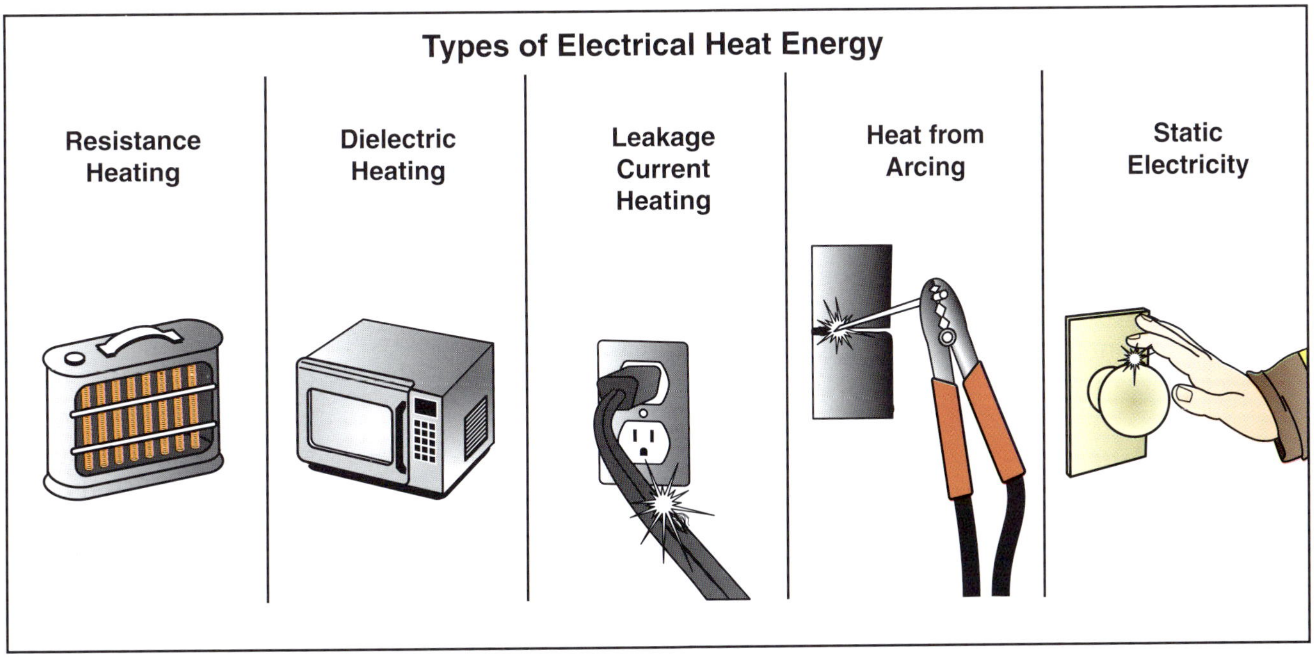

Figure 1.5 Examples of the five types of electric heat energy: resistance heating, dielectric heating, leakage current heating, heat from arcing, and static electricity.

Resistance Heating

Resistance heating refers to the heat generated by passing an electrical current through a conductor such as a wire or an appliance. Resistance heating is increased if the conductor is not large enough in diameter for the amount of current flow. Fires may occur when a single extension cord connecting several pieces of electrical equipment causes an overloaded circuit.

Dielectric Heating

Dielectric heating occurs as a result of the action of pulsating either direct current (DC) or alternating current (AC) at high frequency on a nonconductive material. The material is not heated by the dielectric heating but is heated by being in constant contact with electricity. This is somewhat similar to bombarding an object with many small lightning bolts. Dielectric heating is used in microwave ovens.

Leakage Current Heating

Leakage current heating occurs when a wire is not insulated well enough to contain all the current, and some leaks out into the surrounding material such as behind the wall paneling of a compartment (enclosed space) or room. This current causes heat that may result in a fire.

Heat from Arcing

Heat from arcing is a type of electrical heating that occurs when the current flow is interrupted, and electricity "jumps" across an opening or gap in a circuit. Arc temperatures are extremely high and may even melt the conductor in an uncontrolled situation such as at the site of a loose connection or defective switch. A common arc used in industrial applications is the arc welder. In this case, the welding rod (conductor) melts as metals are joined together.

Static Electricity

Static electricity is the buildup of positive charge on one surface and negative charge on another. The charges are naturally attracted to each other and seek to become evenly charged again. This condition is shown when the two surfaces come close to each other — such as when a person's finger touches a metal doorknob — and an arc occurs, producing a spark. Static electricity may cause fires when flammable liquids (any liquid that can burn) are being transferred between containers that are not properly electrically isolated. For this reason, flammable liquids transport vehicles are equipped with grounding wires to prevent sparking caused by the buildup of static electricity.

Heat generated by lightning is static electricity on a very large scale. The heat generated by the discharge of billions of volts from earth-to-cloud, cloud-to-cloud, or cloud-to-ground can be in excess of 60,000°F (33 316°C).

Mechanical Heat Energy

Mechanical heat is generated in two ways: friction and compression. *Heat of friction* is created by the movement of two surfaces against each other, which generates heat and/or sparks. This is the concept that Boy Scouts learn when they rub sticks together to create a fire for cooking **(Figure 1.6)**. *Heat of compression* is generated when a gas is compressed. This principle explains how diesel engines ignite fuel vapor without a spark plug. It is also the reason that self-contained breathing apparatus (SCBA) cylinders feel warm to the touch after they are filled **(Figure 1.7)**.

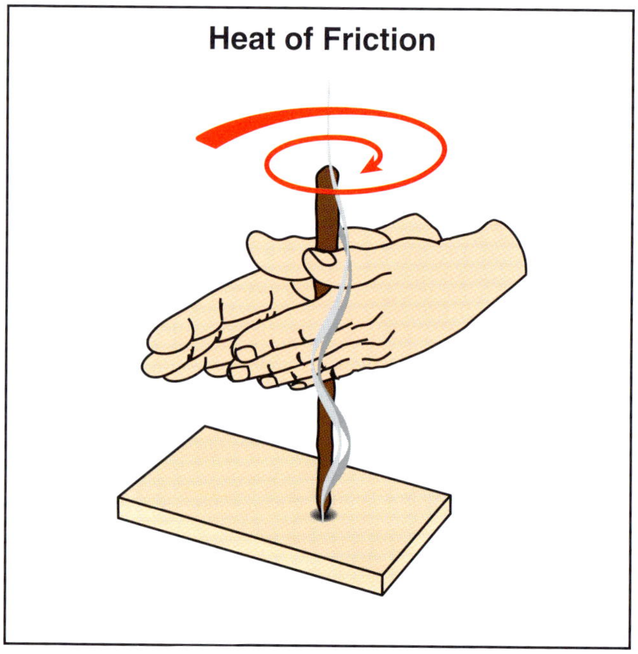

Figure 1.6 Example of the heat of friction: The friction created by rubbing two pieces of wood together can result in the ignition of the wood.

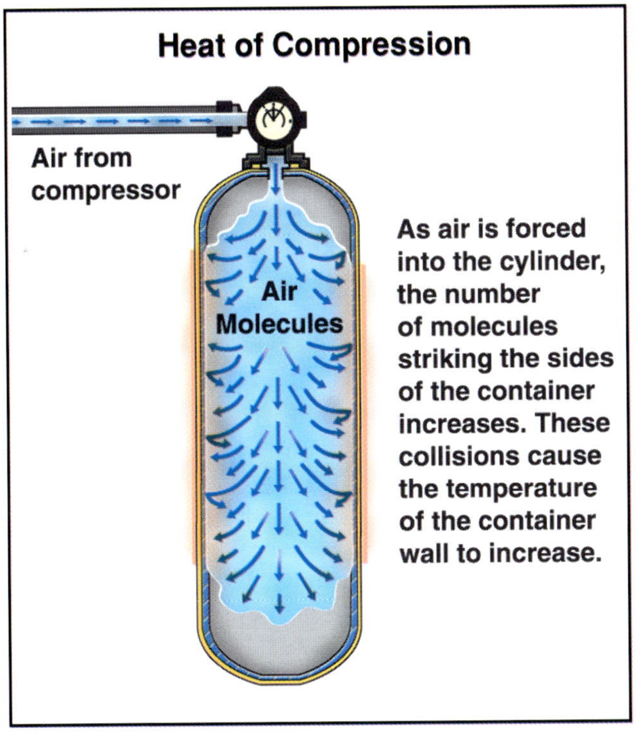

Figure 1.7 Example of heat of compression: The compression of any gas generates heat.

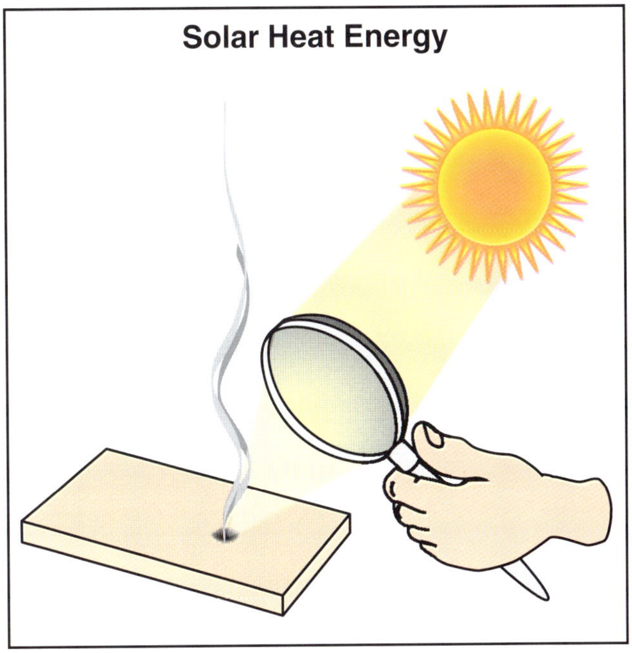

Figure 1.8 Example of solar heat: Solar rays passing through a glass surface can result in the ignition of combustible material at the point the rays are focused.

Nuclear Heat Energy

Nuclear heat energy is generated when atoms are either split apart (fission) or combined (fusion). In a controlled setting, fission is used to heat water to drive steam turbines and produce electricity. Currently, fusion cannot be controlled and has no commercial use.

Solar Heat Energy

Solar heat energy is the energy transmitted from the sun in the form of electromagnetic radiation. Typically, solar energy is distributed fairly evenly over the face of the earth and in itself is not capable of starting a fire. However, when solar energy is concentrated on a particular point such as through a lens, it may ignite combustible materials (**Figure 1.8**).

Heat Transfer Methods

[NFPA 1001: 5.3.12; NFPA 1003: 3-3.8]

A number of the natural laws of physics are involved in the transfer or transmission of heat. For example, the *Law of Heat Flow* specifies that heat tends to flow from a hot substance to a cold substance. The colder of two bodies in contact absorbs heat until both objects are at the same temperature. Heat transfer is measured in British thermal units (Btu) or in joules/kilojoules (J/kJ). A *British thermal unit* is the amount of heat needed to raise the temperature of 1 pound of water 1 degree Fahrenheit in the English or Customary System. The *joule (J),* a unit of work, has taken the place of the calorie in the International System of Units (SI) for heat measurement. A *calorie* is the amount of heat required to raise the temperature of 1 gram of water 1 degree Celsius (1 calorie = 4.19 J) (1 Btu = 1 055 J or 1.055 kJ). Heat can travel from one body to another by one or more of three methods: conduction, convection, and radiation. Direct flame contact (impingement) was once considered a fourth method but is really a combination of convection and radiation. These methods are discussed in the sections that follow.

Conduction

Conduction is the transfer of heat energy from one body to another by direct contact of the two bodies or by an intervening heat-conducting medium (**Figure 1.9**). An example of conduction is when a fire in a compartment heats pipes hot enough to ignite material inside an adjacent compartment. The amount of heat that is transferred and its rate of travel depend upon the conductivity of the

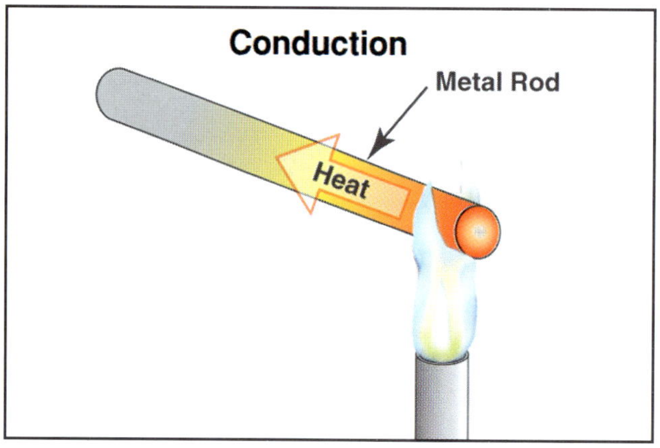

Figure 1.9 Example of the theory of conduction: The temperature along the metal rod rises because of the increased movement of molecules from the heat of the flame.

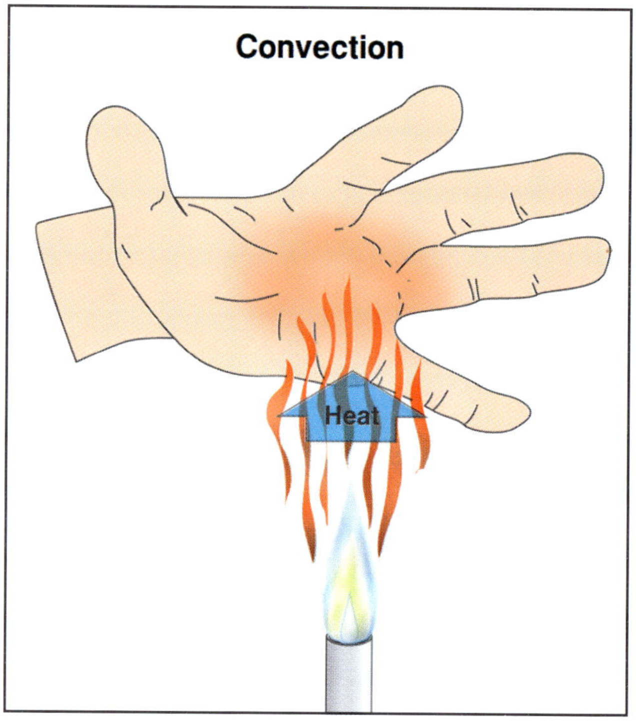

Figure 1.10 Example of the theory of convection: When the air between the flame and hand is heated, it rises and transfers heat to the hand.

material through which the heat is passing. Not all materials have the same heat conductivity. Aluminum, copper, and iron are good conductors. Fibrous materials such as felt, cloth, and paper are poor conductors, thus they are heat insulators (materials that slow heat transfer by separating conducting bodies). Air is a relatively poor conductor, so sandwiched airspaces between interior and exterior walls of a structure provide additional insulation from outside air temperatures. Certain solid materials, such as fiberglass, that are shredded into fibers and packed into batts make good insulation because the material itself is a poor conductor and air pockets exist inside the batting.

Convection

Convection is the transfer of heat by the movement of heated air or liquid (**Figure 1.10**). When water is heated in a glass container, the movement within the container is observed through the glass. If sawdust is added to the water, the movement is more apparent. As the water is heated, it expands and grows lighter; hence, the upward movement. In the same manner, as air near a steam radiator becomes heated by conduction, it expands, becomes lighter, and moves upward. As the heated air moves upward, cooler air takes its place at the lower levels. When liquids and gases are heated, they begin to move within themselves. This movement is responsible for heat transfer by convection. Convection is the reason a hand held over a flame feels heat even though the hand is not in direct contact with the flame.

Because heated air expands and rises, fire spread by convection is mostly in an upward direction; however, air currents can carry heat in any direction. Convection currents are generally the cause of heat movement through structures from floor to floor, room to room, and area to area. If the convecting heat encounters a ceiling or other horizontal barrier that keeps it from rising, it spreads out laterally (sideways) along the ceiling or barrier. This phenomenon is commonly referred to as *mushrooming*. When the heat reaches the walls, it is pushed by more heated air rising behind it and travels down the walls toward the floor.

Radiation

Although air is a poor conductor, it is obvious that heat can travel where matter does not exist. *Radiation* is the transmission of energy as an electromagnetic wave without an intervening medium (**Figure 1.11**). A hand held a few inches (millimeters) to the side of a flame feels heat by radiation. The heat of the sun reaches the earth even though it is not in direct contact with the earth (conduction), nor is it heating gases that travel to the earth (convection). The sun's heat transmits by

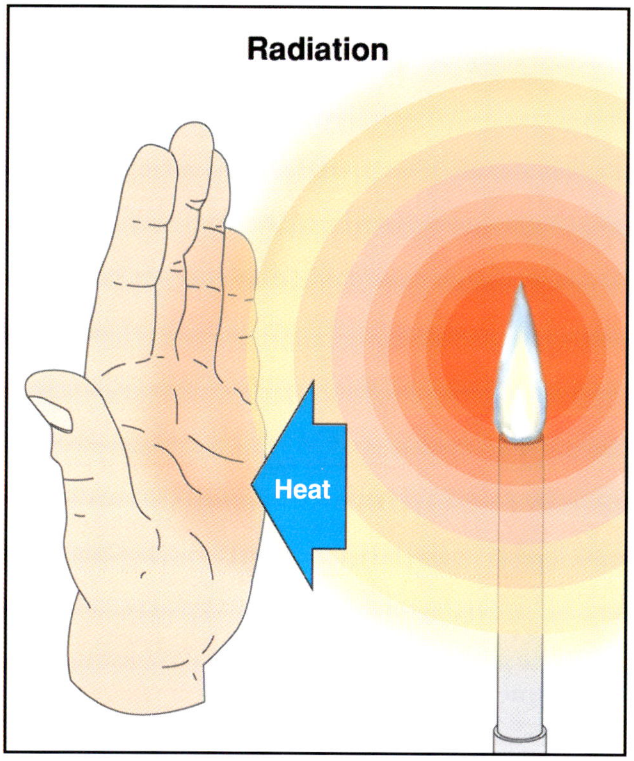

Figure 1.11 Example of the theory of radiation: Heat is transferred as energy in the form of electromagnetic waves without an intervening medium.

the radiation of heat waves. Heat and light waves are similar in nature, but they differ in length per cycle. *Heat waves* (sometimes called *infrared rays*) are longer than light waves. Radiated heat travels through space until it reaches an opaque object. As the object is exposed to heat radiation, it in turn radiates heat from its surface. Radiated heat is one of the major sources of fire spread to exposures (fuel separate from an original fire to which fire could spread). Radiation is important as a source of fire spread and demands immediate attention at locations where radiation exposure is severe.

Although at one time considered a fourth method of heat transfer, *direct flame contact (impingement)* is really a combination of heat transfer by convection and radiation at close range. When a substance is heated to the point where flammable vapors are released, these vapors may ignite, creating a flame. As other flammable materials come into contact with the burning vapors, or flame, they may be heated to a temperature where they, too, ignite and burn.

Fire Behavior Principles

[NFPA 1081: 5.1.2, 8.2.2(A); NFPA 1051: 5.1.1(A), 5.5.3, 5.5.6, 6.5.3, 6.5.7, 7.5.2]

Fuel may be found in any of three states of matter: *solid, liquid,* or *gas* (**Figure 1.12**). Although it appears to the eye that solids and liquids burn, only gases burn. The initiation of combustion of either a liquid or solid fuel requires its conversion into a gaseous state by heating. Fuel gases evolve from solid fuels by a *pyrolysis process (sublimation)* — the

Figure 1.12 Example of water in three states: solid, liquid, and gas. The transition from one state to another is due to the increase or decrease of the temperature of the water.

chemical decomposition of a substance through the action of heat. Fuel gases evolve from liquids by *vaporization* (process of a liquid turning into gas). This process is the same for water evaporating by boiling or water evaporating in sunlight. In both cases, heat causes the liquid to vaporize. Generally, the vaporization process for liquid fuels requires less heat input than does the pyrolysis process for solid fuels. Extinguishment of gas fuel fires is difficult because reignition is much more likely. Gaseous fuels are in the natural state required for ignition; no pyrolysis or vaporization is needed.

Knowledge of fire behavior, fuels, and how fuel affects fire behavior can make the difference between success or failure in fire suppression. The following sections identify and explain fuel characteristics, explain fuel vapor-to-air mixtures, and discuss the burning process.

Fuel Characteristics

The degree of fuel flammability is determined by characteristics of the fuel, including shape and size, position, density, water solubility, reactivity, and volatility. These characteristics are important factors in the combustion process. Material safety data sheets (MSDSs) contain the specific information on these characteristics for materials that are considered hazardous.

Shape and Size

Solid fuels have definite shapes and sizes that significantly affect their ignitability. A primary consideration is the *surface-to-mass ratio:* the ratio of the surface area of the fuel to the mass of the fuel. As fuel particles become smaller and more finely divided (for example, sawdust as opposed to logs), this ratio increases; ignitability also increases tremendously. As the surface area increases, heat transfer is easier and the material heats more rapidly, thus speeding pyrolysis.

A liquid fuel assumes the shape of its container. When a flammable or combustible liquid (any liquid that can burn) spills from a container, it assumes the shape or form of the containment area or the open ground surrounding the spill. Gravity causes the liquid to flow into and fill the lowest spaces first. This fact makes the size and shape of the surface area very unpredictable.

Figure 1.13 Once flammable and combustible liquids are released from a container, they assume the form of the surface upon which they are spilled, seeking the lowest places to pool. Foam is used to cover the spill and extinguish any resulting spill fire as shown in this photograph. *Courtesy of Edward J. Prendergast.*

Therefore, the containment and extinguishment techniques must be tailored to the specific spill (**Figure 1.13**).

Position

The physical position of a solid fuel is also a great concern to emergency services personnel. Fire spread is more rapid when a solid fuel is in a vertical position rather than a horizontal position. The rapidity of fire spread is due to increased heat transfer through convection as well as conduction and radiation. Position is not a consideration when dealing with a liquid fuel because it generally remains in a horizontal plane (**Figure 1.14**).

Density

Density is a measure of how tightly the molecules of a substance are packed together. Dense materials are heavy (for example, saltwater has a density of 1.025, whereas freshwater has a density of 1). The concept of density applies to both solids and liquids. *Vapor density* (the density of gas or vapor in relation to air) is a concern with volatile liquids and gaseous fuels.

The density of a solid determines the rate at which the solid absorbs heat and reaches its *ignition temperature:* the point at which flammable/combustible vapors are released. The more dense the molecules of a solid, the greater the heat required or the longer it takes to ignite. For example, a hardwood like teak requires more heat or time to ignite than a soft wood like balsa.

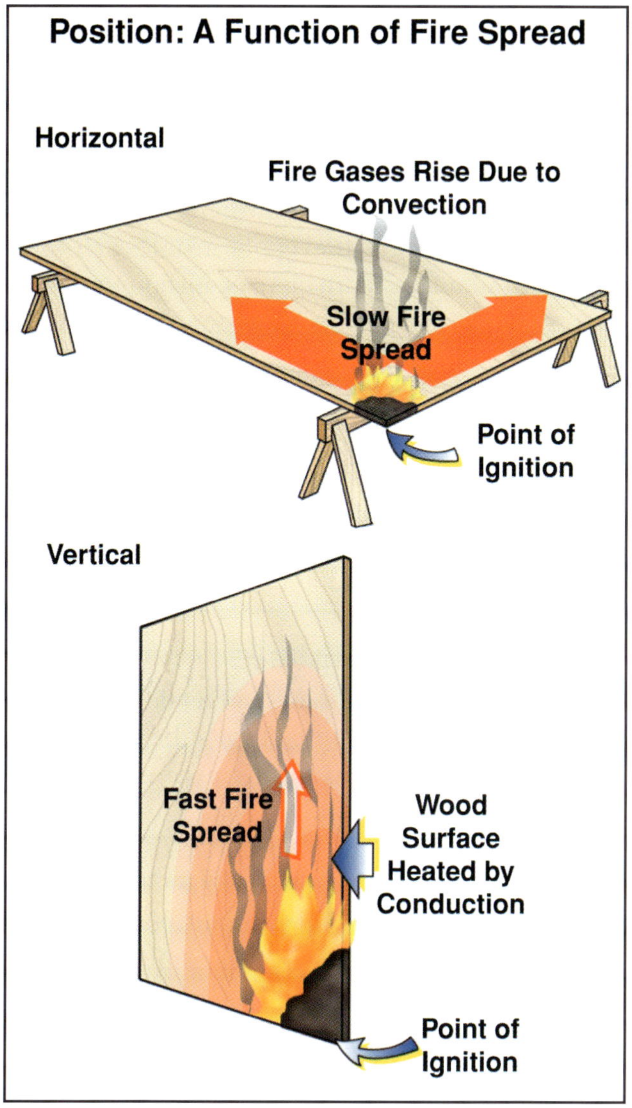

Figure 1.14 The position of a solid fuel affects the rate at which it burns.

The density of liquids is measured in relation to water and referred to as the liquid's *specific gravity*. Water is given a value of 1. Liquids with a specific gravity less than 1 are lighter than water, while those with a specific gravity greater than 1 are heavier than water. If the other liquid also has a density of 1, it mixes evenly with water (which is not the same as dissolving in water). Most flammable liquids have a specific gravity of less than 1. This means that if water is discharged onto a flammable liquids fire, the water may vaporize explosively (thus spreading the fire) or the whole fire can just float away on the flow of water and ignite everything in its path.

Gases tend to assume the shape of their containers but have no specific volume. If a vapor is less dense than air (air is given a value of 1), it rises and tends to dissipate. If a gas or vapor is heavier than air, which is a more common situation, it tends to stay at the lowest level and travel as directed by gravity, terrain configurations, and air convection. Every base hydrocarbon except the lightest one, methane, has a vapor density greater than 1. Common gases such as ethane, propane, and butane are examples of hydrocarbon gases that are heavier than air. Natural gas (composed largely of methane) is lighter than air. A cryogenic liquid (for example, liquefied natural gas [LNG]) that vaporizes may produce a vapor that is initially heavier than air but that becomes lighter than air as it warms.

Water Solubility

The solubility of a liquid fuel in water is also an important factor in combustion. Alcohols and other polar solvents (such as paint thinners and acetone) dissolve in water (soluble). If large volumes of water are used for fire extinguishment or spill control, alcohols and other polar solvents may dilute to the point where they will not burn. As a rule, hydrocarbon liquids (nonpolar solvents) do not dissolve in water (insoluble), which is why water alone cannot wash oil from the hands. Soap must be used with water to dissolve the oil. Consideration must be given to which extinguishing agents are effective on fires involving hydrocarbons (insoluble) and which ones are effective on polar solvents and alcohols (soluble) (**Figure 1.15**).

Figure 1.15 Alcohol-resistant aqueous film forming foam (AR-AFFF) is used to blanket soluble polar solvent fuel spills. *Courtesy of Tom Ruane.*

Reactivity

Solids that are classified as Class D fuels (combustible metals) are reactive with water. (See Class D Fires section for details.) Liquid fuels with flash points (minimum temperature at which a liquid releases enough vapors to form an ignitable mixture) higher than water's boiling point (212°F [100°C]) react violently to water and foam extinguishment applications. (See Flammable Liquids and Combustible Liquids sections.) Since most foam solutions are from 93 to 99 parts water and from 1 to 6 parts foam concentrate (known as *1-percent, 3-percent,* or *6-percent foam solutions*), water contained in the finished foam boils on contact with hot fuel, causing a dangerous frothover condition when fuel-tank fires are encountered (**Figure 1.16**). See Chapter 7, Class B. Foam Fire Fighting: Class B Liquids at Fixed Sites.

Figure 1.16 Water and foam streams directed into open-topped fuel storage tanks can result in the spread of burning liquids due to the reaction with water. *Courtesy of Ron Jeffers.*

Volatility

The *volatility* (ease with which a liquid releases vapors) of a liquid influences fire control and extinguishment. Volatility is based on the vapor pressure of the liquid. *Vapor pressure* is a measure of the pressure that is exerted by the liquid on the atmosphere it is in contact with. The higher the vapor pressure of a liquid, the greater the rate at which it evaporates and the lower its boiling point becomes. As a liquid fuel increases in temperature, its vapor pressure increases. Hot diesel fuel or gasoline can pose significant postfire or postspill security problems as ignitable fuel vapor invades an otherwise secure-looking foam blanket. All liquids release vapors in the form of simple evaporation. Ignitable vapors are released from liquid fuels at different temperatures. In addition, flammable and combustible liquids are more volatile in hot-weather climates than they are in cold-weather climates. Liquids that release large quantities of flammable or combustible vapors can be dangerous because they are easily ignited at almost any temperature.

Fuel Vapor-to-Air Mixture

For combustion to occur after a fuel has been converted into a gaseous state, it must be mixed with the oxygen (oxidizer) in air in the proper ratio — that is, within the flammable limits. The flammable (explosive) range of a fuel is reported using the percent by volume of gas or vapor in air for the *lower flammable limit (LFL)* (the minimum concentration of fuel vapor and air that supports combustion) and for the *upper flammable limit (UFL)* (concentration above which combustion cannot take place). Concentrations that are below the LFL are *too lean* (not enough fuel vapor) to burn. Concentrations that are above the UFL are *too rich* (too much fuel vapor) to burn (**Figure 1.17**).

Table 1.1 presents the flammable ranges for some common materials. The flammable limits for combustible gases are presented in chemical handbooks and documents such as NFPA's *Fire Protection Guide to Hazardous Materials* and *Fire Protection Handbook*. The limits are normally reported at ambient (surrounding air) temperatures (generally 70°F or 21°C) and atmospheric pressures. Variations in temperature and pressure can cause the flammable range to vary considerably.

NOTE: In an open space, concentrations of a substance that are too rich to support combustion near the source of a spill or release can be less rich near the outer edges of the spill or release. These leaner concentrations may allow for ignition in these areas.

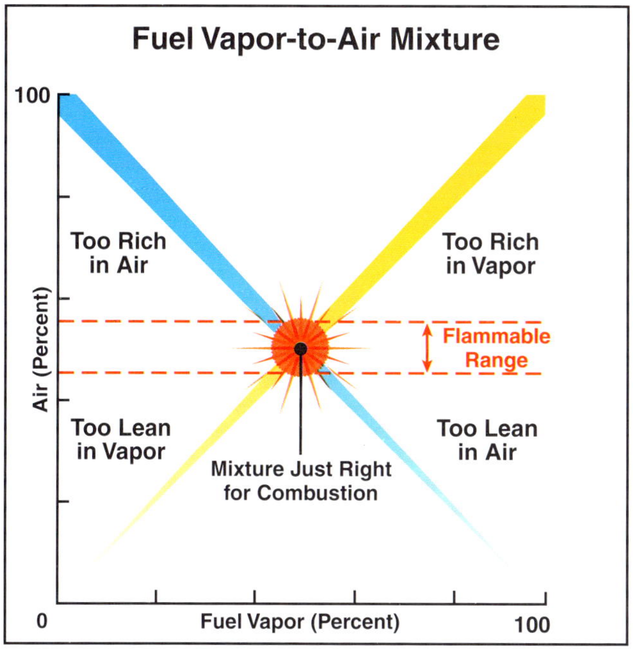

Figure 1.17 Illustration of the fuel vapor-to-air mixture: The flammable range is the point at which the mixture will ignite and sustain burning.

Table 1.1
Flammable Ranges for Selected Materials

Material	Lower Flammable Limit (LFL)	Upper Flammable Limit (UFL)
Acetylene	2.5	100.0
Carbon Monoxide	12.5	74.0
Ethyl Alcohol	3.3	19.0
Fuel Oil No. 1	0.7	5.0
Gasoline	1.4	7.6
Hydrogen	4.0	75.0
Methane	5.0	15.0
Propane	2.1	9.5

Source: *Fire Protection Guide to Hazardous Materials,* 12th edition, 1997, by National Fire Protection Association.

A fuel may be classified as either flammable or combustible depending on its *flash point:* the minimum temperature at which a fuel releases enough vapors to form an ignitable mixture with air near the fuel's surface and refers to the degree of volatility a particular fuel may possess. *Fire point* (or *burning point*) is the temperature at which a fuel produces sufficient vapors to support combustion once it is ignited. The fire point is usually a few degrees above the flash point. The flash point, then, is the LFL of the fuel. NFPA determined the difference between flammable and combustible liquids as whether the flash point is below or above 100°F (38°C).

Flammable and Combustible Liquids Classifications

According to the definition provided in NFPA 11, *Standard for Low-, Medium-, and High-Expansion Foam,* 2002 Edition, flammable liquids are any liquids with flash points below 100°F (37.8°C) and vapor pressures not exceeding 40 psi (276 kPa) at 100°F (37.8°C). Flammable liquids are further divided into the following three subclasses:

- ***Class I*** — Any liquid having a flash point below 100°F (37.8°C)
  - — ***Class IA:*** Liquids having flash points below 73°F (22.8°C) and a boiling point below 100°F (37.8°C)
  - — ***Class IB:*** Liquids having flash points below 73°F (22.8°C) and a boiling point above 100°F (37.8°C)
  - — ***Class IC***: Liquids having flash points at or above 73°F (22.8°C) and a boiling point below 100°F (37.8°C)

Combustible liquids include any liquids having a flash point at or above 100°F (37.8°C) and are further divided into the following three categories:

- ***Class II*** — Liquids having flash points at or above 100°F (37.8°C) and below 140°F (60°C)
- ***Class IIIA*** — Liquids having flash points at or above 140°F (60°C) and below 200°F (93.3°C)
- ***Class IIIB*** — Liquids having flash points at or above 200°F (93.3°C)

Flammable Liquids

Liquids that release enough vapors to support combustion at temperatures less than 100°F (38°C) are *flammable liquids.* These liquids will generally be in a state where vapors are ignitable with a stray spark.

Flammable Liquid Flash Points

- Gasoline — Vaporizes at -45°F (-42.7°C) (flash point)
- Heptane — Vaporizes at 25°F (-3.8°C) (flash point)
- Benzene — Vaporizes at 12°F (-11.1°C) (flash point)
- Naphtha — Vaporizes at 50°F (10°C) (flash point)
- Toluene — Vaporizes at 40°F (4.4°C) (flash point)
- Aviation gasoline (AVGAS) — Vaporizes at -75°F to -85°F (-59.4°C to -65°C) (flash point)
- JP-B/JP-4 (jet engine test fuel) — Vaporizes at -60°F (-51.1°C) (flash point)

Combustible Liquids

Liquids that must be at temperatures greater than 100°F (38°C) to release flammable vapors are *combustible liquids.* These liquids are generally in a state where vapors are not ignitable with a stray spark. Combustible fuel (such as diesel) spilled on a hot road can be as dangerous as gasoline (flammable liquid) in terms of flash point because the spilled fuel warms to the temperature of the surface where it is spilled.

Combustible Liquid Flash Points

- Kerosene — Vaporizes at 100°F (37.7°C) (flash point)
- Diesel fuel — Vaporizes at 126°F (52.2°C) (flash point)
- Home fuel oil — Vaporizes at 126°F (52.2°C) (flash point)
- Jet A/JP-5, JP-6, JP-8 — Vaporizes at 95°F to 165°F (35°C to 73.8°C) (flash point)
- Bunker oil (No. 6) — Vaporizes at 150°F (65.5°C) (flash point)
- Lubricating oil — Vaporizes at 444°F (228.8°C) (flash point)

The Burning Process

In order for burning to occur, not only must the proper fuel vapor-to-air mixture be present, but also the material must be at its ignition temperature or the point where self-sustained combustion continues. However, an accumulation of vapors can occur even below fire or flash points in an enclosed container such as an empty storage tank or transport.

Fire burns in two basic combustion modes: *flaming* and *smoldering*. Educators previously used the fire triangle to represent the smoldering mode of combustion, until it was proven that another factor was involved besides the presence of fuel, heat, and an oxidizer (oxygen). Scientific experimentation showed the presence of an uninhibited chemical chain reaction (flaming mode). To illustrate this more accurate description of the flaming mode of the burning process, the fire tetrahedron was developed. The *fire tetrahedron* includes the three parts of the old fire triangle (fuel, heat, and an oxidizer), but it adds the fourth dimension of an uninhibited chemical chain reaction **(Figure 1.18)**. Remove any one of the four components and combustion will not occur. The four components are briefly discussed in the sections that follow.

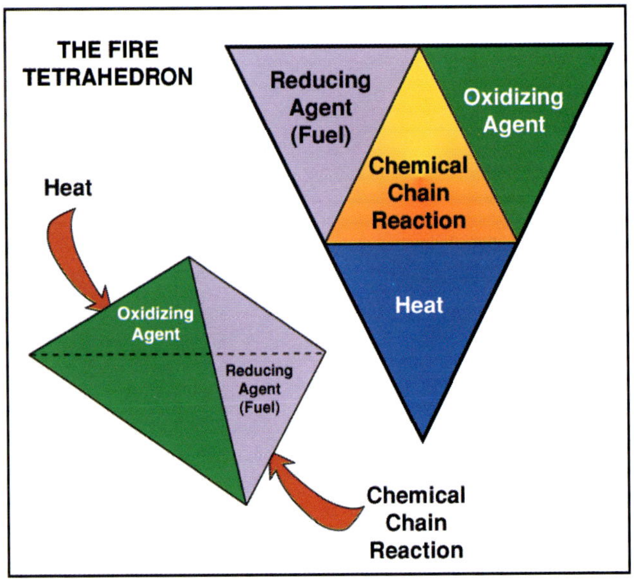

Figure 1.18 Components of the fire tetrahedron: fuel, heat, oxygen, and chemical chain reaction.

Fuel

The fuel segment of the diagram is any solid, liquid, or gas that can combine with oxygen in the oxidation chemical reaction **(Figure 1.19)**. A fuel with a sufficiently high temperature ignites if an oxidizing agent is liberated. Combustion continues as long as enough energy or heat is present. Under most conditions, the oxidizing agent is the oxygen

Oxidizing-Agent Container Labels

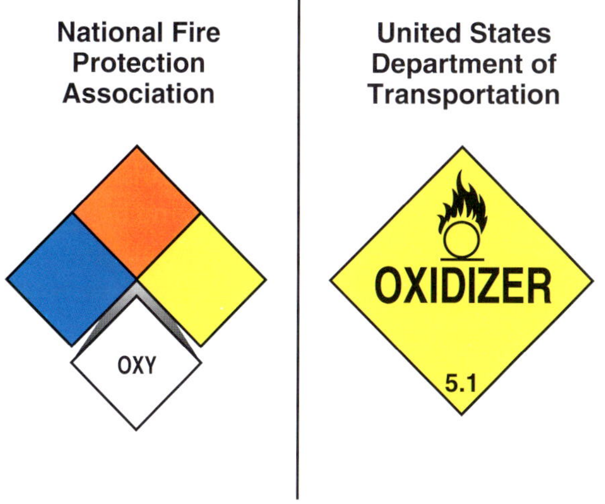

Figure 1.19 Example of labels/placards commonly found on containers of chemicals that may generate their own oxygen and cause ignition or sustain combustion.

in air. Some materials, such as sodium nitrate and potassium chlorate, release their own oxygen during combustion and can cause fuels to burn in an oxygen-free atmosphere.

Heat

A self-sustaining combustion reaction of solids and liquids depends on *radiative feedback:* radiant heat providing energy for continued vapor production. When sufficient heat is present to maintain or increase this feedback, the fire either remains constant or grows, depending on the heat produced. A *positive heat balance* occurs when heat is fed back to the fuel and is required to maintain combustion. If heat is dissipated faster than it is generated, a *negative heat balance* is created.

Oxygen

The amount of oxygen available to support combustion is important. As stated previously, air contains about 21 percent oxygen under normal circumstances. Oxygen concentrations below the normal 21 percent adversely affect both fire production and life safety. From a fire standpoint, the intensity of a fire begins to decrease below an 18-percent oxygen concentration. Oxygen concentrations below 15 percent do not support combustion.

Chemical Chain Reaction

A *chain reaction* is a series of reactions that occur in sequence with the results of each individual reaction being added to the rest. While scientists only partially understand what happens in the combustion chemical chain reaction, they do know that heating a fuel can produce vapors that contain substances that combine with oxygen and burn. Once flaming combustion or fire occurs, it can only continue when enough heat energy is produced to cause the continued development of fuel vapors. The self-sustained chemical reaction and the related rapid growth are the factors that separate fire from slower oxidation reactions.

Fire Development Principles

[NFPA 1021: 2-6.1]

When the four components of the fire tetrahedron come together, ignition occurs. For a fire to grow beyond the first material ignited, heat must be transmitted beyond that first material to additional fuel groups. Early in the development of a fire, heat rises and forms a plume of hot gas. If the fire is in the open, the plume rises unobstructed; air is drawn (entrained) into it as it rises (**Figure 1.20**). Because

Figure 1.20 The unobstructed smoke plume or column of an exterior fire draws fresh air into the plume, cooling the rising gases above the fire.

the air being pulled into the plume is cooler than the fire gases, this action has a cooling effect on the gases above the fire. The spread of fire in an open area is primarily due to heat energy transmitted from the plume to nearby fuels. Fire spread in outside fires is increased by wind and a sloping surface, which allows exposed fuels to preheat.

The development of fire in a compartment is more complex than one in the open. The growth and development of fires is usually controlled by the availability of fuel and oxygen. When the amount of fuel available is limited, a fire is *fuel controlled.* When the amount of available oxygen is limited, the condition is *ventilation controlled.* Researchers have attempted to describe compartment fires in terms of stages or phases that occur as the fire develops. These stages include the following:

- Ignition
- Growth
- Flashover
- Fully developed fire
- Decay

Figure 1.21 shows the development of a compartment fire in terms of time and temperature rise. Describing the stages as a fire develops in a space where no suppression action is taken helps to explain the reaction that occurs. The development of fire is very complex and influenced by many variables. As a result, all fires may not develop through each of the stages described. Fire is a dynamic event that depends on many factors for its growth and development.

Understanding into which stage a fire has developed determines, in part, the tactics used to extinguish it. A small fire in the ignition or growth stages (also called the *incipient phase*) might easily be extinguished by a portable extinguisher, while a fully developed fire may require using fire hoses or fixed fire-suppression systems. In addition, it is important to understand the rate at which a fire develops. This rate is expressed as the time-versus-temperature-rise concept. Because this development rate varies, the choice of extinguishing agents and attack tactics may vary.

The stages of fire development and the time-versus-temperature-rise concept are not the only things that affect the way in which fires develop. Compartment, or interior, fires are subject to the variations in size and number of openings, volume of the space, height of the space, and quantity and location of the fuel in the space. Exterior fires are governed by wind direction and velocity, relative humidity, terrain configuration, quantity and location of the first fuel ignited, and the location and composition of other fuel sources.

Ignition

The *ignition stage* is the period when the four elements of the fire tetrahedron come together and flaming combustion begins. The physical act of ig-

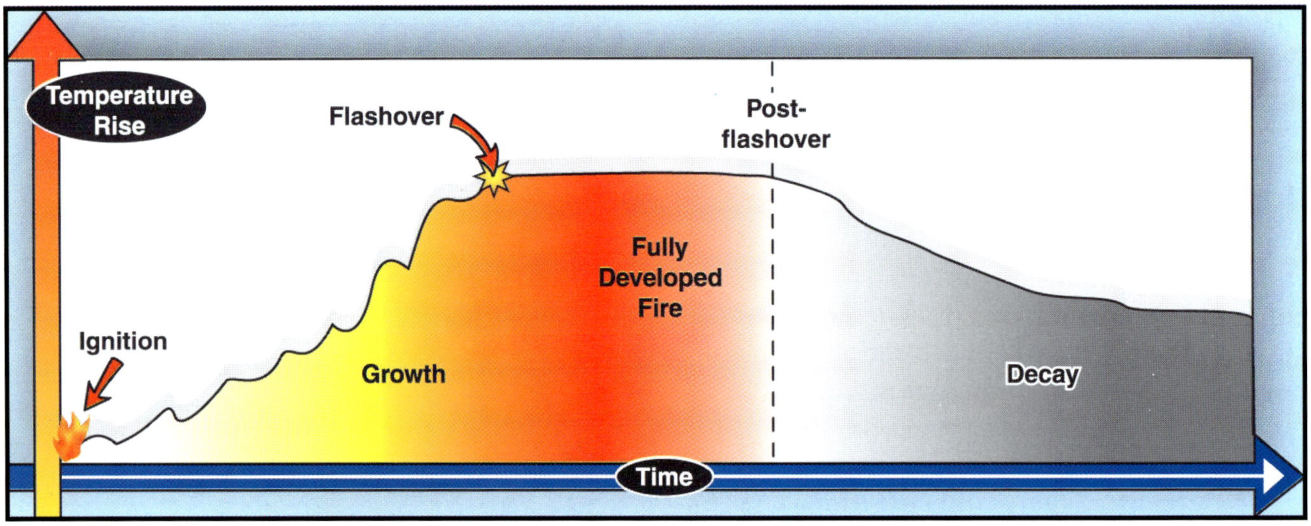

Figure 1.21 Stages of fire development in a compartment: ignition, growth, flashover, fully developed fire, and decay.

nition can be *piloted* (caused by a spark or flame) or *spontaneous* (caused when a material reaches its autoignition temperature as a result of self-heating). At this point, the fire is small and generally confined to the material (fuel) first ignited. All fires — whether in an open area or within a compartment — occur as a result of some type of ignition.

Growth

Shortly after ignition, a fire gas plume begins to form above the burning fuel. Air is drawn into the plume, and convection causes the heated gases to rise. The initial growth is similar to that of an outside unconfined fire. Unlike an unconfined fire, the ceiling and walls of a confined-space affect the plume. The hot gases rise until they reach the ceiling, and then they begin to spread outward (mushrooming). When the gases reach the ceiling, the ceiling covering materials absorb heat energy by conduction. The absorption of heat and the cooling effect of the entrained air cause the temperature at the highest level to decrease as the distance from the centerline of the plume increases. **Figure 1.22** shows plume development, air entrainment, and fire gas temperatures. Fire also spreads because radiated heat from the flame increases the temperature of surrounding fuels to their fire or flash points.

The growth stage continues as long as there is sufficient fuel and oxygen available. Compartment fires in the growth stage are generally fuel controlled. As the fire grows, the temperature in the compartment increases, as does the temperature of the gas layer at the upper level.

Flashover

Flashover is the transition between the growth and the fully developed fire stages but is not a specific event. During flashover, conditions in the compartment change very rapidly: Temperatures are rapidly increasing, and additional fuel groups are becoming involved. The hot-gas layer that develops at the ceiling level during the growth stage causes radiant heating of combustible materials remote from the origin of the fire. This heating causes pyrolysis to take place in these materials. The gases generated during this time are heated to their ignition temperatures by the radiant energy from the gas layer at the ceiling. The combustible materials and gases in the compartment ignite as flashover occurs — the compartment is fully involved in fire.

Scientists define *flashover* in many ways, but most base their definition on the temperature in a confined space that results in the simultaneous ignition of all combustible contents in the space. While there is no exact temperature associated with this occurrence, a range from approximately 900 to 1,200°F (483°C to 649°C) is widely used. This range correlates with the ignition temperature of carbon monoxide (1,128°F or 609°C), one of the most common gases released by pyrolysis. Persons who have not escaped from a compartment before flashover are not likely to survive. Even in full personal protective equipment with respiratory protection, emergency services personnel in a compartment at flashover are at extreme risk **(Figure 1.23)**.

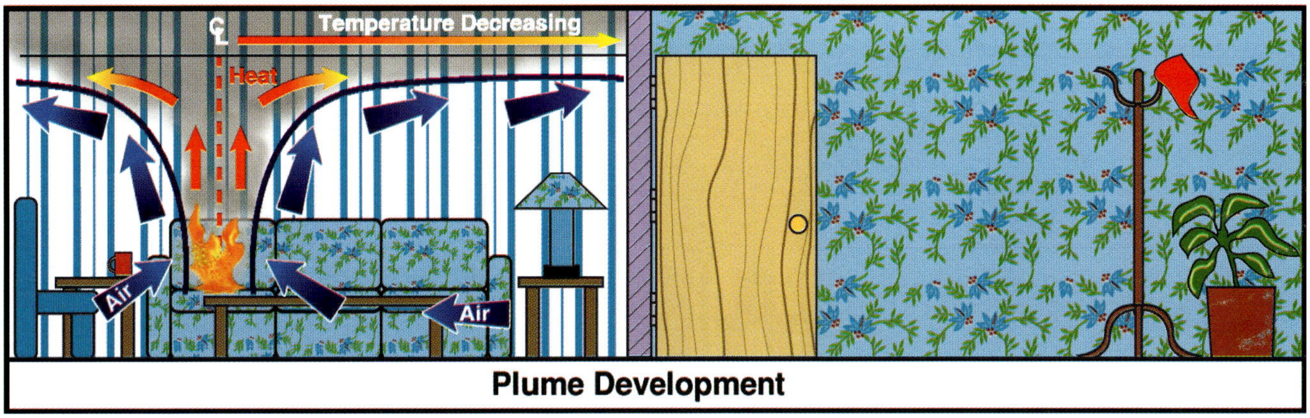

Figure 1.22 In the initial phases of a compartment fire, the temperature of the fire gases decreases as they move away from the centerline of the plume.

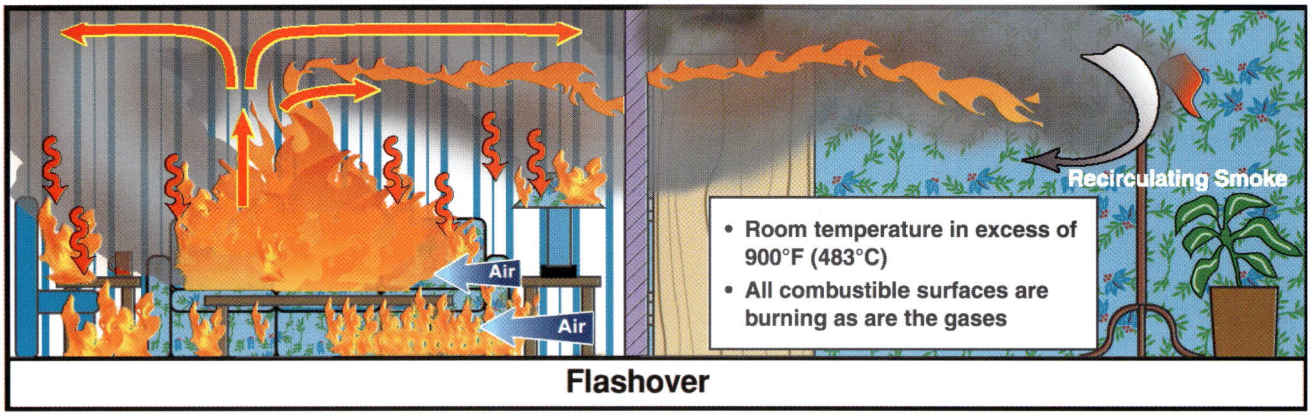

Figure 1.23 Flashover in a compartment fire.

Fully Developed Fire

After the flashover stage, all combustible materials in the compartment are involved in fire. During this period, the burning fuels are releasing the maximum amount of heat possible and producing large volumes of unburned fire gases. The fire can frequently become ventilation controlled, depending on the number and size of ventilation openings in the compartment. If these fire gases flow from the compartment into adjacent spaces, they often ignite when they mix with abundant air.

A condition known as *rollover* (also called *flameover*) occurs when flames move through or across unburned gases during a fire's progression. This condition may occur during the growth stage as the hot-gas layer forms at the ceiling or when unburned fire gases discharge from a compartment during the growth and fully developed stages. As these hot gases flow from the burning compartment into the adjacent space, they mix with oxygen, and flames often become visible in the layer. Rollover is distinguished from flashover because only the gases are burning in rollover, not the entire contents of the compartment (**Figure 1.24**).

Decay

As the available fuel in the compartment is consumed by fire, the amount of heat energy released begins to decline. Once again, the fire becomes fuel controlled, the amount of fire diminishes, and the temperatures within the compartment begin to decline. The remaining mass of glowing embers can, however, result in moderately high temperatures in the compartment for some time. For example, concrete and steel structural members retain heat for a long time following extinguishment.

Time-Versus-Temperature-Rise Concept

The fact that fire follows a time-versus-temperature-rise curve during development is an important concept. Understanding this concept makes it clear just how quickly a fire can develop, spread, and grow out of control. Following a fire through each of the

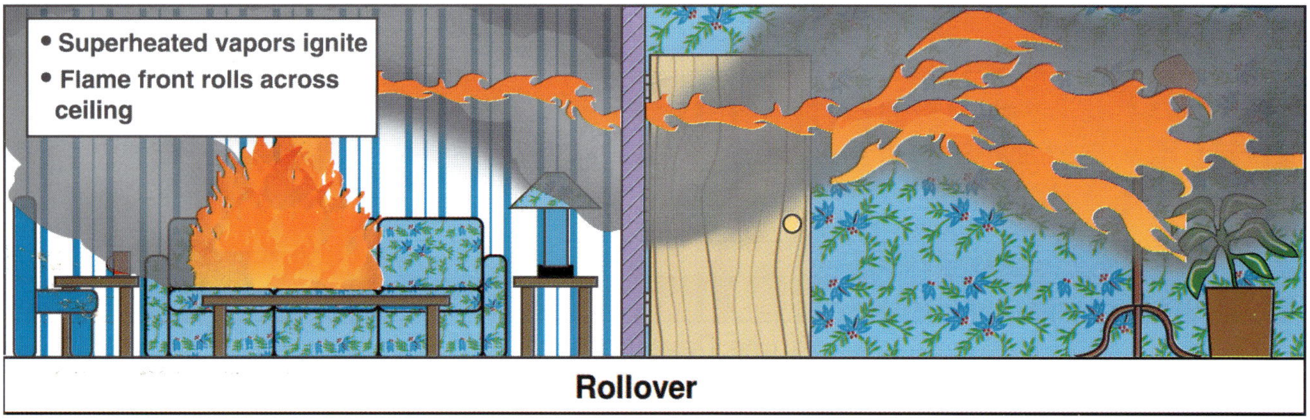

Figure 1.24 Rollover in a compartment fire.

previously described stages, **Figures 1.25** and **1.26** show two scenarios in which fires develop at different rates. The objective of fire fighting is to interrupt a fire's progress as early as possible during its development. The more time delays there are in attacking a fire, the further it develops, the more damage is sustained, and the more difficult it becomes to control the fire and save property.

Fires clearly do not develop at the same rate, given differing factors. But fires can, and often do, develop with amazing speed. Fires have been known to reach the flashover stage in less than 2 minutes. In other situations, the factors affecting fire development may result in flashover being reached much later, after 8 minutes or so. In still other situations such as a small fire with limited fuel in a large, otherwise empty, closed cargo hold, fire may never reach flashover. In the latter case, the fuel would be consumed before the dynamics of the fire carried it through all of the stages. The chances of such a fire occurring are unlikely, and one must not assume the best-case scenario.

Figure 1.25 An example of a small, smoldering fire developing slowly in a compartment wastebasket.

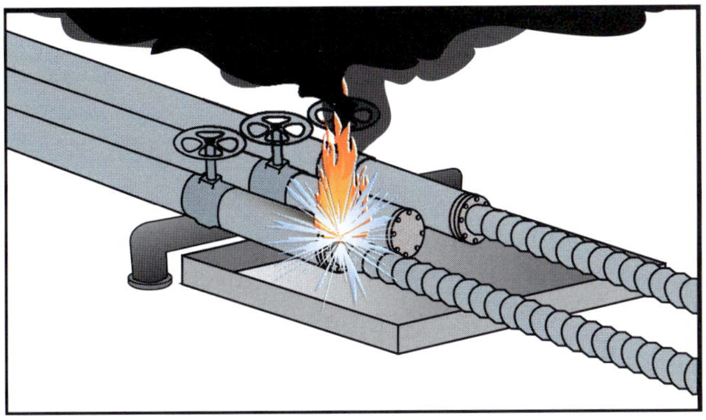

Figure 1.26 An example of sudden, rapidly developing fire in a pipe flange on a flammable-liquids fuel line.

Factors Affecting Fire Development

Fires may occur within a compartment or structure or in the open. The factors that determine the development of a fire vary depending on whether it is contained within a structure or compartment or in the open. Fire in a confined compartment or structure has two particularly important characteristics. The first characteristic is that there is a limited amount of oxygen. This fire differs from an exterior fire where the oxygen supply is unlimited. The second characteristic is that the fire gases released are trapped inside the structure and build up, unlike outdoors where they can dissipate.

Compartment Fire

Factors that impact the development of a fire in a compartment are as follows:

- Size and number of ventilation openings
- Volume of the compartment
- Height of the compartment
- Size and location of the first fuel group ignited and total fuel load in the compartment

For a fire to develop, sufficient air must be available to support burning beyond the ignition stage. The size and number of ventilation openings determines how the fire develops within the space. The compartment's size, shape, and height determine whether a significant hot-gas layer forms. The size and location of the initial fuel group are also very important in the development of the hot-gas layer. The plumes of burning fuel in the center of a compartment entrain more air and thus are cooler than those against the ceiling or in corners of the compartment (**Figure 1.27**).

The temperatures that develop in a burning compartment are the direct result of the energy released as the fuels burn. In a fire, the resulting energy is in the form of heat and light. The amount of heat energy released in a fire (heat release rate) is measured in Btu per second (Btu/s) or kilowatts (kW). This heat energy is directly related to the

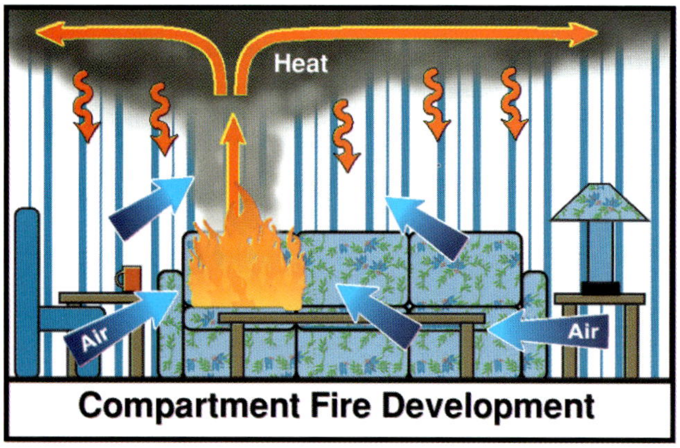

Figure 1.27 An example of the growth of a fire within a compartment as the overall temperature within the space rises and gases form layers at the ceiling.

amount of fuel consumed over time and that fuel's heat of combustion (amount of heat a specific mass of a substance gives off when burned). Materials with high heat release rates (foam-padded furniture or polyurethane foam mattresses, for example) would burn rapidly once ignition occurs. Fires in materials with lower heat release rates are expected to take longer to develop.

The heat generated in a compartment fire is transmitted from the initial fuel group to other fuels in the space by all modes of heat transfer. The heat rising in the initial fire plume is transported by convection and ventilation. As the hot gases travel over surfaces of other fuels in the compartment, heat is transferred to them by conduction. Radiation and direct flame contact play significant roles in the transition from a growing fire to a fully developed fire in a space. As the hot-gas layer forms at the ceiling, hot particles in the smoke begin to radiate energy to the other fuel groups in the compartment. As the radiant energy increases, the other fuel groups pyrolyze and release ignitable gases. When the temperature in the compartment reaches the ignition temperature of these gases, the entire space becomes involved in fire (flashover).

Exterior Fire

Factors that effect the development of fire outside a compartment are as follows:

- Wind direction and velocity
- Relative humidity
- Terrain and slope
- Size, composition, and location of the fuel package that is first ignited
- Availability and locations of additional fuel packages (target fuels)

Fuels involved in exterior fires may be solid, liquid, or gaseous. They may take the form of trees, grass, lumber, waste products, spilled flammable or combustible liquids, or spilled gases. Exposure hazards may consist of similar fuels such as unburned grass or underbrush adjacent to a brush fire or nonsimilar fuels such as structures in the path of a brush fire. The materials burning or the potential for fire created by spills determine the selection of fire-suppression tactics and extinguishing agents. The quantity, type, and physical makeup of the primary fuel source determine the amount of heat generated.

The wind direction and velocity are factors in plume dispersion, fire spread, and exposure hazards. Wind continues to force oxygen into a fire, mixes the products of combustion (heat, light, smoke, fire gases, and flame) with fresh oxygen, and spreads burning embers onto unburned fuel packages **(Figure 1.28)**. (See Products of Combustion section.) Wind direction and velocity also determine fire-suppression attack tactics and extinguishing agent choices to extinguish an exterior fire. Depending on wind direction and velocity, convection or radiation usually transfers heat outside a structure.

The *relative humidity* is a measure of the moisture content (water quantity expressed in a percentage) in both the air and solid fuels in the vicinity of a fire (that is, the percentage of moisture in the air or solid compared to the maximum amount the air or solid can hold). The higher the humidity, the less likely a material is to ignite and burn.

Special Considerations for Fire Suppression

Several conditions or situations that impact fire-suppression efforts occur during a fire's growth and development within a compartment or confined space. Special considerations for exterior fires are specific to wildland and forested areas and are covered in detail in the IFSTA **Wildland Fire Fighting for Structural Firefighters** manual. The following

Figure 1.28 Wind direction and velocity can affect the spread of exterior fires rapidly, spreading fire into unburned areas.

sections provide an overview of thermal layering of gases, backdraft, and products of combustion along with the potential safety concerns for each.

Thermal Layering of Gases

[NFPA 1001: 5.3.12]

Thermal layering of gases is the tendency of gases to form layers according to temperature. Other terms sometimes used to describe this layering of gases by heat are *heat stratification* and *thermal balance*. The hottest gases tend to be in the top layer, while the cooler ones form the bottom layer (**Figure 1.29**). Smoke is a heated mixture of air, gases, and particles, and it rises. If a hole is made in a roof, smoke will rise from the room or compartment to the outside.

Thermal layering is critical to fire-fighting activities, especially within confined spaces. As long as the hottest air and gases are allowed to rise, the lower levels will be safe for fire crews. If water is

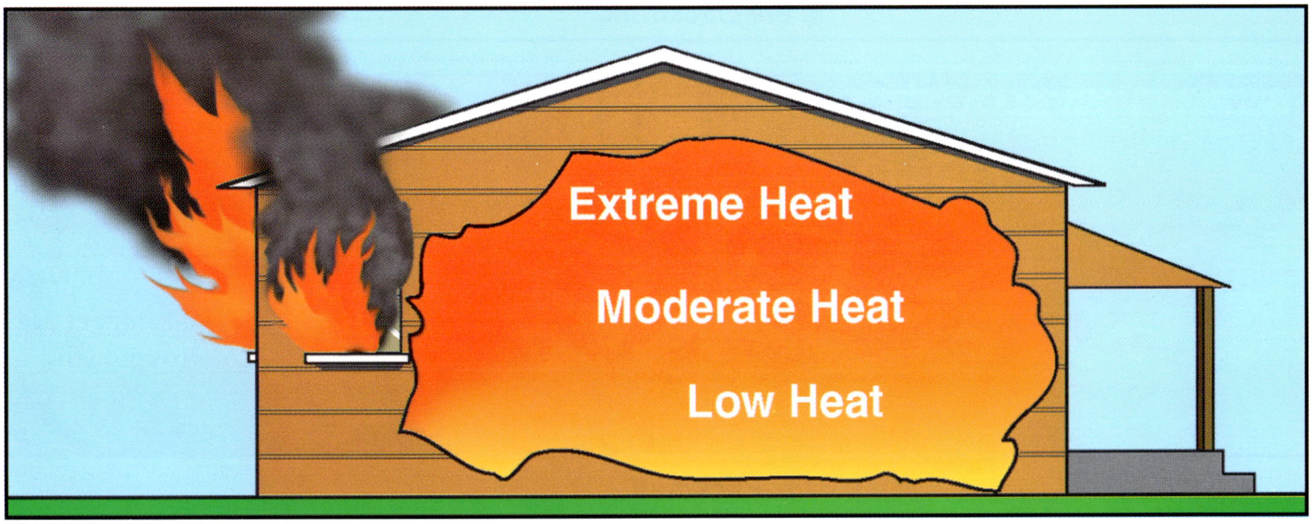

Figure 1.29 Under normal fire conditions in a closed structure, the highest levels of heat are at the ceiling level and the lowest level of heat is at the floor level.

applied continuously to the upper level of the layer where the temperatures are highest, the rapid conversion to steam can cause the gases to mix rapidly. This swirling of smoke and steam disrupts normal thermal layering and causes hot gases to mix throughout the compartment. This mixing *disrupts the thermal balance* or *creates a thermal imbalance* that can burn fire and emergency responders. Once the normal layering is disrupted, forced ventilation procedures (such as using fans) must be used to clear the area. Otherwise fire and emergency responders will be forced to retreat, and their attack becomes ineffective.

The proper interior attack procedure is to ventilate the compartment (if possible), allow the hot gases to escape, and direct the fire stream to the ceiling in short bursts to maintain the thermal balance. After the initial application of water, direct the fire stream at the base of the fire, keeping it out of the upper layers of gases. Using this same technique in an unventilated space causes the least amount of thermal disruption. See Chapter 5, Class A Foam Fire Fighting, for attack techniques involving the use of Class A foam and compressed air foam systems (CAFSs) on interior structural fires.

Backdraft

Fire and emergency responders operating at fires in compartments or other confined areas must use care when opening doors, windows, or other potential ventilation openings. During the growth, fully developed, and decay phases of fire development, large volumes of hot, unburned fire gases can collect in unventilated spaces. These gases are at or above their ignition temperatures, but insufficient oxygen is available for them to ignite. Any action during fire-fighting operations that allows air to enter this space and mix with these hot gases can result in *backdraft* — an explosive ignition of the gases. Backdrafts can injure or kill fire and emergency responders attempting to enter these spaces **(Figures 1.30 a and b)**. The potential for backdraft

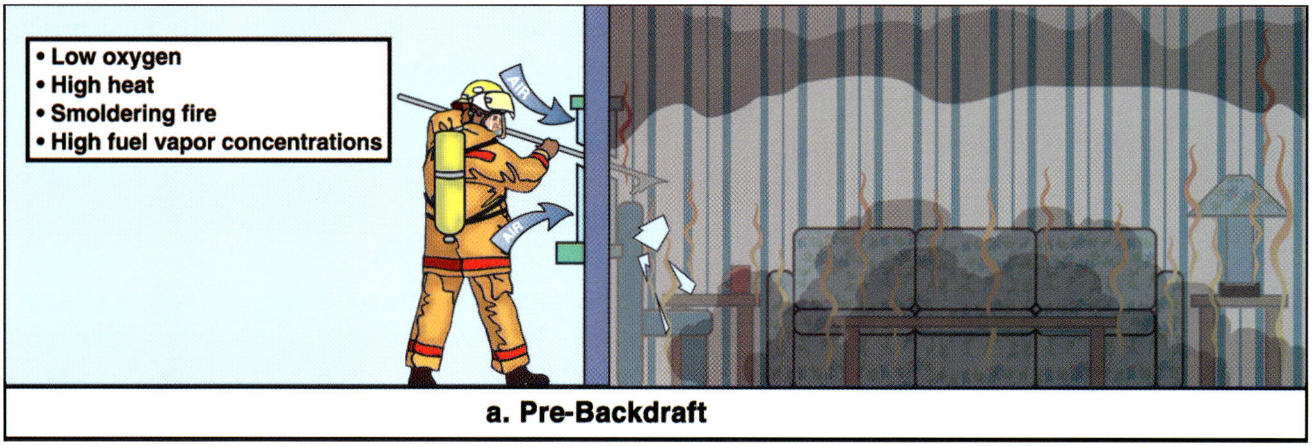

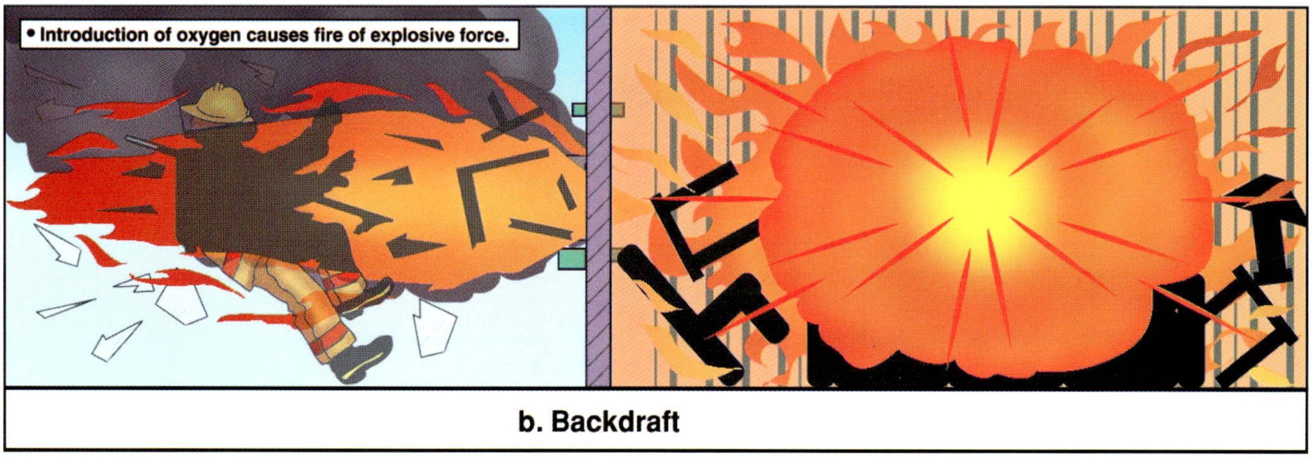

Figures 1.30 a and b Backdraft occurs when fresh air is introduced to a smoldering compartment fire, creating a rapid reignition of the heated fire gases in the space.

is reduced with vertical ventilation (opening at the highest point) because the unburned gases rise. Opening the space at the highest possible point allows gases to escape before fire and emergency responders enter. The following conditions may indicate the potential for backdraft (**Figure 1.31**):

- Pressurized smoke exiting small openings
- Black smoke becoming dense gray yellow
- Confinement and excessive heat
- Little or no visible flame
- Smoke leaving the space in puffs or at intervals (appearance of breathing)
- Smoke-stained window glass

Products of Combustion

As a fuel burns, the chemical composition of the material changes. This change results in the production of new substances (products of combustion: heat, light, smoke, fire gases, and flame) and the generation of energy (**Figure 1.32**). As a fuel burns, some of it is actually consumed. The

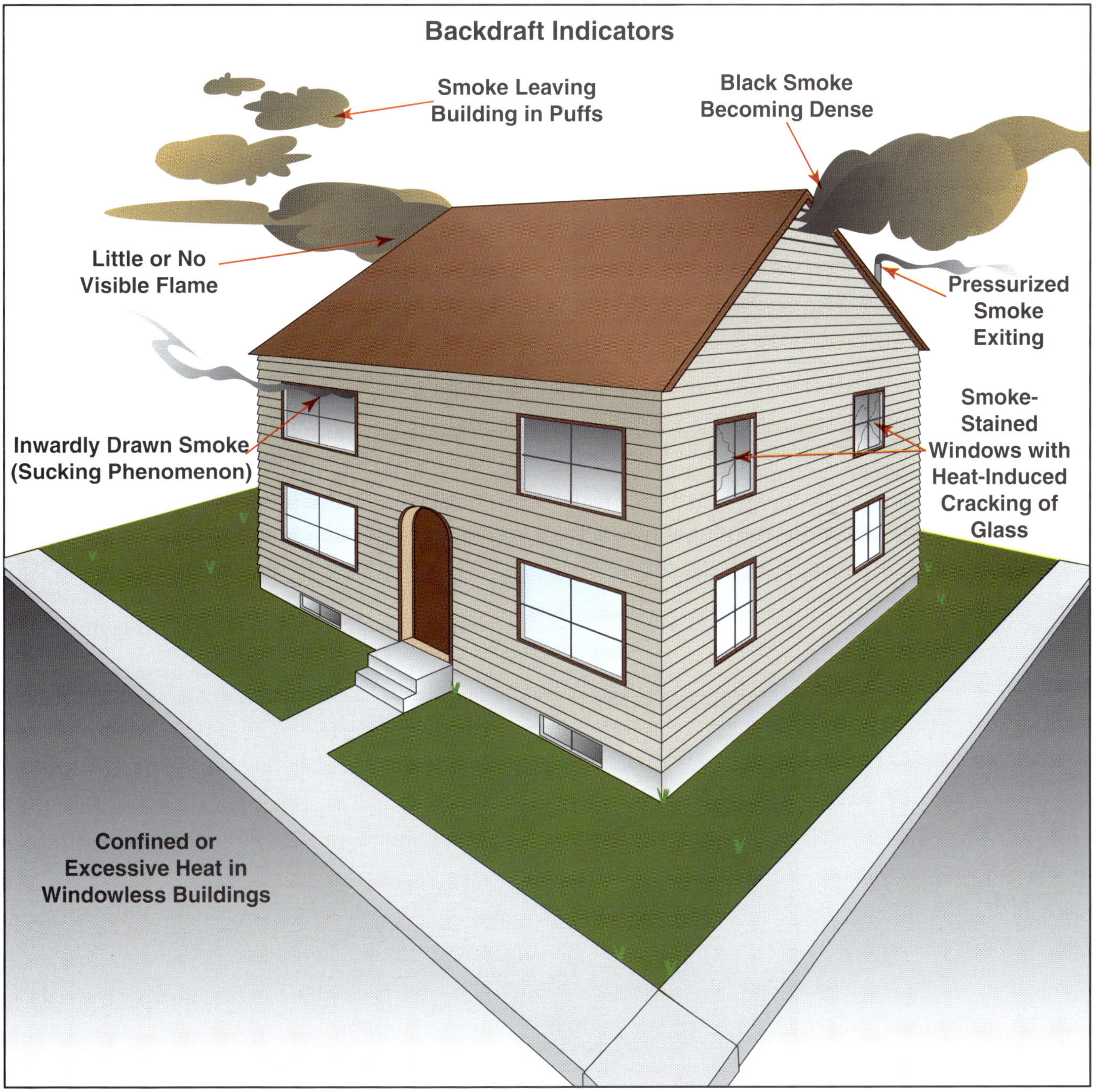

Figure 1.31 Fire and emergency services personnel must be able to recognize the indicators of a potential backdraft situation.

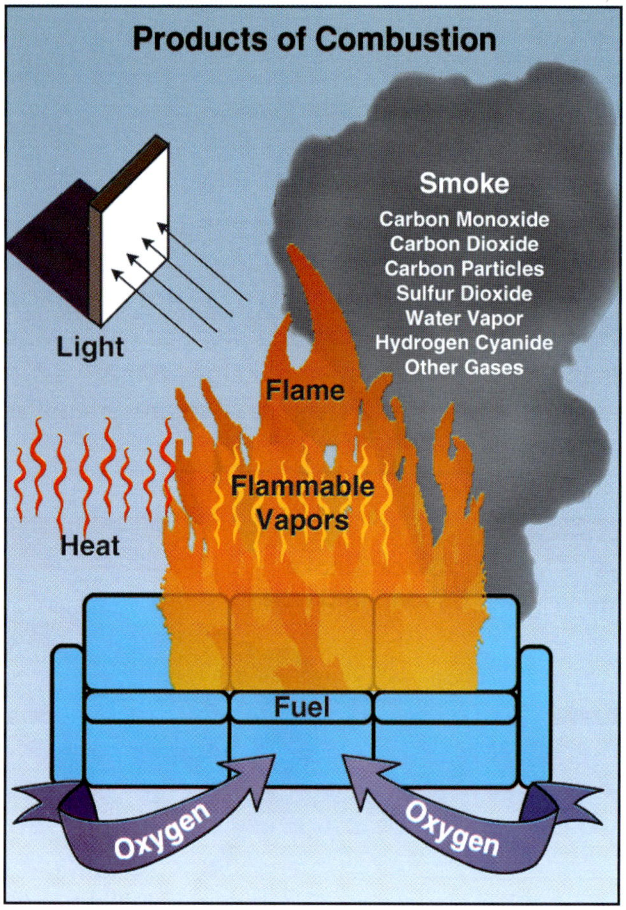

Figure 1.32 Products of combustion: heat, light, smoke, fire gases, and flame.

Law of Conservation of Mass explains that any mass lost converts to energy. This energy is in the form of *heat* and *light* in the case of fire. Burning also results in the generation of *smoke* — airborne fire gases, particles, and liquids. Heat is responsible for the spread of fire.

Flame is the visible, luminous body of a burning gas. When a burning gas is mixed with the proper amount of oxygen, the flame becomes hotter and less luminous. The loss of luminosity is caused by a more complete combustion of carbon. For these reasons, flame is considered to be a product of combustion. Of course, it is not present in those types of combustion that do not produce a flame such as smoldering fires.

Fire Classifications

[NFPA 1001: 3-3.15; NFPA 1081: 5.2.1(A), 6.2.3(A), 6.3.4(A)]

A description of how fires are classified is important when discussing extinguishment. The four broad classes of fire (A, B, C, and D) are based on the type of fuel that is involved. Each class has its own requirements for extinguishment (see Extinguishing Properties section). Examples and descriptions of fuels in each class are as follows:

- *Class A* — Ordinary combustible materials such as wood, cloth, paper, rubber, and many plastics. The majority of fires involve Class A fuels in the form of building materials, interior finishes, furnishings, consumer goods, and wildland vegetation (**Figure 1.33**).
- *Class B* — Flammable and combustible liquids and gases such as gasoline, oil, lacquers, paints, mineral spirits, alcohols, and natural gas. Gaseous fuels are already in the natural state required for ignition and the most difficult to contain. Neither pyrolysis nor vaporization is needed to

Figure 1.33 Class A fuel examples.

prepare them for ignition. Flammable liquids are further divided into hydrocarbon fuels, polar solvent fuels, acids, and bases (**Figure 1.34**).

— *Hydrocarbon fuels* are petroleum-based organic compounds that contain only hydrogen and carbon. They have a specific gravity that is less than 1; therefore, they float on water. Hydrocarbon fuels are immiscible; that is, they do not mix with water. Hydrocarbon fuels include crude oil, fuel oil, gasoline, benzene, naphtha, jet fuel, petroleum gas, and kerosene.

— *Polar solvent fuels* are flammable liquids that have an attraction for water and are miscible; that is, they dissolve in water. Some examples of polar solvents are alcohol, acetone, ketone, ester, aldehyde, and ether.

— *Acids* are corrosive chemicals that react with water to produce hydrogen ions. Acids can be found in gas, liquid, or solid forms, but most are liquids. In general, organic acids (such as amino acids, ascorbic acids, picric acid, and acetic acid), which contain carbon atoms and are obtained from living matter, will burn. Inorganic acids (such as hydrochloric acid, sulfuric acid, and nitric acid), which do not contain carbon and are usually stronger than organic acids, will not. However, both organic and inorganic acids can create combustible situations when combined with certain materials.

— *Bases* are corrosive water-soluble compounds that react with acids to form salts. Bases come in both solid and liquid forms. Although they are not considered a fuel source, some bases such as sodium hydroxide and potassium hydroxide generate great amounts of heat when they are dissolved in water and may be considered a type of ignition source.

- ***Class C*** — Energized electrical equipment such as household electrical appliances, transformers (found both inside and outside structures), power station switchgears, power transfer lines, and computerized equipment. Interior transformers are small and air insulated while units located outside are large, oil-cooled types used for main generator step-up. The oil capacities of transformers may range from 20,000 to 25,000 gallons (76 000 L to 95 000 L) for the large types to 300 to 20,000 gallons (1 140 L to 76 000 liters) for smaller vaulted or subterranean transformers (**Figure 1.35**).

Figure 1.34 Class B fuel examples.

Figure 1.35 Class C fuel examples.

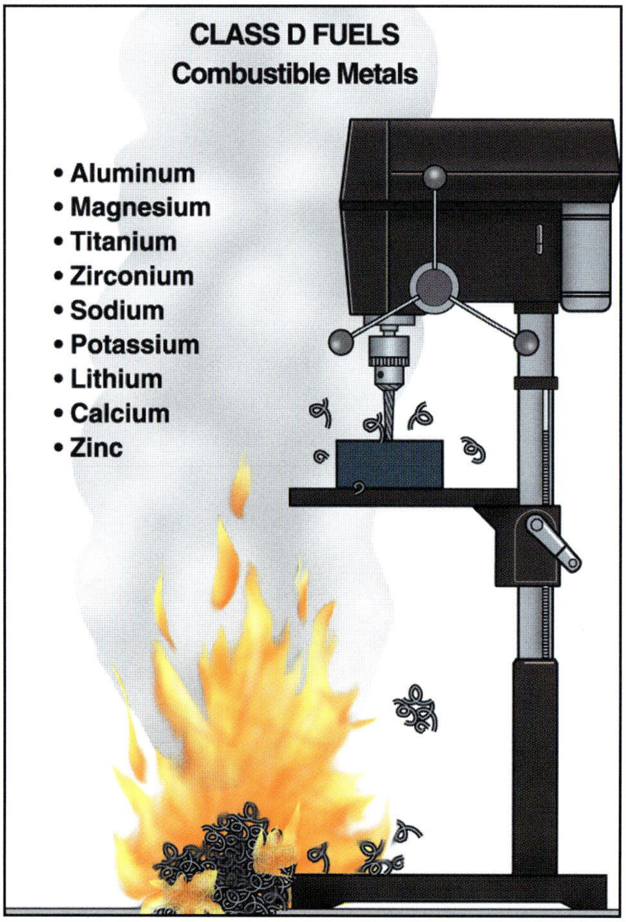

Figure 1.36 Class D fuel examples.

- *Class D* — Combustible metals such as aluminum, magnesium, titanium, zirconium, sodium, and potassium. These materials are particularly hazardous in their powdered forms. Given a suitable ignition source, airborne concentrations of metal dusts can cause powerful explosions (**Figure 1.36**).

Fire Extinguishment Theory

Limiting or interrupting one or more of the essential elements in the combustion process (fire tetrahedron) extinguishes fire. Thus, a fire may be extinguished by four methods: (1) reducing its temperature, (2) eliminating available fuel, (3) excluding oxygen, or (4) stopping the uninhibited chemical chain reaction.

Fuel Temperature Reduction

Fire extinguishment depends on reducing a fuel's temperature to the point where it does not produce sufficient vapor to burn. Both solid and liquid fuels with high flash points can be extinguished by cooling. One of the most common methods of cooling is to use water. However, fires involving low flash-point liquids and flammable gases cannot be extinguished by cooling with water because vapor production cannot be sufficiently reduced. Reduction of temperature depends on the application of an adequate flow of water to establish a negative heat balance (**Figure 1.37**).

Figure 1.37 Water fog can be an effective method for cooling burning materials or containers of flammable or combustible liquids.

Fuel Removal

In some cases, removing the fuel source effectively extinguishes a fire. For example, stopping the flow of liquid or gaseous fuel and removing solid fuel in the fire's path are ways of removing the fuel source. Another method is to allow a fire to burn until all fuel is consumed (**Figure 1.38**).

Figure 1.38 One of the most effective ways of controlling a pipeline, loading rack, or storage tank fire is to turn off the flow of fuel to the fire, thereby removing the fuel source. Fixed sites have control valves located throughout the facility for the purpose of controlling the flow of the liquids. *Courtesy of Jim Rackl.*

Oxygen Exclusion

Reducing the oxygen available to the combustion process reduces a fire's growth and may totally extinguish it over time. In its simplest form, this method is used to extinguish cooking stove fires when a cover is placed over a pan of burning food. Flooding an area with an inert gas such as carbon dioxide, which displaces the oxygen and disrupts the combustion process, can reduce oxygen content. Blanketing (covering) a fuel with fire-fighting foam also separates oxygen from the fuel. Neither of these latter methods works on those rare fuels that are self-oxidizing (**Figure 1.39**).

Chemical Flame Inhibition

Extinguishing agents such as some dry chemicals and halogenated agents (halons) interrupt the flame-producing chemical reaction and stop flaming combustion. This method of extinguishment is effective on gas and liquid fuels because they must flame to burn (**Figure 1.40**). These agents do not easily extinguish smoldering fires. Very high agent concentrations and extended periods of application are necessary to extinguish smoldering fires, making these agents impractical. Thus, cooling is the only practical way to extinguish a smoldering fire.

Extinguishing Properties

The theory of fire extinguishment is translated into action through the use of extinguishing agents. Those agents consist of water, foam solutions, dry chemical, carbon dioxide, and halon substitutes. Water and foam agents are discussed in the sections that follow.

Water

Water has always been the primary extinguishing agent for fires involving Class A fuels. Water extinguishes fire either through cooling (or quenching) or smothering, but the primary way is by cooling the fuel. Basically, water removes the heat from the fire. Water has advantages that make it valuable for fire extinguishment. It also has disadvantages that affect its value as an extinguishing agent. The advantages of water that are extremely valuable for fire extinguishment are as follows:

- Water is readily available and inexpensive.
- Water has a greater heat-absorbing capacity than other common extinguishing agents.
- A relatively large amount of heat is required to change water into steam; therefore, more heat will be absorbed by a given quantity of water.
- The greater the surface area of the water exposed, the more rapidly heat is absorbed.

To understand how the cooling action occurs, it is necessary to take a brief look at what water is and how it reacts to and with heat. *Water* is a compound formed of hydrogen and oxygen when two parts of hydrogen combine with one part of oxygen. At normal temperatures (32 to 212°F [0°C to 100°C]), water exists in a liquid state. Below 32°F (0°C) (the freezing point of water), water

Figure 1.39 Oxygen is excluded from a fire by blanketing the burning material with foam.

Figure 1.40 Dry chemical extinguishing agents are used to interrupt the flame-producing chemical reaction and stop the flaming combustion.

converts to a solid state in the form of ice. Above 212°F (100°C) (the boiling point of water), it converts into a gas called *water vapor* or *steam*. Water cannot be seen in this vapor form. When steam starts to cool, however, its visible form is called *condensed condensate*.

Complete vaporization does not happen the instant water reaches the boiling point because each pound (0.5 kg) of water requires approximately 970 Btu of additional heat to completely turn into steam. When a fire stream is broken into small particles, it absorbs heat and converts into steam more rapidly than it would in a compact form because more of the water surface is exposed to the heat. For example, if a single cube of ice is dropped into a glass of water, it takes some time for it to absorb its capacity of heat and cool the water because only the surface area of 6 sides of ice is exposed to the water (**Figure 1.41**). However, if that cube of ice is crushed into multiple particles, it will cool the water much faster. The finely divided particles of ice absorb heat more rapidly. This same principle applies to water in the liquid state.

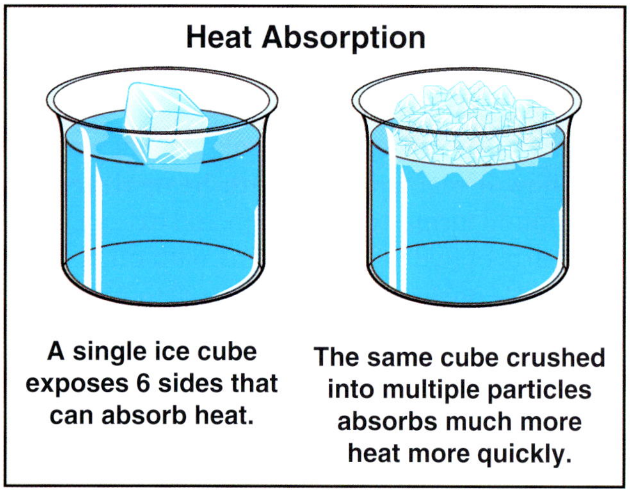

Figure 1.41 Example of the heat absorption capacity of ice in various forms.

Steam expansion helps cool a fire area by driving heat and smoke from it. Expanding steam can displace superheated fire gases that can cause burn injuries to fire and emergency responders and occupants in spaces over or adjacent to where the steam is generated. This fact is why venting is so important in the early stages of structural fire fighting. The amount of expansion varies with the temperatures of the fire area. At 212°F (100°C), a cubic foot (0.028 m^3) of water expands approximately 1,700 times its original volume (**Figure 1.42**). The higher the temperature, the higher the amount of expansion becomes. For example, at 500°F (260°C) the amount of expansion is approximately 2,400 times its original volume; and at 1,200°F (649°C) the amount of expansion is approximately 4,200 times its original volume.

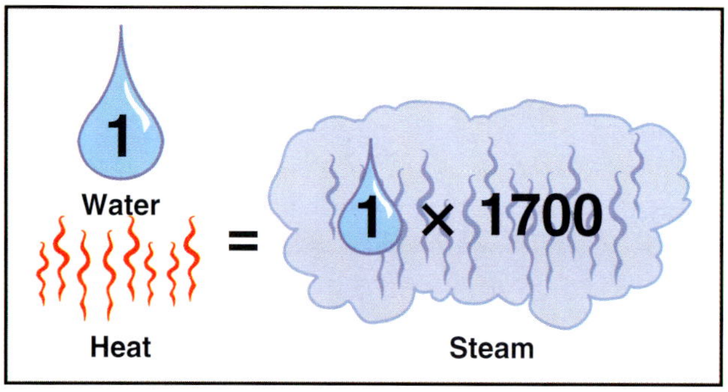

Figure 1.42 Water expands to 1,700 times its original volume when it converts to steam.

Steam expansion is not gradual but rapid. If a compartment is already full of smoke and gases, the steam that is generated displaces these gases when adequate ventilation openings are provided (**Figure 1.43**). As the compartment cools, the steam condenses and allows it to refill with cooler air. The steam produced also aids in fire extinguishment by smothering the gases from certain types of burning materials. Smothering is accomplished when the expansion of steam reduces oxygen in a confined space.

Water is the extinguishing agent of choice for the majority of emergency service organizations around the world. While water continues to be the primary weapon against fire, it does have the following limitations or disadvantages:

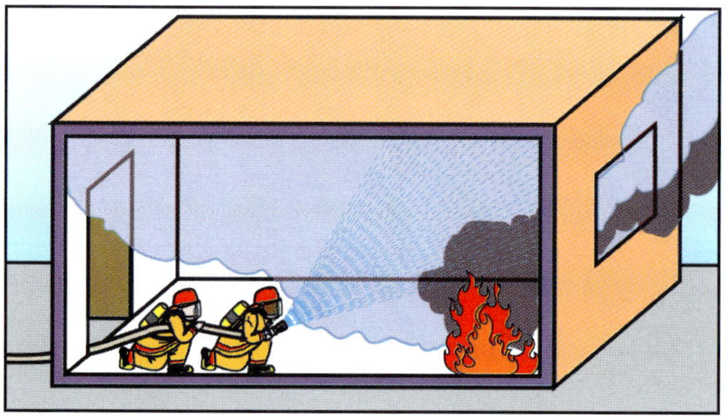

Figure 1.43 Water fog is used to disperse the products of combustion within a compartment.

- Does not provide an effective barrier against reignition
- Will not adhere to vertical surfaces (has low viscosity)
- Has difficulty being absorbed by some materials
- May react with some burning materials
- May cause water damage to unburned property
- May cause a structure to collapse because of water's weight
- May spread burning flammable or combustible liquids when they float on the water's surface
- May cause pollution of bodies of water by carrying contaminated runoff into waterways
- May create an electrocution hazard when used on energized equipment
- Freezes at 32°F (0°C), which requires the addition of antifreeze solutions in cold climates to prevent water freezing in hoselines, water tanks, and pumps

Water can be made into a more effective extinguishing agent for some fuels by adding foam concentrate and entrained air. These additions allow water to be used on fuels where water alone would not be effective. For example, fires involving Class B flammable and combustible liquids that are either miscible or lighter than water can be extinguished with water if a foam concentrate is added. In some cases, water can be used to dilute a miscible liquid enough that its temperature will be lower than its flash point or ignition temperature. This dilution can only occur when the container holding the liquid is large enough to hold all of the fuel and the water.

Foam Agents

To overcome some of the limitations of water as an extinguishing agent, foam concentrates are added. The type of foam agent used depends on the class of material that is involved in fire. Foam concentrates are designed for each specific class of material. Attempting to use a Class A foam solution on a Class B fire, for instance, can result in fire spread, reignition, or even injuries to emergency personnel. Fire-fighting foam concentrates and solutions are described in detail in Chapter 2, Foam Concentrate Technology.

Class A Fires

The addition of Class A foam concentrates to water enhances the ability of water to extinguish fires involving Class A materials by reducing water's surface tension (force minimizing a liquid surface's area), allowing it to penetrate more easily into dense or piled Class A materials. Class A finished foam can also cling to or cover the material and release (drain) water into the fuel as the foam blanket dissolves (breaks down or drains). Because of its quenching (cooling) characteristics, Class A finished foam is very effective when smoldering Class A fuels are encountered **(Figure 1.44)**.

Class A materials contain a certain amount of moisture to begin with. *Moisture content* is the quantity of water in a mass expressed in a percentage. It depends on the material itself and its size and shape, relative humidity of the atmosphere, temperature, and wind velocity. The higher the moisture content, the lower the oxygen content, and the greater the fuel vapor reduction will be. Materials with high moisture contents generate less fuel vapors when heated and are slow to ignite.

Figure 1.44 Class A foam is very effective for extinguishing wildland fires. *Courtesy of National Interagency Fire Center.*

When Class A finished foam is applied to the material, it penetrates, keeps the moisture content high, and reduces the release of fuel vapors.

Class B Fires

The smothering or blanketing effect of oxygen exclusion is most effective for extinguishing fires involving Class B fuels. Class B foam concentrates are effective as a fire-extinguishing agent and vapor suppressant on Class B liquids because finished foam can float on their surfaces. Class B finished foam also allows water to float on top of burning liquids, thus cooling the surface of the liquid as well as preventing the fuel from mixing with air (oxygen) **(Figure 1.45)**. Flammable gas fires are best extinguished by turning off the fuel source. Extinguishment occurs when the flow is stopped and the previously released fuel (residual fuel) is consumed. Foam is generally not used in such applications.

Figure 1.45 Class B foams provide an effective extinguishing agent for most flammable and combustible liquid fires by cooling and smothering the burning liquids.

Specialty foams have been created and used to successfully manage fuming spills of acids and base materials. Fuming spills result in a rapid release of vapors. Extinguishing fires involving water-miscible Class B fuels or blends of hydrocarbon and polar solvent fuels requires special alcohol-resistant foam agents that do not readily mix with the fuels.

In addition to foam concentrates, some extinguishing agents such as dry chemicals, halogenated hydrocarbons, and halon replacement agents interrupt the flame-producing chemical reaction and stop flaming. This method of extinguishment is effective on gas and liquid fuels because the fuels must flame to burn. These agents may be used in conjunction with Class B foams to improve extinguishment times.

Class C Fires

Foam agents are not recommended for Class C fires because of the conductivity of the agent, which results from the high water content. Nonconducting extinguishing agents such as halon, dry chemical, or carbon dioxide can sometimes control Class C fires. The safest extinguishment procedure is to first turn off the power and de-energize the involved circuits and then treat the fire as a Class A or Class B fire depending upon the fuel involved. After electrical equipment is de-energized, foam agents may be used.

When large oil-cooled electrical transformers or high-energy vaults are exposed to fire, a violent eruption may result, scattering oil over a wide area. Once this eruption occurs and the oil ignites, the fire may be considered a Class B fire. Typical smaller vaulted or subterranean transformer fires may be handled as subsurface in-depth fires. See Chapter 7, Class B. Foam Fire Fighting: Class B Liquids at Fixed Sites, for extinguishing systems used in these types of fires.

Class D Fires

Class D materials are particularly hazardous in their powdered forms. Given a suitable ignition source, airborne concentrations of metal dusts mixed in the correct proportion with oxygen can cause powerful explosions. The extremely high temperatures of some burning metals make water, foam solutions, and other common extinguishing agents ineffective and even dangerous in some cases. No single agent is available that effectively controls fires in all combustible metals. Special extinguishing agents are available for control of fire in each of the metals, and they are marked specifically for that metal. These agents cover the burning material and smother the fire.

Summary

A thorough understanding of the basic principles of heat energy sources, heat transfer, fire behavior, and fire development provide the emergency services member with the background necessary for selecting the appropriate agent for controlling or extinguishing a fire. Extinguishing agents are available for virtually all types of fuels and classifications of fires. While water is acceptable for the majority of Class A fires that an emergency services organization may encounter, it is not always suitable for other types of fuels. Fire-fighting foam concentrates, when used properly and according to manufacturers' recommendations, provide the necessary extinguishing agents for Class B fires and for the containment of nonignited Class B materials. In addition, Class A foam concentrates can be used to improve the extinguishing abilities of water on wildland, structure, and other types of Class A fires.

Chapter 2

Foam Concentrate Technology

Job Performance Requirements

This chapter provides information that will assist the reader in meeting the following job performance requirements from NFPA 1001, *Standard for Fire Fighter Professional Qualifications*, 2002 edition; NFPA 1002, *Standard for Fire Apparatus Driver/Operator Professional Qualifications*, 1998 edition; NFPA 1081, *Standard for Industrial Fire Brigade Member Professional Qualifications*, 2001 edition; NFPA 1003, *Standard for Airport Fire Fighter Professional Qualifications*, 2000 edition; NFPA 472, *Standard for Professional Competence of Responders to Hazardous Materials Incidents*, 2002 edition; and NFPA 1051, *Standard for Wildland Fire Fighter Professional Qualifications*, 2002 edition. Colored portions of the standard are specifically addressed in this chapter.

NFPA 1001

6.3.1 Extinguish an ignitable liquid fire, operating as a member of a team, given an assignment, an attack line, personal protective equipment, a foam proportioning device, a nozzle, foam concentrates, and a water supply, so that the correct type of foam concentrate is selected for the given fuel and conditions, a properly proportioned foam stream is applied to the surface of the fuel to create and maintain a foam blanket, fire is extinguished, reignition is prevented, team protection is maintained with a foam stream, and the hazard is faced until retreat to safe haven is reached.

(A) *Requisite Knowledge:* **Methods by which foam prevents or controls a hazard; principles by which foam is generated; causes for poor foam generation and corrective measures; difference between hydrocarbon and polar solvent fuels and the concentrates that work on each; the characteristics, uses, and limitations of firefighting foams; the advantages and disadvantages of using fog nozzles versus foam nozzles for foam application; foam stream application techniques; hazards associated with foam usage; and methods to reduce or avoid hazards.**

(B) *Requisite Skills:* The ability to prepare a foam concentrate supply for use, assemble foam stream components, master various foam application techniques, and approach and retreat from spills as part of a coordinated team.

NFPA 1002

3-2.3 Produce a foam fire stream, given foam-producing equipment, so that properly proportioned foam is provided.

(a) *Requisite Knowledge:* **Proportioning rates and concentrations, equipment assembly procedures, foam systems limitations**, and manufacturer specifications.

(b) *Requisite Skills:* The ability to operate foam proportioning equipment and connect foam stream equipment.

NFPA 1081

5.3.2 Activate a fixed fire protection system, given a fixed fire protection system, a procedure, and an assignment, so that the steps are followed and the system operates.

(A) Requisite Knowledge. Types of extinguishing agents, hazards associated with system operation, how the system operates, sequence of operation, system overrides and manual intervention procedures, and shutdown procedures to prevent damage to the operated system or to those systems associated with the operated system.

(B) Requisite Skills. The ability to operate fixed fire protection systems via electrical or mechanical means.

6.1.2.1 Utilize a pre-incident plan, given pre-incident plans and an assignment, so that the industrial fire brigade member implements the responses detailed by the plan.

(A) Requisite Knowledge. The sources of water supply for fire protection or other fire-extinguishing agents, site-specific hazards, **the fundamentals of fire suppression and detection systems including specialized agents,** and common symbols used in diagramming construction features, utilities, hazards, and fire protection systems.

(B) Requisite Skills. The ability to identify the components of the pre-fire plan such as fire suppression and detection systems, structural features, site-specific hazards, and response considerations.

7.3.4 Operating as a member of a team, extinguish an ignitable liquid fire, given an assignment, an attack line, personal protective equipment, a foam proportioning device, a nozzle, foam concentrates, and a water supply, so that the correct type of foam concentrate is selected for the given fuel and conditions, a correctly proportioned foam stream is applied to the surface of the fuel to create and maintain a foam blanket, fire is extinguished, re-ignition is prevented, and team protection is maintained.

(A) Requisite Knowledge. Methods by which foam prevents or controls a hazard; principles by which foam is generated; causes for poor foam generation and corrective measures; difference between hydrocarbon and polar solvent fuels and the concentrates that work on each; the characteristics, uses, and limitations of firefighting foams; the advantages and disadvantages of using fog nozzles versus foam nozzles for foam application; **foam stream application techniques; hazards associated with foam usage; and methods to reduce or avoid hazards.**

(B) Requisite Skills. The ability to prepare a foam concentrate supply for use, assemble foam stream components, master various foam application techniques, and approach and retreat from fires/spills as part of a coordinated team.

NFPA 1003

3-3.2 Extinguish an aircraft fuel spill fire, given PrPPE, an assignment, an ARFF vehicle hand line flowing a minimum of 95 gpm (359 L/min) of AFFF extinguishing agent, and a fire sized to the AFFF gpm flow rate divided by 0.13 (gpm/0.13 = fire square footage) (L/min/0.492 = 0.304 m^2), so that the agent is applied using the proper techniques and the fire is extinguished in 90 seconds.

(a) *Requisite Knowledge:* The fire behavior of aircraft fuels in pools, physical properties and characteristics of aircraft fuel, **agent application rates and densities.**

(b) *Requisite Skills:* Operate fire streams and apply agent.

3-3.3 Extinguish an aircraft fuel spill fire, given PrPPE, an ARFF vehicle turret, and a fire sized to the AFFF flow rate of 0.13 gpm (0.492 L/min) divided by the square feet of fire area, so that the agent is applied using the proper technique and the fire is extinguished in 90 seconds.

(a) *Requisite Knowledge:* Operation of ARFF vehicle agent delivery systems, the fire behavior of aircraft fuels in pools, physical properties and characteristics of aircraft fuel, **agent application rates and densities.**

(b) *Requisite Skills:* Apply fire-fighting agents and streams using ARFF vehicle turrets.

3-3.4 Extinguish a three-dimensional aircraft fuel fire, given PrPPE, an assignment, and ARFF vehicle hand line(s) using primary and secondary agents, so that a dual agent attack is used, the agent is applied using the proper technique, the fire is extinguished, and the fuel source is secured.

(a) *Requisite Knowledge:* The fire behavior of aircraft fuels in three-dimensional and atomized states, physical properties and characteristics of aircraft fuel, **agent application rates and densities,** and methods of controlling fuel sources.

(b) *Requisite Skills:* Operate fire streams and apply agents, secure fuel sources.

3-3.9 Replenish extinguishing agents while operating as a member of a team, given an assignment, an ARFF vehicle, a fixed or mobile water source, a supply of agent, and supply lines and fittings, so that agents are available for application by the ARFF vehicle within the time established by the authority having jurisdiction (AHJ).

(a) *Requisite Knowledge:* **Resupply procedures,** operation procedures for ARFF vehicle replenishment.

(b) *Requisite Skills:* Connect hose lines, operate valves.

3-3.11 Overhaul the accident scene, given PrPPE, an assignment, hand lines, and property conservation equipment, so that all fires are extinguished and all property is protected from further damage.

(a) *Requisite Knowledge:* **Methods of complete extinguishment and prevention of re-ignition,** purpose for conservation, operating procedures for property conservation equipment.

(b) *Requisite Skills:* Use property conservation equipment.

NFPA 1051

5.1.1 The Wildland Fire Fighter I shall meet the job performance requirements defined in Sections 5.1 through 5.5.

(A) Requisite Knowledge. Fireline safety, use, and limitations of personal protective equipment, agency policy on fire shelter use, basic wildland fire behavior, **fire suppression techniques,** basic wildland fire tactics, the fire fighter's role within the local incident management system, and first aid.

(B) Requisite Skills. Basic verbal communications and the use of required personal protective equipment.

5.5.6 Describe the methods to reduce the threat of fire exposure to improved properties given a wildland or urban/interface fire, suppression tools, and equipment so that improvements are protected.

(A) Requisite Knowledge. Wildland fire behavior, wildland fuel removal, **structure protection methods,** and equipment and personnel capabilities.

(B) Requisite Skills. The application of requisite knowledge to protect structures.

6.5.2 Select fireline construction methods, given a wildland fire and line construction standards, so that the technique used is compatible with the conditions and meets agency standards.

(A) Requisite Knowledge. Resource capabilities and limitations, fireline construction methods, and agency standards.

(B) Requisite Skills. None specified.

7.5.5 Deploy resources to suppress a wildland fire, given an assignment, personnel, equipment, and agency policies and procedures, so that appropriate suppression actions are taken, and safety of personnel is ensured.

(A) Requisite Knowledge. Fireline location and construction techniques, burning out procedures, capabilities of fire-fighting equipment and personnel, radio communications capabilities and protocols, and **techniques of the proper and safe deployment of the assigned resources.**

(B) Requisite Skills. Capabilities of assigned personnel and equipment.

NFPA 472

5.4.4 **Performing Defensive Control Actions.** Given a plan of action for a hazardous materials incident within their capabilities, the first responder at the operational level shall demonstrate defensive control actions set out in the plan and shall meet the following related requirements:

(1) Using the type of fire-fighting foam or vapor suppressing agent and foam equipment furnished by the authority having jurisdiction, **demonstrate the effective application of the fire-fighting foam(s)** or vapor suppressing agent(s) on a spill or fire involving hazardous materials.

(2) **Identify the characteristics and applicability of the following foams:**

(a) **Protein**

(b) **Fluoroprotein**

(c) **Special purpose**

i. **Polar solvent alcohol-resistant concentrates**

ii. **Hazardous materials concentrates**

(d) **Aqueous film-forming foam (AFFF)**

(e) **High expansion**

Chapter 2
Foam Concentrate Technology

Having covered the basics of fire behavior, extinguishing properties of water and foam agents, and potential fuels that fire and emergency responders face in Chapter 1, Fire Behavior and Extinguishment, attention now turns to mechanical fire-fighting foam concentrate technology. Although the majority of fires that municipal and industrial fire and emergency responders face can be successfully extinguished with plain water, there are incidents that cannot be safely handled by water alone. These incidents involve flammable or combustible liquids, deep-seated below-grade fires that are inaccessible by standard fire-fighting techniques, or wildland fires that are complicated by rugged terrain, dry ground cover, or high winds. On these occasions, emergency responders who have a thorough knowledge of foam concentrate technology can make the difference in bringing incidents to safe and expedient conclusions. Even Class A fires that can be successfully extinguished using plain water may be extinguished more effectively when a Class A foam concentrate is added to the water.

This chapter explains mechanical foam components, their characteristics, how they are proportioned (mixed) for use, and how they are stored. The basics of how foam concentrates work and specific foam concentrate types are also discussed. Information on characteristics, proportioning, application rate, and operations using both Class A and Class B foam concentrates are given.

How Foam Concentrates Work

[NFPA 1001: 6.3.1; NFPA 1051: 6.5.2(A); NFPA 1081: 7.3.4]

As discussed in the previous chapter, water alone is not always effective as a fire-extinguishing agent. Under certain circumstances, foam is needed as an extinguishing agent. The resulting finished foam extinguishes and/or prevents fire by the following methods (**Figure 2.1**):

- ***Separating*** — Creates a barrier between the fuel and burning vapors
- ***Cooling*** — Lowers the temperature of the fuel and adjacent surfaces
- ***Suppressing*** (sometimes referred to as *smothering*) — Prevents the release of additional flammable vapors, access to oxygen in the atmosphere, and therefore reduces the possibility of ignition or reignition

Through a water/concentrate proportioning process, a foam solution creates a finished foam product that forms a smothering blanket over the burning fuel (see Foam Proportioning Methods). The finished foam blanket excludes oxygen and inhibits the burning process. As water in the finished foam blanket releases (drains), the foam blanket becomes less effective as a vapor barrier and/or heat shield. The rate at which a foam blanket reduces in effectiveness is proportional to the rate at which its water drains. See Drainage section for more information. A good foam blanket is required

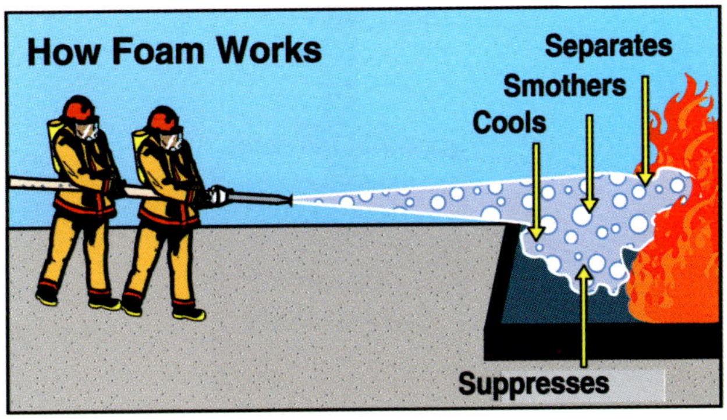

Figure 2.1 Foam prevents ignition and extinguishes fire by cooling, smothering, separating, and suppressing vapors.

to maintain an effective cover over either Class A or Class B fuels for the period of time desired. Uniform-sized bubbles provide long lasting foam blankets that are highly desirable in postfire or unignited spill events.

Mechanical Foam Components

Mechanical foam concentrates are divided into two general categories: those intended for use on Class A fuels (ordinary combustibles) and those for use on Class B fuels (flammable and combustible liquids). The mechanical-type foam concentrates in use today must be proportioned (mixed with water) and aerated (mixed with air) before they can be used. Foam concentrate, water, air, and mechanical aeration are needed to produce a foam blanket (**Figure 2.2**). These elements must be present and blended in the correct ratios. Removing any element results in either no foam production or poor-quality foam. Before discussing the foam-making process, it is important to understand the following terms (**Figure 2.3**):

- ***Foam concentrate*** — Liquid found in a foam storage container before the introduction of water
- ***Foam solution*** — Mixture in the proper ratio of foam concentrate and water before the introduction of air

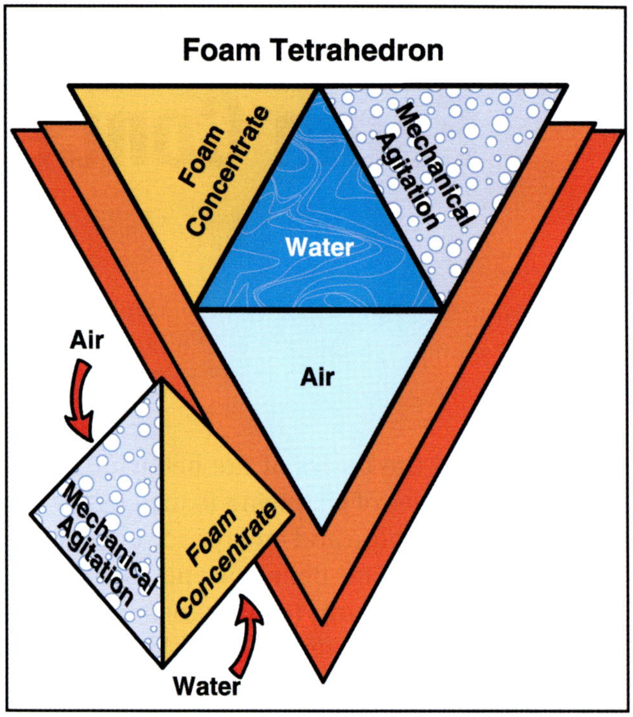

Figure 2.2 Foam tetrahedron.

- ***Foam proportioner*** — Device that mixes foam concentrate in the proper ratio with water
- ***Finished foam*** — Completed product after air is introduced into the foam solution and after it leaves the nozzle or aerator

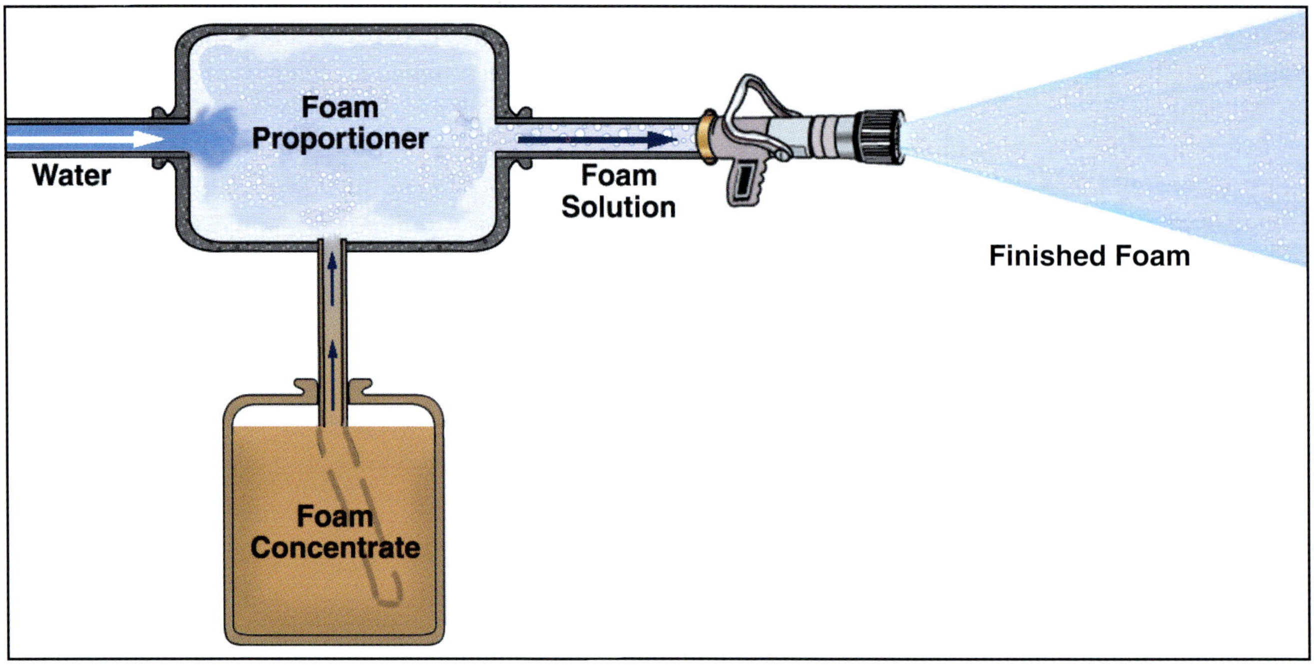

Figure 2.3 The foam process includes water, foam concentrate, foam proportioner, foam solution, and finished foam.

Foam concentrates must match the fuel to which they are applied to be effective. Class A foam concentrates are not designed to extinguish Class B fires. Class B foam concentrates designed solely for hydrocarbon fires (such as regular fluoroprotein and regular aqueous film forming foam [AFFF]) will not extinguish polar solvent (alcohol-type fuel) fires regardless of the concentration at which they are used. However, foam concentrates that are intended for polar solvents may be used on hydrocarbon fires. See the Specific Foam Concentrates section for a discussion of these concentrates.

CAUTION

Failure to match the proper foam concentrate with the fuel may result in an unsuccessful extinguishing attempt and could endanger fire and emergency responders.

Foam Concentrate Characteristics

The general characteristics of foam concentrates must be understood before continuing the discussion of foam concentrate use in the following chapters. Foam expansion, shelf or storage life, corrosiveness, health risks, compatibility with other extinguishing agents, environmental impact, and drainage characteristics are discussed in the sections that follow.

Foam Expansion

Foam expansion refers to the increase in volume of a foam solution when it is aerated. Expansion is a key characteristic to consider when choosing a foam concentrate for a specific application. The method of aerating a foam solution results in varying degrees of expansion, which depends on the following factors:

- Type of foam concentrate used
- Accurate proportioning of the foam concentrate and water
- Quality of the foam concentrate (how much foaming chemical it contains)
- Method of aspiration (fog [water spray] nozzle or aerating nozzle)

Depending on its purpose, foam concentrates can be described as one of three types: low-expansion, medium-expansion, and high-expansion. NFPA 11, *Standard for Low-, Medium-, and High-Expansion Foam* (2002) states that low-expansion foam has an air/solution ratio up to 20 parts finished foam for every part of foam solution (20:1 ratio). Medium-expansion foam is most commonly used at the rate of 20:1 to 200:1 through hydraulically operated nozzle-style delivery devices. In the high-expansion foam category, the rate is 200:1 to 1,000:1 (**Figure 2.4**).

Figure 2.4 High-expansion foam is typically used in total flooding applications such as this ship's engine room.

Shelf or Storage Life

Most modern foam concentrates may be stored for long periods of time (in excess of 10 years). Optimum storage life can be achieved by following manufacturers' recommendations on foam concentrate storage containers and their related environmental conditions. UL-listed foam concentrates (those listed by Underwriters Laboratories Inc., *Fire Protection Directory*) have storage temperature requirements of not less than 35°F (1.6°C) and not higher than 120°F (48.8°C). Freezing and thawing shall have no effect on UL-listed foam concentrate performance. UL-listed foam concentrates are shipped in "standard" containers that must pass drop-and-impact tests with their contents frozen. Some concentrates are available in freeze-protected formulations for unheated storage spaces. This factor allows them to be stored and used in cold climates (less than 35°F [1.6°C]). Appendix B, Foam Concentrate Certification/Regulatory Organizations and Testing Methods, contains information on testing foam concentrates.

Corrosiveness

Class A and Class B foam concentrates may be mildly corrosive. NFPA has established maximum allowable corrosion rates for Class A foam concentrates in NFPA 1150, *Standard on Fire-fighting Foam Chemicals for Class A Fuels in Rural, Suburban, and Vegetated Areas* (1999). To prevent corrosion on metal parts, pumps, hoselines, proportioners, and nozzles, flush with clean water following each use of foam **(Figure 2.5)**. Always lubricate moving parts of proportioners, pumps, and nozzles after using foams. Appliances and devices that require occasional cleaning and lubrication may need attention after prolonged foam use. Always follow the foam concentrate manufacturer's recommendations for cleanup procedures and/or concentrate disposal.

Figure 2.5 Because of the mild corrosive effects of foam, thoroughly flush hoselines, proportioners, and nozzles following each use. *Courtesy of Tom Ruane.*

Health Risks

Foam concentrates, either at full strength or in diluted form, pose minimal health risks to fire and emergency services responders. In both forms, concentrates may be mildly irritating to the skin and eyes, so flush affected areas with water. Concentrates, or their vapors, may be harmful if ingested or inhaled. Consult the manufacturer's material safety data sheet (MSDS) for information on a specific foam concentrate. Many foam manufacturers maintain up-to-date material safety data on their websites.

Compatibility with Other Extinguishing Agents

Aqueous film forming foam (AFFF), film forming fluoroprotein (FFFP) foam, and fluoroprotein foams are all compatible with carbon dioxide, halon substitutes, and dry-chemical agents and may be simultaneously discharged with them. (See Specific Foam Concentrates section for details.) Using foam, a halon substitute, and dry-chemical agents together is commonly referred to as a *multiagent attack* **(Figure 2.6)**. Several different types of special apparatus are used for multiagent applications, ranging from small cart-mounted units commonly found in industrial facilities to large aircraft rescue and fire fighting (ARFF) vehicles.

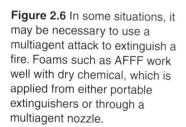

Figure 2.6 In some situations, it may be necessary to use a multiagent attack to extinguish a fire. Foams such as AFFF work well with dry chemical, which is applied from either portable extinguishers or through a multiagent nozzle.

Environmental Impact

The environmental impact of foam concentrates and foam solutions varies. The primary concern is the impact of finished foam after it has been applied to a fire or spill. The biodegradability of foam in either solution or concentrate form is determined by the rate at which environmental bacterias dissolve or degrade (break down) the foam. This decomposition process results in the consumption of oxygen. A reduction in oxygen from surrounding water can cause damage in waterways by killing water-inhabiting creatures and vegetation. The less oxygen that is required to degrade a particular foam, the better or more environmentally friendly the foam is when it enters a body of water. Studies by the U.S. Forest Service and the National Biological Survey Office show that Class A foams can have a lethal affect on fish. Therefore, avoid the direct discharge of foam concentrates or finished foam into rivers, streams, lakes, and other bodies of water if possible.

Where AFFF concentrate is concerned, environmental issues revolve around glycol ethers (also known as *butylcarbitol*) and perfluorooctylsulfonates (PFOS). See the textbox for additional information. Some AFFF foam concentrates use butylcarbitol as a water-dispersing solvent and refractive index tracker, which is used to determine the concentration of the finished foam when it has been applied to a fuel. The U.S. Environmental Protection Agency (EPA) determined that both glycol ethers and PFOS might be hazardous to the environment. Many foam concentrate manufacturers have substituted more environmentally responsible solvents for butylcarbitol. Military specification (commonly called *milspec*) AFFF foam concentrate (U.S. Navy, military specification [milspec MIL-F-24385F]) formulations are not environmentally friendly solvents. In May, 2001, the U.S. Navy began seeking alternative products to replace PFOS and glycol ether foam concentrates with those that are both compatible with saltwater and the environment.

Each foam concentrate manufacturer provides information in a material safety data sheet for each of its products. In the United States, Class A foam concentrates require U.S. Forest Service approvals from the U.S. Department of Agriculture (USDA) where environmental issues are concerns. NFPA 1150 addresses environmental issues associated with Class A foam concentrates.

Perfluorooctylsulfonates

Perfluorooctylsulfonates (PFOS) are used as textile stain-preventing coatings; oil-, grease-, and water-resistant coatings for paperboard and food contact paper; acid-mist suppressants for metal plating and electronic etching baths; and alkaline cleaners. They are also used in the manufacture of film forming fire fighting foam concentrates. The Minnesota Mining and Manufacturing Company (3M) Company has primarily manufactured these products using an electrochemical process since the 1950s.

According to the U.S. Environmental Protection Agency (EPA), this electrochemical process creates a compound that is a persistent, bioaccumulative, and toxic (PBT) chemical. Although the EPA ***does not*** consider these chemicals an imminent health risk to the general population, there is concern for future health risks if these chemicals continue to accumulate in the environment. On May 16, 2000, the 3M Company publicly announced its voluntary withdrawal from manufacturing such chemicals. The 3M Company committed to this withdrawal by discontinuing the manufacture of certain chemical substances on a global basis for their most widespread uses. It will discontinue the manufacture of all PFOS chemicals by December 31, 2002. Alternative foam concentrates that do not contain PFOS are available from various manufacturers.

Foam Drainage

Drainage (also referred to as *quarter-drain time, 25-percent drainage time,* or *quarter-life*) plays a role in finished foam effectiveness. The bubbles formed in the foaming process start to break down once they are applied (**Figure 2.7**). Class A foams continue to wet the fuel as they drain. Aqueous film forming foams form a film over the surface of the fuel as they drain (see more details in Specific Foam Concentrates section). One of the measurements for determining the stability of finished foam is the rate at which the foam blanket drains (breaks down). A short drain time means rapid wetting. A long drain time means that the foam blanket holds water and provides an insulating foam layer for an extended period before the water releases. Several elements affect the drainage process: fuel temperature, heat of the fire, size of the flame front, and to a lesser extent, ambient air temperature and wind.

Figure 2.7 The effectiveness of a foam blanket can be determined by its drain time; that is, the time it takes the blanket to dissolve or break down. In this photograph, the foam blanket within the detention pit has started to dissolve, exposing the surface of the fuel.

Foam Concentrate Types

[NFPA 1001: 6.3.1; NFPA 1002: 3-2.3; NFPA 1003: 3-3.2, 3-3.3, 3-3.4, 3-3.11; NFPA 1051: 5.1.1(A), 5.5.6(A), 7.5.5(A); NFPA 1081: 6.1.2.1, 7.3.2, 7.3.4]

Numerous types of foam concentrates have been developed for specific applications according to their properties and performance. Some foam concentrates are thick and viscous. These concentrates are typically alcohol-resistant aqueous film forming foams (AR-AFFF, 3 to 6 percent or 1 to 3 percent) and produce finished foam that can form tough, heat-resistant blankets that drain slowly when proportioned at higher ratios of water and concentrate. Thinner concentrates are generally found in the nonalcohol-resistant (regular AFFF, FFFP, and fluoroprotein foam) or Class A foams. AFFF and FFFP tend to be fast drainers that spread rapidly across liquid fuel surfaces, sending a quick-spreading film ahead of the actual foam blanket. The draining AFFF foam blanket becomes the reservoir for continued water-filming action. Others, such as high-expansion foam concentrates, are used from 1 to 2.5 percent concentrations to produce large volumes of finished foam for flooding confined spaces. The sections that follow highlight common types of foam concentrates in the two general categories of fuel: Class A and Class B. Information on characteristics, proportioning, fire-fighting uses, and application rates are also given. Appendix C, Foam Properties, gives a matrix view of the properties of foam concentrates.

Accidental Foam Release Measures

In the event of an accidental release of foam concentrates, foam solutions, or finished foam, specific steps must be taken to mitigate the incident: Stop the flow of concentrate, solution, or finished foam as soon as possible. Use appropriate personal protective equipment during cleanup. Collect spilled concentrate, solution, or finished foam with absorbent material. Flush the area with water until foam bubbles are no longer produced. Exercise caution while walking in the area because surfaces may be slippery. Prevent discharge of concentrate, solution, or finished foam into waterways. Do not discharge into biological sewer treatment systems without prior approval. Dispose of recovered concentrate, solution, or finished foam in accordance with federal, state/provincial, and local regulations.

Information concerning the collection and disposal of concentrates, solutions, or finished foam and references to U.S. Environmental Protection Agency (EPA) and some state-regulated chemicals will be found on manufacturers' MSDS documentation under headings that refer to Regulatory Information. U.S. federal regulations that impact the release of foam concentrates, solutions, or finished foam include the following:

- U.S. EPA's Extremely Hazardous Substances (EHS) list — Lists components in foam concentrates with known Controlled Atmosphere(s) (CA[s]) numbers (CA[s] are controlled atmospheres used to extend the life of fresh and durable products and are typically low in oxygen content and high in carbon dioxide such as dry ice.)
- Toxic Substances Control Act (TSCA)
- Superfund Amendments and Reauthorization Act of 1988 (SARA), Title III, Section 302/304
- Comprehensive Environmental Response, Compensation, and Liability Act (CERCLA), Section 313

CAUTION

Many foam concentrates are currently on the market. Not all of them meet NFPA standards; some claim to work on both Class A and Class B fuels. To ensure that the product meets the requirements of the fire or emergency services organization, certify the product by a third-party testing organization.

Class A Foam Concentrates

It has long been known that wetting agents (nonfoaming, ionic, and nonionic surfactants) added to water will improve its absorption into Class A materials. In fact, wetting agents have been in use in the fire-suppression industry since the 1940s. However, Class A products have only been in use since the late 1970s. During the 1980s and 1990s, proportioning technology improvements increased the popularity and effectiveness of both Class A foams and wetting agents. These agents have proven effective in fighting fires in structures, wildland settings, coal mines, tire storage sites, and other incident sites involving similar deep-seated fuels.

WARNING

Use Class A foam only on Class A fuels. It is not formulated for fighting Class B fires.

Class A foam concentrate is a formulation of hydrocarbon surfactants. These ionic and nonionic surfactants help reduce water's surface tension in the foam solution. Reducing water's surface tension provides better fuel penetration of the water, thereby increasing its effectiveness. When used in conjunction with a compressed-air foam system (CAFS), Class A foam has outstanding reach and insulating qualities. CAFSs typically entrain one or more parts of air with one part water. Compressed-air foam proportioning systems are discussed in Chapter 3, Foam Proportioning and Delivery Equipment.

Class A foam may be used with fog nozzles, aerating foam nozzles, medium- and high-expansion devices, and compressed-air foam systems using almost any nozzle, including solid stream nozzles (**Figure 2.8**). The shelf life of Class A foam concentrates is indefinite. Because this type of foam concentrate is used in very small percentages in the foam solution, harm to the environment is not a concern under ordinary fire-suppression conditions. Nevertheless, take care to prevent direct discharge of concentrates into public or private waterways and sanitary sewer systems.

Figure 2.8 Class A foam has proven effective in the extinguishment of structure fires as well as wildland fires. *Courtesy of Tom Ruane.*

Characteristics

To be effective, Class A foam must contain the right blend of physical characteristics. The following terms explain the characteristics that affect Class A foam:

- ***Surface tension reduction*** — The effect of a surfactant on the water/concentrate foam solution allows water to spread more rapidly over the surface of Class A fuels and penetrate organic fuels.
- ***Expansion*** — Aeration of the water/concentrate foam solution into a volume of bubbles that is greater than the volume of the solution is referred to as *expansion ratio* and is expressed in air/solution ratios such as 2:1, 5:1, 10:1, etc. Water quality, aeration method (nozzle, compressed air, airdrop altitude), concentrate ratio, and concentrate formulation all affect the obtainable expansion ratio.
- ***Drainage*** — The amount of liquid that drains from the bubble mass to wet and/or penetrate the Class A fuel is affected by the expansion ratio.

For example, Class A foam with a low-expansion ratio drains faster than does Class A foam with a high-expansion ratio. Generally, finished foam with a low-expansion ratio produces a quicker knockdown than does foam with a high-expansion ratio.

- *Consistency* — Small bubbles of a consistent size are best for the development of long-lasting finished Class A foam that can adhere to surfaces (horizontal, vertical, and overhead). They also provide an insulation barrier and minimize the oxygen supply to Class A fuels. Large bubbles tend to break down more rapidly and release more liquid to wet and penetrate the Class A fuel (**Figure 2.9**). Consistency is an important consideration when establishing Class A foam delivery to a fire. Class A foam does not adhere to hot, vertical fuel surfaces that are in excess of 212°F (100°C).
- *Retention* — The characteristic of Class A foam to remain on and in the fuel, reduce the fuel temperature, and increase the fuel moisture content is known as its *retention*. Air temperature, wind movement, the fuel's latent heat, and the amount of heat present affect retention time.
- *Viscosity* — Viscosity refers to a liquid's thickness or ability to flow. Highly viscous liquid concentrates are very thick and may be difficult to proportion through some types of foam concentrate pumps. Liquid concentrates with low viscosity are very thin and pour easily. Temperature affects viscosity. Class A foam concentrates become more viscous when cold and less viscous when heated.

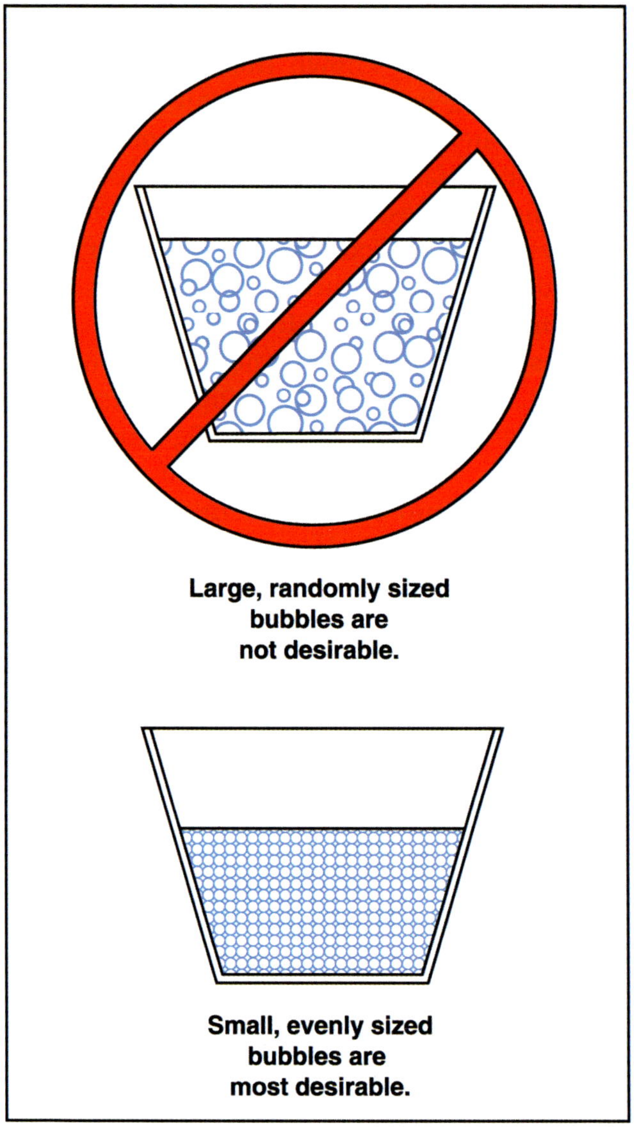

Figure 2.9 For effective foam blankets, small bubbles of consistent size are more desirable than large bubbles that break down rapidly.

Proportioning

Class A foam concentrates are mixed in proportions of 0.1 to 1 percent (**Figure 2.10**). Class A foam drain times increase with an increase in the proportioning ratio. A rich concentrate/water mixture ratio (using a higher proportion of foam concentrate than normal) produces thicker, richer foam that drains slower than a lean ratio (using a lower proportion of foam concentrate than normal). See Chapter 3, Foam Proportioning and Delivery Systems, for additional information on rich/lean ratios. Most foam nozzles produce more stable finished foams at a 1-percent concentration than they do at 0.4- to 0.5-percent concentrations. Employing percentages

Figure 2.10 Class A foam is typically mixed in proportions of 0.1 to 1 percent, which results in a thick foam blanket. *Courtesy of Tom Ruane.*

greater than 0.5 percent with standard fog nozzles does not appear to increase fire-fighting performance. Exposure protection may be enhanced with applying Class A foam through fog nozzles at 1-percent or greater proportioning settings. When performing the following operations, use the common guidelines listed for Class A foam concentrate proportioning:

- ***Fire attack and overhaul with standard fog nozzles*** — 0.2 to 0.5 percent foam concentrate
- ***Exposure protection with standard fog nozzles*** — 0.5 to 1 percent foam concentrate
- ***Most operations with air-aspirating foam nozzles*** — 0.3 to 1 percent foam concentrate
- ***Most operations with a compressed-air foam system*** — 0.2 to 0.5 percent foam concentrate

Application Rates

Application rate refers to the amount of finished foam that must be applied to a fire, per minute, per square foot (square meter) of fire. In structural fire-fighting combat, the application rate for Class A foam is the same as the minimum flow rate for water. Knowing the correct application rates is important during emergency operations. Determine these rates during pre-incident planning. Do not reduce initial attack flow rates when using Class A foam.

Operations

Class A foam can be tailored to meet certain needs, depending on the following situations:

- ***Areas requiring maximum penetration*** — Wet finished foam is very fluid and desirable for areas requiring maximum penetration. Wet finished foam has high water content and a very fast drainage rate.
- ***Vertical surfaces*** — Dry finished foam is a rigid coat that adheres well. Its slow drainage rate allows it to cling to these surfaces for extended periods. Dry finished foam has very low water content and high air content. It resembles shaving cream. Foam will not cling to very hot metal (**Figure 2.11**).
- ***Areas requiring a balance of penetration and clinging ability*** — Medium finished foam has the ability to blanket and wet the fuel equally well. With medium finished foam, a strong need to adjust the air and concentrate ratio is present.

Figure 2.11 Class A foam applied in a thick blanket makes an effective exposure barrier because it adheres well to vertical surfaces.

The importance of the wetting characteristic of foam becomes evident with a better understanding of the following types of Class A fires:

- Structures
- Wildlands
- Deeply concentrated fuels
- Civil disturbances
- Limited or disrupted water supply incidents
- Limited staffing incidents

Structures. Historically, water has been the most effective and most commonly used extinguishing agent on structure fires. However, the strong surface tension of water may inhibit its ability to penetrate and cool Class A fuels. What further complicates matters is the fact that many home furnishings and structural finishes in use today are made of synthetic materials that may not absorb water; whereas materials made of fibrous elements can absorb water. The addition of Class A foam concentrate into a fire stream helps water work against these obstacles by reducing its surface tension. The nature of finished foam also permits it to coat materials such as plastics that do not allow penetration. These properties offer benefits such as faster knockdown, reduced rekindles, minimal water damage, and reduced overhaul time (**Figure 2.12**).

Wildlands. Fuel density, fuel composition, climatic conditions, and fuel and moisture content strongly influence the course of a wildland fire. A common problem with fires involving heavy brush is the difficulty fire and emergency responders have in applying water to the underside of vegetation.

Figure 2.12 Class A foam can coat various manmade materials such as synthetic and aluminum siding as well as natural materials such as vegetation. *Courtesy of Mount Shasta (CA) Fire Protection District.*

Figure 2.13 Foam can create wet line barriers to halt the spread of fire on open grasslands. *Courtesy of National Interagency Fire Center.*

This problem slows fire-suppression efforts and creates long mop-up operations. The addition of Class A foam concentrates into a wildland fire attack aids the absorption of water to the concealed areas of the fuel. The soaking characteristics of Class A foam also allow water to seep more quickly into vegetation with a low-moisture content. Raising moisture content buys time because the fire has to spend time heating the water before the fuel releases its flammable components. A fire front will rapidly weaken in the face of rehydrated fuel (**Figure 2.13**).

Deeply concentrated fuels. Deeply concentrated fuels come in a variety of forms and are found in a variety of settings such as tire storage sites, rolled paper storage facilities, hay barns or open stacks of baled hay, lumber storage, landfills, auto salvage yards, junkyards, open storage sites, construction sites, vehicles, Dumpster® or other waste receptacles, and other densely stored fuel sites (**Figure 2.14**). These types of concentrated storage tend to be barriers for fire-extinguishing efforts because little space is available between materials for water penetration. The use of Class A foam in finished or plain solution forms can play a major role in extinguishing these types of fires. The characteristics of

Figure 2.14 Large area fires, such as this used tire storage site, create unique problems for fire and emergency services personnel. Foam can be affective in extinguishing this type of fire by penetrating into the fuels. *Courtesy of Bill Willcox.*

Class A finished foam that were previously described in terms of extinguishing wildland fires and structure fires also apply to deeply concentrated fuel fires. The reduction in surface tension allows the finished foam to more thoroughly penetrate the fuel. This action allows for more effective fire extinguishment and less chance of a rekindle.

Civil disturbances. Civil disturbances create special tactical problems for fire and emergency responders. The incident priority of life safety is of greatest importance. Class A finished foam application allows emergency responders to make a rapid attack for rescue purposes and then withdraw from the incident scene in a timely manner. Rapid attacks limit emergency responders to brief exposure to personal attacks from unruly crowds and demonstrators.

Limited or disrupted water supply incidents. Class A finished foam is beneficial for rapid fire extinguishment when water supplies are limited or become disrupted during the operation. It increases the effectiveness of the water that is available to responders.

Limited staffing incidents. Volunteer, combination, and some full-time paid fire and emergency services organizations are often faced with limited staffing during the initial response to an incident. Because a fire can grow rapidly, the quicker an extinguishing agent is applied, the less damage the fire causes. Class A finished foam and compressed-air foam systems provide the advantage that is lost by limited personnel during an initial response. The Class A foam extinguishing agent reduces the fire more rapidly, cools the interior of the structure, and prevents reignition within the structure. A CAFS has the advantage of providing quick fire extinguishment from great distances, while allowing for the maneuverability of the hoseline by fewer people. CAFS-charged hoselines are approximately half the weight of water-filled hoselines.

Class B Foam Concentrates

Class B foam extinguishes fires involving flammable and combustible liquids (**Figure 2.15**). It also suppresses vapors from unignited spills of these liquids. Several types of Class B foam concentrates are available. The following sections examine Class B foam concentrates and their characteristics. Class B finished foam may be proportioned into the fire stream via a fixed system, an apparatus-mounted system, or portable foam proportioning equipment. Proportioning equipment is discussed in Chapter 3, Foam Proportioning and Delivery Equipment, and Chapter 4, Foam Delivery Systems. The foam may be applied either with standard fog nozzles (AFFF and FFFP concentrates) or with air-aspirating foam nozzles (all types). (See Specific Foam concentrates section for details.) See Chapter 6, Class B Foam Use: Unignited Class B Liquid Spills, Chapter 7, Class B Foam Fire Fighting: Class B Liquids at Fixed Sites, and Chapter 8, Class B Foam Fire Fighting: Transportation Incidents, for discussions of Class B fire and nonfire incident operations.

Class B foam concentrates are manufactured from either a synthetic or protein base. Protein-based foam concentrates are derived from animal protein. Synthetic foam concentrate is made from a mixture of fluorosurfactants. Some foam concentrates are made from a combination of synthetic and protein bases. The various types of protein-based, synthetic-based, and combination foam concentrates are discussed in the Specific Foam Concentrates section.

Figure 2.15 Class B foam is used to combat large-scale flammable and combustible liquid fires. *Courtesy of Harvey Eisner.*

Individual foam concentrates may or may not be subject to third-party assurance testing or military specifications. Each manufacturer determines whether or not to seek these listings. Some manufacturers offer both listed and nonlisted concentrates. Foam concentrates that have not been third-party tested (by Underwriters Laboratories Inc. [UL] for instance) are often called *nonlisted* concentrates. These concentrates may or may not be as effective as those that have UL or milspec listings **(Figure 2.16)**. Milspec-listed AFFF concentrates must pass more stringent fire-fighting tests than UL-listed AFFF concentrates. As a result, milspec-listed concentrates contain more film-forming and foaming chemicals than do normal UL-listed AFFF. Milspec-listed AFFF is only suitable for use on simple hydrocarbon fuels. No alchohol-resistant versions of milspec-listed foams are available. Milspec-listed AFFF is produced by approved manufacturers and formulated using the U.S. Navy's formulation (milspec MIL-F-24385F). Therefore, milspec-listed AFFF is storage compatible regardless of manufacturer.

Foam concentrates that are stored in accordance with manufacturers' recommendations have an indefinite shelf life; twenty years is not an unreasonable expectation. This expectation is also true of protein-based foam concentrates. Do not expose original packaging (pails, drums, or intermediate bulk container [IBC] totes) to direct sunlight for extended periods **(Figure 2.17)**.

According to NFPA 11, different manufacturer's foam concentrates should not be mixed together in storage because they may be chemically incompatible. The exception to this recommendation would be milspec concentrates. The milspec MIL-F-24385F is written so that mixing can be done with no adverse affects. On the emergency scene, con-

Figure 2.16 Foam concentrates that have been tested and certified by Underwriters Laboratories Inc. carry the UL label on all containers.

Figure 2.17 To improve the shelf life of foam concentrates, properly store containers in cool areas. When possible, store foam containers in easily accessible compartments on apparatus.

centrates of a similar type (all AFFFs, all fluoroproteins, etc.) but from different manufacturers may be mixed together immediately before application.

Class B foam concentrates are available in freeze-protected formulations. Antifreeze is added to standard concentrate to prevent the water in the concentrate from freezing. Freeze-protected AR-AFFF concentrates may not be UL-listed because the antifreeze additives tend to reduce the effectiveness of finished foams.

The chemical properties of Class B foam concentrates and their environmental impact may vary depending on the type of concentrate and the manufacturer. While protein-based foam concentrates degrade faster after application than do synthetic foams (usually in 2 to 8 days), the rapid biodegradation may have a negative impact on aquatic life. The process of biodegradation removes oxygen from the water thereby suffocating fish and plants. Synthetic-foam concentrates degrade in 20 to 40 days and do not reduce the oxygen levels in the water at the same speed that protein-foam concentrates do. Consult the material safety data sheets provided by the manufacturer for information on specific concentrates. Full-disclosure MSDS documentation provides information on a particular foam's chemical or biological oxygen demands in use concentrations. Generally speaking, the lower the oxygen demand, the less environmental damage the foam causes.

Characteristics

To be effective, good Class B finished foam must contain the right blend of the following characteristics:

- ***Water retention*** — Water-retention properties of finished foam are the keys to long-life vapor suppression. The faster a foam blanket drains its water, the sooner the foam blanket becomes vulnerable to heat attack and vapor migration.
- ***Finished foam life*** — The finished foam life refers to the period of time that the foam blanket remains in place until more finished foam must be applied to replenish it. The life of finished foam is affected by heat, type of concentrate, expansion ratio, fuel involved, and environmental conditions.
- ***Heat resistance*** — Heat resistance refers to the ability of finished foam to resist the actual heat of the liquid surface and containment vessel on which it is applied. Most finished foam breaks down when the fuel's temperature is near or over 212°F (100°C). Foam breaks down rapidly as the fuel temperature rises and will perhaps become ineffective when the fuel's temperature is near or over 150°F (66°C).

CAUTION

Applying foam to fuels with temperatures in excess of 212°F (100°C) can be dangerous. When the foam breaks down, the remaining water causes a violent reaction with the fuel allowing the burning fuel to spread on top of the water.

- ***Multipurpose use*** — Some foam concentrates can be used on more than one type of hazard. An example is one that can be used on both hydrocarbon and polar solvent liquids. Only foam concentrates that have been specifically designed for multipurpose use should be used in this manner. Typical examples of these foams are AR-AFFF and AR-FFFP.
- ***Viscosity*** — Viscosity refers to a liquid's thickness or ability to flow. Highly viscous liquid concentrates are very thick and may be difficult to proportion through some types of foam concentrate pumps. Typically these viscous foams are AR-FFFP. Foam concentrates with low viscosity are very thin and pour easily. Typically these foams are regular AFFF, fluoroprotein, Class A, and wetting agent concentrates. Viscosity is affected by temperature. Foam concentrates become more viscous when they are cold and less viscous when they are heated. UL-listed foam concentrates must be pourable and easily proportioned at temperatures as low as 35°F (1.6°C).
- ***Knockdown speed and flow characteristics*** — This characteristic refers to the time required for a foam blanket to spread across a fuel surface or around obstacles and wreckage to achieve complete fire extinguishment.

- ***Fuel resistance*** — Finished foam must be able to resist liquid fuel contamination through absorption. Foams with high fuel resistance can be used for fixed-system subsurface injection into storage tanks. AFFF, FFFP, and fluoroprotein foams possess a fuel resistance that prevents contamination of the foam. Regular protein, high-expansion, and Class A foams have little or no fuel resistance, which is the primary reason why these agents are not recommended for use on Class B fuel fires.
- ***Vapor suppression*** — A vapor-tight foam blanket or an aqueous film floating on a hydrocarbon fuel is capable of preventing ignition or reignition of flammable/combustible liquid vapors. Alcohol-resistant foams are capable of producing a vapor blanket on both hydrocarbon and polar solvent fuels.
- ***Alcohol resistance*** — Polar solvent fuels (alcohols, ketones, and esters) create special problems because they are mixable (miscible) with water. These fuels draw water out of nonalcohol-resistant foams as fast as they are applied. Alcohol-resistant foam concentrates are formulated to protect the finished foam by providing a barrier (membrane) between the fuel and the finished foam. These foams are often referred to as *smart foams* because they tend to adjust their chemistry when they detect polar (alcohol) or hydrocarbon fuels (**Figure 2.18**).

  NOTE: Gasoline blended with more than 10 percent oxygen compound or ethanol must be treated with alcohol-resistant foams (see NFPA 11).
- ***Quarter-life*** — Quarter-life is the time required in minutes for one-fourth of the total liquid solution to drain from the finished foam (also referred to as the *25-percent drainage time* or *quarter-drain time*). For example, fire-fighting foams behave the same way as beer in terms of drainage characteristics: Weak beer drains its head quickly, while strong brews hold their heads longer.

Proportioning

Today's Class B foam concentrates are mixed in proportions from 1 to 6 percent (1:99 or 6:94 ratio of water to concentrate). The proper proportion for any particular concentrate is listed on the outside of the foam concentrate container (**Figure 2.19**). Some multipurpose foam concentrates designed for use on both hydrocarbon and polar solvent fuels can be used at different concentrations, depending on which of the two fuels they are used on. These concentrates are normally used at a 3-percent rate (3:97 ratio) on hydrocarbons and a 6-percent rate (6:94 ratio) on polar solvents. New multipurpose

Figure 2.18 Alcohol-resistant foam concentrates are available for use on polar solvent spills and fires. *Courtesy of Kidde Fire Fighting.*

Figure 2.19 Foam container labels provide the manufacturer's recommended proportioning ratio such as this 3- to 6-percent concentrate. *Courtesy of Conoco, Inc.*

foam concentrates may be used at 3-percent concentrations regardless of which type of fuel they are used on. High-expansion foam concentrates are typically used between 1- or 2.5-percent concentrations. In addition, 1-percent AR foam concentrates are available for use on hydrocarbon spill fires using Type III application devices and procedures. Type III devices may include foam cannon or handheld nozzles. See Chapter 4, Foam Delivery Systems, for a broader discussion of Type III foam application devices.

Application Rates

The dynamics of ignited fuels cause them to consume finished foam as it is applied. Therefore, the rate at which finished foam is applied must exceed its rate of consumption. For example, when employing AFFF, FFFP, or AR-AFFF solution on a hydrocarbon spill fire, the discharge rate should equal 10 percent of the area of a spill. Thus if a spill measures 1,000 square feet (93 m^2), the solution application rate is 200 gpm (379 L/min). This rate varies with fuel type, fuel depth, finished foam type, and method of foam application (**Figure 2.20**). Once the fire is extinguished, finished foam consumption is only effected by the latent heat of the fuel, weather conditions, and finished foam's natural drainage rate. Specific application rates can be found in Chapter 7, Class B Foam Fire Fighting: Class B Liquids at Fixed Sites. Also see Appendix C, Foam Properties.

Specific Foam Concentrates

[NFPA 472: 5.4.4(2); NFPA 1081: 5.3.2]

As stated earlier, numerous types of foam concentrates are selected for specific applications according to their properties and performance. Some foam concentrates are thick and viscous and form tough, heat-resistant finished foam blankets

Figure 2.20 Foam application rates are determined by the fuel type, fuel depth, foam type, and application method. Flammable liquid fires such as this simulated shipboard fire may require more than one attack line flowing foam.

over burning liquid surfaces; other foam concentrates are thinner and spread more rapidly in their finished forms. Some foam concentrates produce a vapor-sealing film of surface-active water solution on a liquid surface. Others such as medium- and high-expansion foam concentrates are used in large volumes for vapor suppression or to flood confined spaces. The sections that follow highlight each of the common foam concentrate types.

Regular Protein Foam

The use of regular protein foam concentrates started before World War II. These foams are virtually nonexistent in today's municipal, industrial, or military fire service. Although rare, regular protein foam concentrate may still be found in a fixed fire-suppression system. However, modern derivatives such as film forming fluoroprotein foam and aqueous film forming foam are in service worldwide. Regular protein foam concentrate is derived from naturally occurring sources of protein such as hoof, horn, or feather meal. The protein meal is hydrolyzed in the presence of lime and converted to a protein liquid. Other components such as foam stabilizers, corrosion inhibitors, antimicrobial agents, and freezing-point depressants are then added (**Figure 2.21**). Regular protein foam generally has very good heat stability, but it is not as mobile or fluid on the fuel surface as synthetic-based foam concentrates or modern fluoroprotein derivatives. Regular protein foam concentrate is very susceptible to fuel pickup; consequently, care should be taken to minimize submergence through plunging of the foam stream into the fuel. Regular protein foam concentrate conforms to the following characteristics:

- Available in 3- and 6-percent concentrations
- Excellent water-retention capabilities
- High heat resistance
- Performance can be affected by freezing and thawing
- Stores at temperatures ranging from 35 to 120°F (2°C to 49°C)
- Compounded for freeze protection using a nonflammable antifreeze solution
- Not compatible with dry-chemical extinguishing agents
- Only used on hydrocarbon fuels

Figure 2.21 Protein foam concentrates have been available since World War II and are available from numerous manufacturers. Containers are marked for proportioning ratio and the intended hazard type (in this case hydrocarbons). *Courtesy of Kidde Fire Fighting.*

Fluoroprotein Foam

Fluoroprotein foam concentrate, a combination protein-based and synthetic-based foam concentrate, is derived from protein foam concentrates to which fluorochemical surfactants are added. The fluorochemical surfactants are similar to those developed for AFFF concentrates (described later) but are used in much lower concentrations. The addition of these chemicals produces finished foam that flows across fuel surfaces rapidly. Because of these surfactants, fluoroprotein foam concentrates are oleophobic (oil shedding) and well suited for subsurface injection: a process by which foam solution is pumped into the bottom of a burning petroleum tank and then floats to the top to form a fire-extinguishing foam blanket (**Figure 2.22**). This process is discussed in more detail in Chapter 4, Foam Delivery Systems. Fluoroprotein foam concentrates exhibit the following characteristics:

- Available in 3- and 6-percent concentrations
- Stores at temperatures ranging from 35 to 120°F (2°C to 49°C); however, can be freeze-protected with nonflammable antifreeze solution

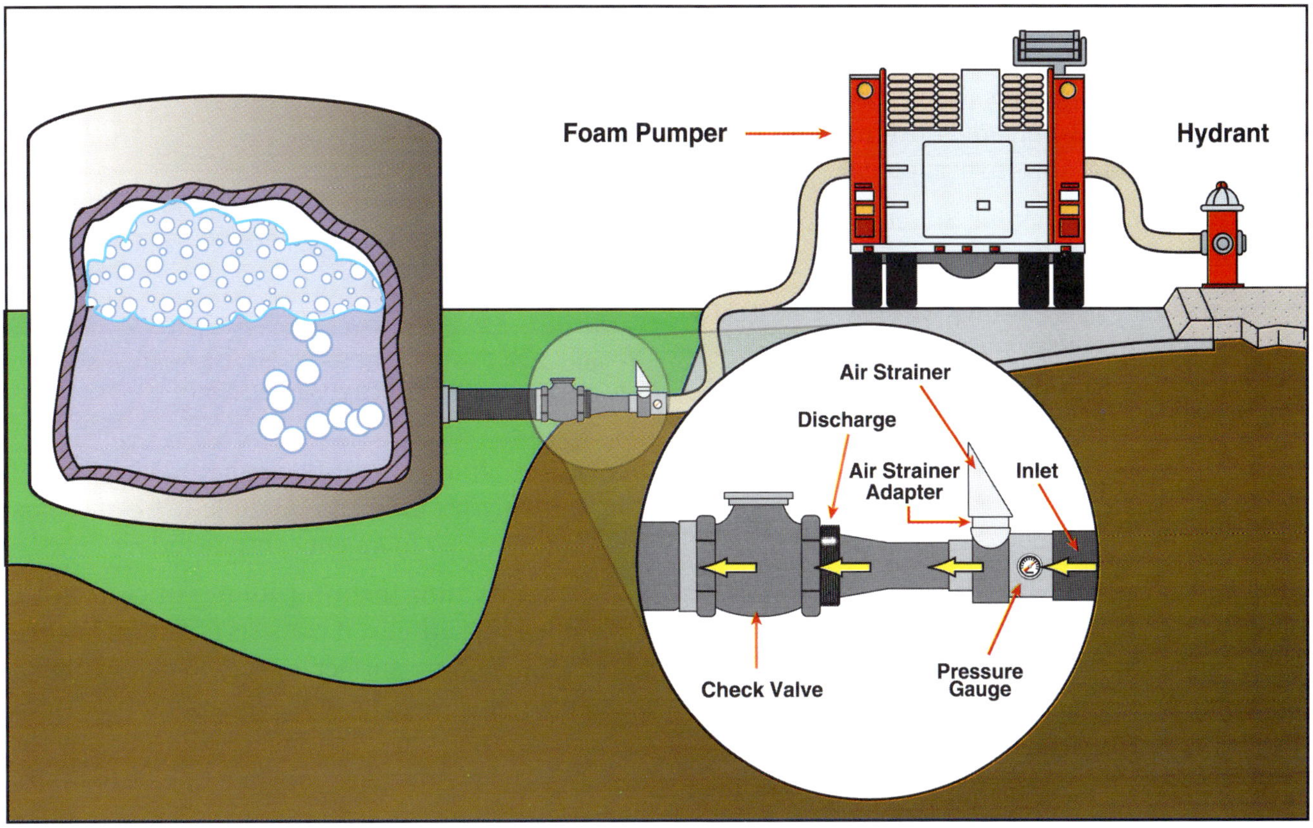

Figure 2.22 Fluoroprotein foam is very effective for subsurface injection into flammable liquids storage tanks. This illustration shows the method generally used for this application.

- Performance not affected by freezing and thawing
- Premixable for short periods of time (based on the manufacturer's recommendations)
- Maintains rather low viscosity at low temperatures
- Compatible with simultaneous application of dry-chemical extinguishing agents
- Delivered through air-aspirating equipment
- Suitable for use on gasoline that has been blended with oxygen additives

Fluoroprotein foam concentrates can be formulated to be alcohol resistant by adding ammonia salts that are suspended in organic solvents. Alcohol-resistant fluoroprotein foam concentrate maintains its alcohol-resistive properties for about 15 minutes. Alcohol-resistant fluoroprotein foam concentrates have a very high degree of heat resistance and water retention.

Film Forming Fluoroprotein Foam

Film forming fluoroprotein (FFFP) foam concentrate is based on fluoroprotein foam technology with aqueous film forming foam (AFFF) concentrate capabilities. This film forming fluoroprotein foam concentrate incorporates the benefits of AFFF concentrate for fast fire knockdown and the benefits of fluoroprotein foam concentrate for long-lasting heat resistance. FFFP conforms to the following characteristics:

- Available in 3- and 6-percent concentrations
- Stores at temperatures ranging from 35 to 120°F (2°C to 49°C) with fair low-temperature viscosity
- Stores premixed in portable fire extinguishers and fire apparatus water tanks (Refer to the manufacturer for long-term storage recommendations.)
- Compatible with simultaneous application of dry-chemical fire-fighting agents
- Performance not affected by freezing and thawing
- Use with either freshwater or saltwater

FFFP concentrate is available in an alcohol-resistant formulation. Alcohol-resistant FFFP concentrate has all the fire-fighting capabilities of regular FFFP concentrate, including some of the following advantages:

- ***Multipurpose*** — Can be used on polar solvent fuels at 6-percent and on hydrocarbon fuels at 3-percent concentrations (New concentrates that can be used at 3-percent concentrations on either type of fuel are also available.)
- ***Storage*** — Can be stored at temperatures ranging from 35 to 120°F (2°C to 49°C)
- ***Premixable*** — Can be mixed into a solution and stored ready for use
- ***Subsurface injection*** — Can be used for subsurface injection applications
- ***Plunge into fuel*** — Can be plunged into the fuel during application

Aqueous Film Forming Foam

With the introduction of high-expansion detergent-based foam, the U.S. Navy became interested in using that technology to improve their flammable liquid fire-fighting capabilities on aircraft carriers (**Figure 2.23**). The resulting research work was called *Project Light Water*. The 3M Company later trademarked the term *Light Water*™. The U.S. Navy discovered that when a fluorinated surfactant was added to detergent foam concentrates, the water that drained from the foam blanket actually floated on jet fuel spills. This film is known as an *aqueous film* (**Figure 2.24**). In order to identify the class of foam concentrate providing a film, the NFPA foam committee introduced the term aqueous film forming foam (AFFF) concentrate. AFFF (commonly pronounced "A-triple-F") is the most commonly used foam concentrate today.

Figure 2.23 Aqueous film forming foam (AFFF) is used by the United States Navy for controlling fires on ships. This flight deck fire on the *U.S.S. Forrestal* occurred in July 1967 while the ship was operating off the coast of Viet Nam. *Courtesy of United States Navy.*

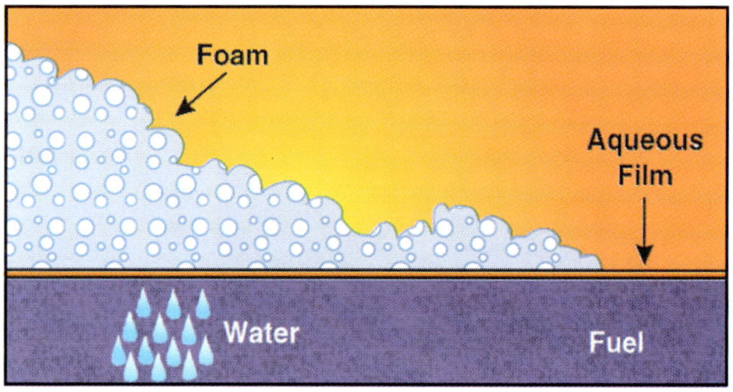

Figure 2.24 The film of AFFF floats on the surface of the fuel, spreading ahead of the foam blanket.

Aqueous film forming foam concentrate is completely synthetic. It consists of fluorochemical and hydrocarbon surfactants combined with high boiling point solvents and water. Fluorochemical surfactants reduce the surface tension of water to a degree less than the surface tension of the hydrocarbon so that a thin aqueous film can spread across the fuel. AFFF foams have the following characteristics:

- Available in 1-, 3-, and 6-percent concentrations for use with either freshwater or saltwater
- Premixable in portable fire extinguishers and apparatus water tanks (**Figure 2.25**)
- Stores at temperatures ranging from 25 to 120°F (-5°C to 49°C) (AFFF concentrates are not adversely affected by freezing and thawing, but consult the manufacturer for details.)
- Freeze-protective with a nonflammable antifreeze solution
- Good low-temperature viscosity
- Suitable for subsurface injection
- Fair penetrating capabilities in baled storage fuels or high surface-tension fuels such as treated wood
- Compatible with dry-chemical extinguishing agents
- Rather fast draining (Reapply AFFF finished foam often to maintain hot-spill security.)
- Film-forming characteristics adversely affected by fuels in excess of 140°F (60°C)

Figure 2.25 Premixed AFFF can be stored in and applied from portable multiagent extinguishing units such as this small towed cart equipped with a hose reel.

- Oxygen additives (mandated by the U.S. EPA) in blended gasoline can adversely affect film-forming characteristics; solvent-based performance additives in reformulated gasoline can also hamper AFFF's performance

When AFFF (as well as the previously mentioned FFFP) finished foam is applied to a hydrocarbon fire, the following three things occur:

- An air/vapor-excluding film is released ahead of the foam blanket.
- The fast-moving foam blanket then moves across the surface and around objects, adding further insulation.
- As the aerated (7:1 to 20:1) foam blanket continues to drain its water, more film is released. This gives AFFF finished foam the ability to "heal" over areas where the foam blanket is disturbed.

Alcohol-Resistant AFFF

Another class of AFFF concentrates is composed of alcohol-resistant concentrates (AR-AFFF). They are available from most foam manufacturers. On most polar solvents, alcohol-resistant AFFF concentrates are used at 3- or 6-percent (3:97 or 6:94 water/concentrate ratios) use concentrations, depending on the particular brand selected. Stronger polar solvents require application rates that are higher than those required for weaker solvents or hydrocarbons. Alcohol-resistant AFFF solutions can also be used on hydrocarbon fires at 3-percent (3:97 ratio) proportions. Concentrates designed to be proportioned at 3 percent on hydrocarbon fuels and 6 percent (6:94 ratio) on polar solvent fuels are commonly referred to as 3 by 6 concentrates. Concentrates proportioned at 3 percent on both types of fuels are called 3 by 3 concentrates. Recently developed AR-AFFF concentrates are available in 1-by-3-percent formulations. Use at 1 percent (1:99) for hydrocarbon spills (1 inch [25 mm] or less in depth) and 3 percent (3:97) for hydrocarbons in depth (more than 1 inch [25 mm]) or polar solvents.

Figure 2.26 Never plunge AR-AFFF into the surface of the fuel. Instead, spray it over the top of the fuel in a rain-down method.

When alcohol-resistant AFFF concentrates are applied to polar solvent fuels, they create a membrane rather than a film over the fuel. This membrane separates the water in the foam blanket from the attack of the solvent. Apply alcohol-resistant AFFF finished foam gently to the fuel so that the membrane can form first. Do not plunge alcohol-resistant AFFF finished foam into the polar fuel (**Figure 2.26**).

Alcohol-resistant AFFF foam may be used in subsurface injection applications on certain light hydrocarbons such as gasoline, kerosene, and jet propulsion fuels. AR-AFFF concentrate is not designed for premix applications.

Vapor-Mitigating Foam

Technological advances in the area of hazardous-material foam concentrates have recently produced AR-AFFF foam concentrates that can help stop acidic or alkaline reactions. Vapor-mitigating foam concentrates are designed solely for use on unignited spills of hazardous liquids. Some are not effective in fire-fighting situations. See Chapter 6, Class B Foam Use: Unignited Class B Liquid Spills, for operational uses of vapor-mitigating foam concentrates.

High-Expansion Foam

High-expansion foam concentrates are special-purpose foam concentrates that are similar to Class A foams. Because they have a low-water content, they minimize water damage. Their low-water content is also useful when runoff is undesirable. The use of high-expansion finished foam outside is generally not recommended because the slightest breeze may remove the foam blanket in sheets and re-expose the hazard to stray ignition sources. High-expansion foam concentrates have three basic applications:

- Concealed spaces such as shipboard compartments, basements, coal mines, and other subterranean spaces **(Figure 2.27)**
- Fixed-extinguishing systems for specific industrial uses such as rolled or bulk paper storage
- Class A fire applications (slow draining) **(Figure 2.28)**

High-expansion foam concentrates have the following characteristics:

- Stores at temperatures ranging from 35 to 120°F (2°C to 49°C)
- Not affected by freezing and thawing
- Poor heat resistance because the air-to-water ratio is very high
- Expansion ratios of 200:1 to 1,000:1 for high-expansion uses and 20:1 to 200:1 for medium-expansion uses (Whether the finished foam is used in either a medium- or high-expansion capacity is determined by the type of application device used.)

Figure 2.28 Class A foam has the advantage of slow draining, which allows the foam to penetrate thoroughly into the material to which it has been applied. *Courtesy of Mark Stuckey.*

Figure 2.27 High-expansion foam is excellent for fires in confined spaces such as basements, vaults, and ship's engine rooms.

Emulsifiers

Emulsifiers are a type of foam concentrate that are intended for use with either Class A or Class B fires. Unlike finished foam that blankets the fuel, the emulsifier is designed to mix with the fuel, breaking it into small droplets and encapsulating them. The resulting emulsion is rendered nonflammable (**Figure 2.29**).

A number of drawbacks exist to the use of emulsifiers, however. First, they should only be used with fuels that are 1 inch (25 m) or less in depth. If the fuel is deeper, it is almost impossible to mix the emulsifier thoroughly with the fuel. Second, once the emulsifier is thoroughly mixed with the fuel, it renders the fuel unsalvageable. Third, emulsifiers do not work effectively with water-soluble or water-miscible fuels because an emulsion cannot be formed between the concentrate and the fuel. Finally, emulsifiers can have a negative effect on fish, aquatic life, and bodies of water. When used with Class A fires and Class B spills, the effects of run-off must be taken into consideration.

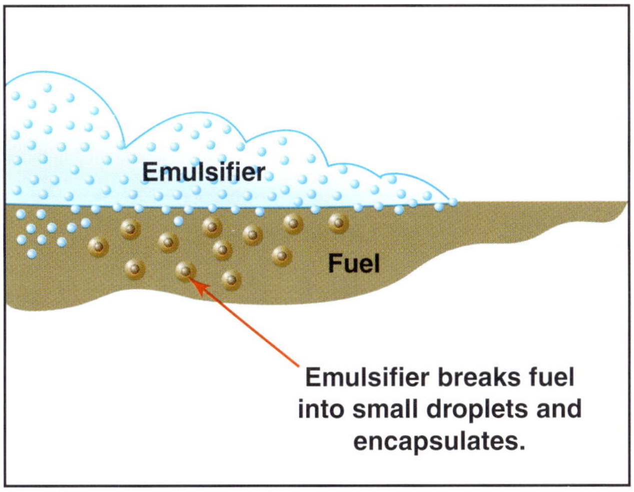

Figure 2.29 Illustration of how emulsifiers work to prevent ignition and stop combustion: They mix with the fuel, making it nonflammable.

Foam Proportioning Methods

[NFPA 1001: 6.3.1; NFPA 1081: 7.3.4]

Proportioning is the mixing of water with foam concentrate to form a foam solution (**Figure 2.30**). Fire-fighting foam concentrates are mixed with 94 to 99.9 percent water. For example, a 3-percent foam concentrate is 97 parts water mixed with 3 parts foam concentrate, which equals 100 parts of foam solution (**Figure 2.31**). For a 6-percent foam concentrate, 94 parts water is mixed with 6 parts foam concentrate to equal 100 parts of foam solution. Either freshwater or saltwater may be used, although impurities within the water source may have a negative effect on the finished foam and clog the distribution system.

For maximum effectiveness, foam concentrates must be proportioned at the specific percentages for which they are designed. The percentage rate for the intended fuel is clearly marked on the outside of every foam concentrate container. Failure to follow this procedure such as trying to mix 6-percent

Figure 2.30 Proportioning takes place when water is mixed with foam concentrate to create a foam solution. This firefighter is monitoring a portable foam proportioner as it draws foam concentrate from a container and mixes it with water in the hoseline.

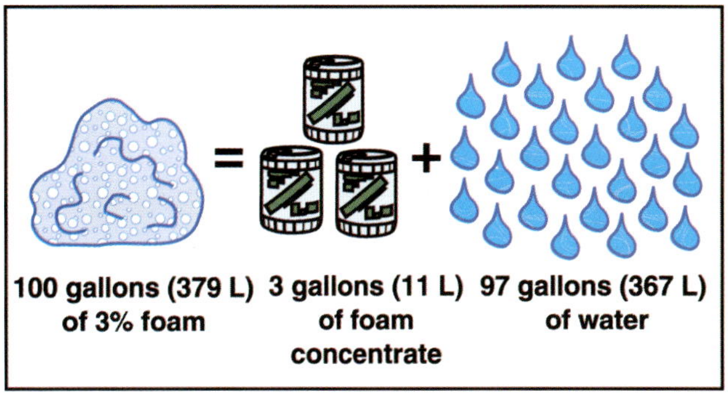

Figure 2.31 Foam solution generated by using a 3-percent foam concentrate consists of 3 parts foam concentrate to 97 parts water.

foam at a 3-percent foam concentrate rate results in wet, poor-quality foam that may not perform as desired. However, under certain circumstances, 3-percent foam concentrates can be mixed at 6 percent (known as *doubling up*), which results in a slow-spreading and slow-draining foam blanket. Doubling the proportioning rate is a good way to slow foam drainage when dealing with postfire fuel remains or unignited liquid fuel spills. Slow-draining foam blankets are desirable when long-term vapor suppression is desired.

Class A foam concentrates are subject to variables in proportioning ratios. The proportioning percentage for Class A foam concentrates can be adjusted (within limits recommended by the manufacturer) to achieve specific objectives. To produce dry (thick) finished foam suitable for exposure protection and firebreaks, the foam concentrate can be adjusted to a higher percentage. To produce wet, fast-soaking foam that rapidly sinks into a fuel's surface, the foam-concentrate proportioning ratio can be adjusted to a lower percentage.

The selection of a proportioner depends on the foam solution flow requirements, available water pressure, cost, type of delivery system (apparatus, fixed, or portable), and the foam concentrate. Proportioners and delivery devices (foam nozzle, foam proportioner, CAFS, etc.) are engineered to work together **(Figure 2.32)**. Using a foam proportioner that is not compatible with the delivery device (even if the two are made by the same manufacturer) can result in unsatisfactory finished foam or none at all. A variety of equipment is used to proportion foam. Some types are designed for mobile apparatus and others are designed for fixed fire-suppression systems. Each of the common types of foam proportioners is covered in detail in Chapter 3, Foam Proportioning and Delivery Equipment. Four basic methods by which foam may be proportioned are listed and discussed in the sections that follow:

- ***Eduction (induction)*** — Uses a foam eductor, inductor, line proportioner, or ratio controller
- ***Injection*** — Uses a foam-concentrate pump or bladder proportioner to pump foam concentrate at the appropriate ratio into a water stream
- ***Batch mixing*** — Uses a "dump and pump" or one-time-use process
- ***Premixing*** — Stores in a solution (such as in a fire extinguisher) ready to use

Eduction (Induction)

The eduction (induction) method of proportioning foam uses the pressure energy in the stream of water to induct (draft) foam concentrate into the fire stream. Drafting is achieved by passing the stream of water through a device called an *eductor* that has a restricted diameter **(Figure 2.33)**. Within the restricted area is a separate orifice that is attached via a hose to the foam concentrate container. The pressure differential created by the water going through

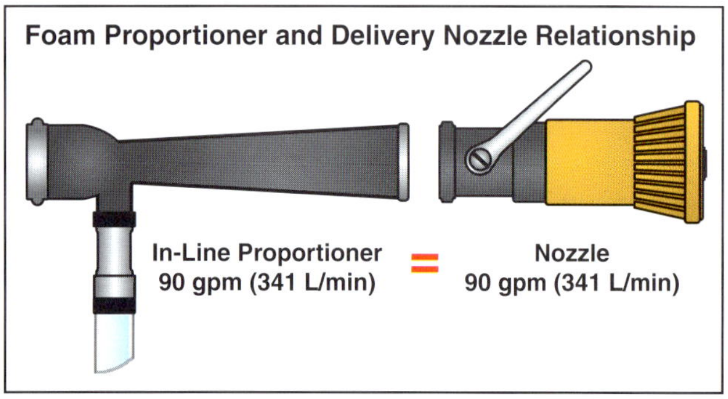

Figure 2.32 Relationship between the foam proportioner and the delivery nozzle: In-line proportioners must have the same flow rate as the nozzles used with them.

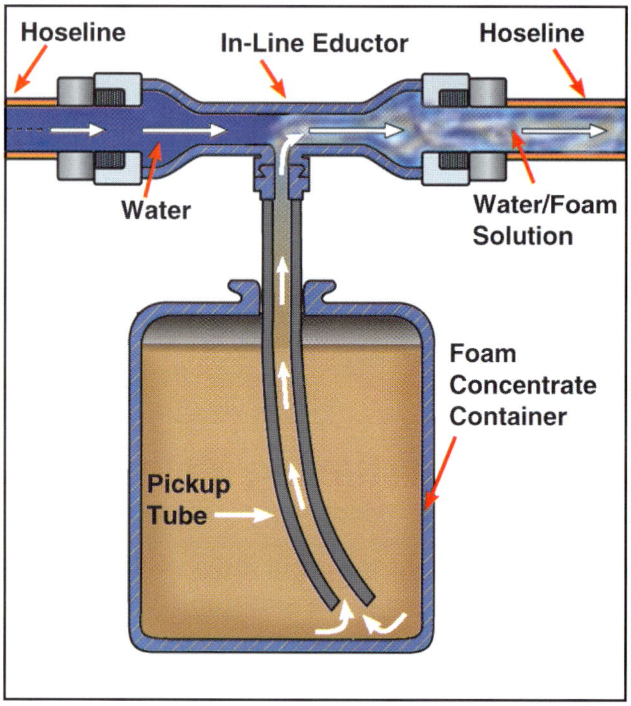

Figure 2.33 Foam concentrate is drawn into the eductor using the Venturi principle.

the restricted area and over the orifice creates a suction that draws the foam concentrate into the fire stream (Venturi process). In-line eductors and foam nozzle eductors are examples of foam proportioners that work by this method. These devices are covered in more detail in Chapter 3, Foam Proportioning and Delivery Equipment.

Injection

The injection method of proportioning foam uses an external pump or head pressure to force foam concentrate into the fire stream at the correct ratio for the flow desired. These systems are commonly employed in apparatus-mounted or fixed fire-suppression system applications. These proportioners are discussed in Chapter 3, Foam Proportioning and Delivery Equipment (**Figure 2.34**).

Batch Mixing

Batch mixing (the simplest method of mixing foam concentrate and water) is commonly used to mix foam concentrates within a fire apparatus water tank or a portable water tank (**Figure 2.35**). It also allows for accurate proportioning of foam concentrates. Batch mixing is commonly practiced with Class A foam concentrates and can be done with all concentrates except alcohol-resistant AFFF. Batch mixing is also commonly referred to as the *dump-and-pump method.*

CAUTION

Because dissimilar types of foam congeal when mixed together, only mix foam concentrates from similar families together. Drain, flush, and refill tanks and proportioners before adding a different type foam concentrate to the water.

Premixing

The premixing method (one of the more commonly used methods of proportioning) mixes premeasured portions of water and foam concentrate in a container. Typically, the premix method is used with portable fire extinguishers, wheeled fire extinguishers, skid-mounted multiagent extinguisher units, and vehicle-mounted tank

Figure 2.34 Direct injection systems are usually mounted on the pump on fire apparatus. This photo shows a compressed air foam system (CAFS), shown in yellow, mounted on top of the pump.

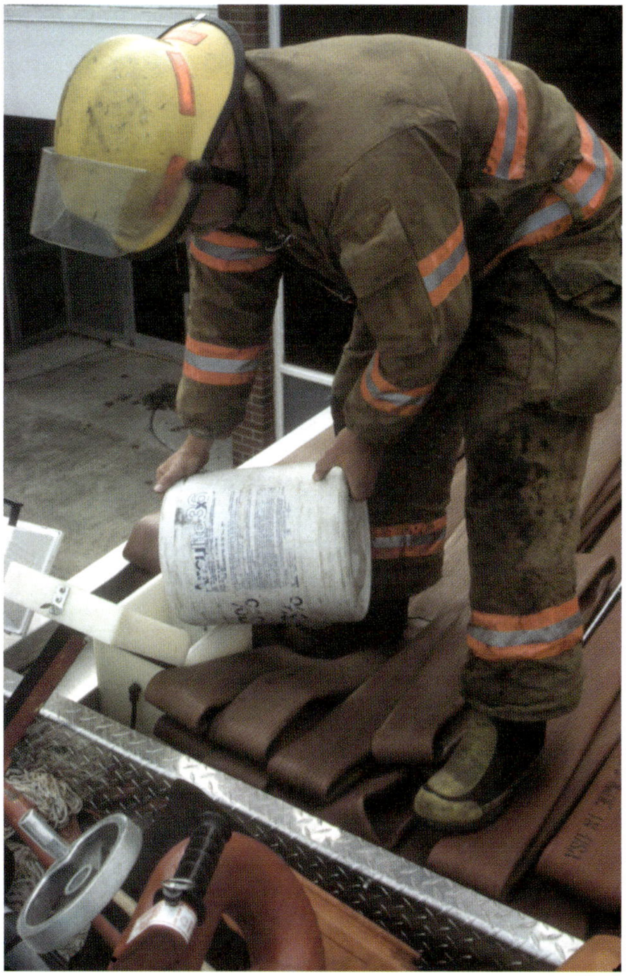

Figure 2.35 Batch mixing foam in the apparatus water tank requires that the entire tank load be used or that the unused portion be disposed of after the incident. The tank, piping, and pump also have to be thoroughly flushed following the incident.

extinguishing systems **(Figures 2.36)**. In most cases, premixed solutions are discharged from a pressure-rated tank using either a compressed inert gas or air. An alternative method of discharge uses a pump and a nonpressure-rated atmospheric storage tank. The pump discharges the foam solution through piping or hose to the discharge devices. Premix systems are limited to a one-time application. When used, the containers must be completely emptied and refilled before they can be used again.

Foam Concentrate Storage

[NFPA 1003: 3-3.9]

Foam concentrate is stored in a variety of containers; the type used in any situation depends on how the foam concentrate is delivered. The most common foam concentrate storage methods are as follows:

- Pails (5 gallons [19 L])
- Barrels (55-gallon [209 L] drums)
- Intermediate bulk containers (tote containers commonly 250 to 450 gallons [950 L to 1 710 L] in size)
- Foam tenders (large-capacity tank trucks)
- Apparatus tanks
- Fixed fire-suppression system tanks (industrial applications and shipboard storage)
- On-site storage tanks

Regardless of the storage method, the storage container must be constructed of a material that is chemically compatible to the concentrate. Failure to use a receptacle made of a concentrate-compatible material may result in a container/tank that eventually fails, either because of container corrosion or foam concentrate contamination. The manufacturers of foam concentrates provide information on the materials that are compatible with their concentrates. Poly tanks, fiberglass tanks, and stainless steel tanks are generally "foam friendly." Do not use mild steel, galvanized steel, or aluminum for foam tank construction or in foam concentrate plumbing. Foam concentrate tanks must not be left open to the atmosphere; keep tanks sealed. Use pressure vacuum vents as required in NFPA 11 and NFPA 1901, *Standard for Automotive Fire Apparatus* (1999).

Figure 2.36 Refineries and fixed-site facilities may have premixed extinguishing units (like this one) that slide rapidly into the beds of pickup trucks. *Courtesy of Conoco, Inc.*

Pails

Five-gallon (19 L) plastic pails are perhaps the most common containers used by fire and emergency services to ship and store foam concentrate **(Figure 2.37)**. Pails tested and certified by Underwriters Laboratories Inc. (UL-listed) and bearing its label are very durable. They are designed to pass several 4-foot (1.2 m) drop tests with their contents frozen. Pails may be carried on the apparatus in compartments, side storage areas, or topside storage areas. In-line proportioners may be inserted into the top of the pail and the foam concentrate educted di-

Figure 2.37 Foam concentrates are normally purchased in 5-gallon (19 L) containers. *Courtesy of Conoco, Inc.*

rectly from the pail. The containers must be airtight to prevent the concentrate from evaporating. AR-AFFF concentrates are particularly prone to evaporation.

Barrels

If small pails are not convenient storage containers, foam concentrate may be shipped and stored in 55-gallon (209 L) plastic or plastic-lined barrels (drums) **(Figure 2.38)**. Foam concentrate weighs slightly more than water. A general guideline for the weight of foam concentrate is to calculate about 9 pounds (4.05 kg) per gallon (3.8 L), which includes the weight of the container. Therefore, a 55-gallon (209 L) drum of foam concentrate weighs 495 pounds (223 kg) (9 × 55 = 495 pounds [223 kg]). Foam concentrate stored in drums listed by Underwriters Laboratories Inc. will withstand falls from 5 feet (1.5 m) with their contents frozen. This fact is important when foam concentrate deliveries must be dropped by the truckload at an emergency scene. Foam concentrate can then be transferred from the barrels to pails or apparatus tanks using specially designed transfer pumps. Some emergency services organizations have apparatus that are designed to carry these barrels directly to an emergency scene for deployment. Foam concentrate is educted or pumped directly from barrels in the same manner that it is educted from pails.

Intermediate Bulk Containers

Intermediate bulk containers (sometimes referred to as *IBC totes*) are used in much the same way as barrels **(Figure 2.39)**. The primary differences in the two containers are their sizes and the manner in which they are handled. Intermediate bulk containers range in capacity from 250 to 450 gallons (950 L to 1 710 L), depending on the manufacturer. Handling is provided by heavy lifting equipment such as forklifts. These containers are designed to fit comfortably onto a forklift pallet. Some airport and hazardous materials response organizations locate these bulk containers on trailers at their facilities to increase the amount of foam concentrate available for refilling apparatus quickly. Portable transfer pumps are used in the refilling operation.

Figure 2.38 Plastic 55-gallon (209 L) drums are available and can be used for refilling apparatus foam tanks. *Courtesy of Conoco, Inc.*

Figure 2.39 The largest portable foam concentrate containers are the 250-gallon (950 L) intermediate bulk containers (IBCs).

Foam Tenders

Foam tenders or tankers are specialty apparatus that are similar in design to large-capacity water tenders found in rural fire departments. Foam tenders and some industrial foam pumpers have large foam-concentrate tanks and no water tanks **(Figure 2.40)**. The vehicles may be built on a straight frame with tandem rear axles or as tractor/trailer combinations. Tank capacities can range from 1,500 gallons (5 678 L) to as much as 8,000 gallons (30 283 L). Units may be equipped with transfer pumps to connect them to pumper apparatus.

Figure 2.40 Some industrial fire brigades have foam concentrate tankers (tenders) available for large-scale foam operations. *Courtesy of Shell Oil Co., Deer Park, Texas.*

Some fire and emergency services organizations have modified old pumpers by replacing the steel water tank with a poly, fiberglass, or stainless steel foam tank. These apparatus can then be placed in service as foam tenders with large-caliber, self-inducting foam cannons or simple eductor systems and foam concentrate transfer pumps.

Apparatus Tanks

Fire apparatus equipped with integral, onboard foam-proportioning systems usually have foam concentrate tanks piped directly to the foam delivery system. This system eliminates the need to use separate pails or barrels. Foam concentrate tanks are found on municipal and industrial fire and emergency services pumpers, foam tenders, and ARFF apparatus. Foam concentrate tanks on fire and emergency services pumpers range from 20 to 200 gallons (80 L to 800 L) **(Figure 2.41)**. ARFF apparatus may carry concentrate in excess of 600 gallons (2 400 L) **(Figure 2.42)**. Apparatus foam tanks must be fitted with airtight tank lids and a vacuum/pressure vent to prevent evaporation **(Figure 2.43)**. Standard vented atmospheric storage tanks are not acceptable for foam concentrate storage because they expose the concentrate in the filler tank to the atmosphere. Long exposure to air causes foam concentrate to dry, which is a particular concern in the storage of AR-AFFF concentrates. This foam has water rather than a dispersing solvent in

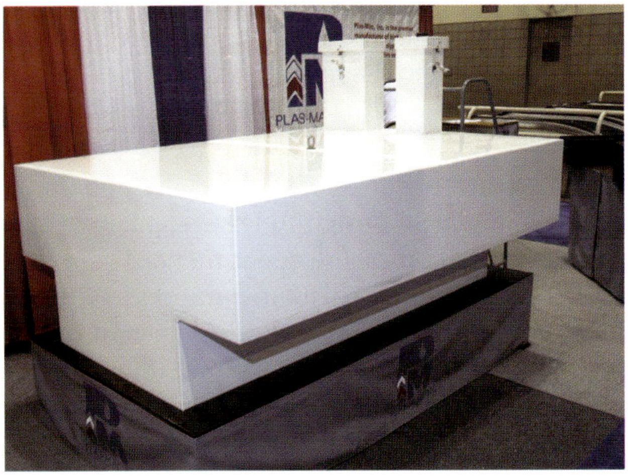

Figure 2.41 Apparatus foam tanks are generally constructed from polypropylene plastic. They are available in a wide variety of capacities.

Figure 2.42 Aircraft rescue and fire fighting apparatus carry large quantities of foam and water and are operated by one person in the cab of the vehicle.

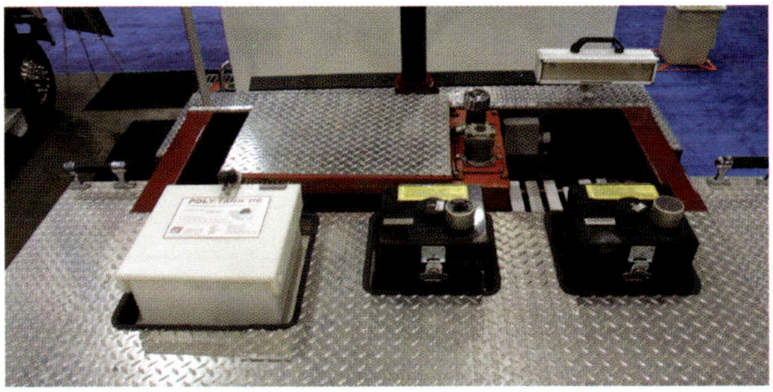

Figure 2.43 Foam tanks are equipped with vacuum/pressure vents on the airtight filler lids to prevent evaporation of the foam concentrate. This photograph shows an apparatus equipped with two foam tanks (right) as well as the water tank (left).

its formulation and is prone to evaporation. Over long time periods (months to years), foam concentrate thickens and becomes difficult, if not impossible, to proportion.

NOTE: Keep mobile apparatus foam tanks full. Failure to do so causes sloshing that will aerate foam concentrate in the foam tank. Aerated foam concentrate will proportion lean. Lean foam solutions can be ineffective or totally useless. AR-AFFF concentrates are of particular concern because they may take weeks to settle after being agitated or improperly pumped during transfer operations.

The type, location, and design of foam concentrate tanks vary depending on the apparatus design. Small foam concentrate tanks are located directly above the fire pump area. New designs incorporate foam concentrate tanks as integral cells within apparatus water tanks. Large foam concentrate tanks such as those on ARFF apparatus may be directly adjacent to the apparatus water tank. Many fire and emergency services organizations that do not use foam concentrates on a regular basis but have the potential to need a large quantity of concentrate have special foam concentrate trailers that can be pulled behind other apparatus when needed **(Figure 2.44)**. NFPA standards provide requirements for foam concentrate storage-tank design. Consult foam concentrate manufacturers for specific storage-tank requirements.

It is advisable to have the ability to fill and drain foam concentrate tanks from ground level rather than from the top of the apparatus. An auxiliary foam pipe with a shutoff valve running from the concentrate tank to the apparatus running board or pump panel can serve as a supply source for portable foam eductors fastened to discharge outlets. This pipe can also drain the tank when maintenance is necessary. Fill foam concentrate tanks from ground level with hand pumps. Pouring foam concentrate into tanks from 5-gallon (19 L) pails is very labor-intensive and dangerous. Inadvertently spilling slippery foam concentrate on boots or the supply hose bed can cause severe slipping and falling hazards. The exception to this advice is the use of overhead fill pipes that are part of an on-site storage tank system (see On-Site Storage Tanks section for more information).

Fixed Fire-Suppression System Tanks

Industrial and shipboard facilities equipped with fixed fire-suppression foam systems have large concentrate tanks that supply these systems. These tanks are usually located in the vicinity of the foam proportioning equipment **(Figure 2.45)**. The sizes of foam concentrate tanks vary, depending on several variables, although tanks in excess of 3,000-gallon (11 400 L) capacity are common. Storage tank-size variables include the following:

- Size or capacity of the foam fire-suppression system
- Hazard being protected
- Discharge duration required for extinguishment

Fire and emergency services organizations should become familiar with fixed fire-suppression systems within their area of responsibility. Proper knowledge of these systems and the appropriate methods of supporting them can be critical to a successful fire-suppression operation.

Figure 2.44 Foam trailers can be used to transport additional foam to an incident. *Courtesy of Fire Wagons, Inc., Stillwater, OK.*

Figure 2.45 Refineries and chemical processing plants may be equipped with bulk foam concentrates located near foam system pumping stations such as this one. *Courtesy of Conoco, Inc.*

On-Site Storage Tanks

Storage tanks for large quantities of foam concentrate may also be found at airport and hazardous materials response facilities. These tanks are designed to hold a specific quantity of foam concentrate, depending on the organization's needs. They usually have permanent transfer pumps and overhead piping that permit the refilling of apparatus tanks from above. This type of system takes the place of ground-level filling. The overhead fill hose is placed into the fill hatch on the top of the apparatus, and the filling is monitored to prevent spillage or overfilling.

Summary

Foam fire-fighting concentrates and delivery systems have been available since the early 1900s. They have provided fire and emergency services personnel with methods of controlling flammable liquid spills and fires, hazardous chemical incidents, maritime and aviation incidents, and transportation incidents. In recent years, advancements in the development of foam concentrates for use with Class A fires have given fire and emergency services personnel a new tool to effectively and efficiently combat structural and wildland fires. These advances have not come without costs. Environmental issues have caused some fire and emergency services organizations to rethink the use of such concentrates that may damage sensitive ecosystems. Fire and emergency services personnel must continue to work with manufacturers and standards-writing organizations to create the best foam fire-fighting concentrates possible.

Fire and emergency services responders who are required to use foam fire-fighting concentrates and equipment must have a basic understanding of mechanical foam components, how they work, and their characteristics, proportioning methods, and storage limitations. From this basic knowledge, a full and complete understanding of foam concentrates and their uses can be developed. The chapters that follow expand on the information in this chapter to provide a detailed understanding.

Chapter 3

Foam Proportioning and Delivery Equipment

Job Performance Requirements

This chapter provides information that will assist the reader in meeting the following job performance requirements from NFPA 1001, *Standard for Fire Fighter Professional Qualifications*, 2002 edition; NFPA 1002, *Standard for Fire Apparatus Driver/Operator Professional Qualifications,* 1998 edition; NFPA 1021, *Standard for Fire Officer Professional Qualifications,* 1997 edition; NFPA 1081, *Standard for Industrial Fire Brigade Member Professional Qualifications,* 2001 edition; NFPA 1003, *Standard for Airport Fire Fighter Professional Qualifications,* 2000 edition; NFPA 472, *Standard for Professional Competence of Responders to Hazardous Materials Incidents,* 2002 edition; and NFPA 1051, *Standard for Wildland Fire Fighter Professional Qualifications,* 2002 edition. Colored portions of the standard are specifically addressed in this chapter.

NFPA 1001

6.3.1 Extinguish an ignitable liquid fire, operating as a member of a team, given an assignment, an attack line, personal protective equipment, a foam proportioning device, a nozzle, foam concentrates, and a water supply, so that the correct type of foam concentrate is selected for the given fuel and conditions, a properly proportioned foam stream is applied to the surface of the fuel to create and maintain a foam blanket, fire is extinguished, reignition is prevented, team protection is maintained with a foam stream, and the hazard is faced until retreat to safe haven is reached.

(A) *Requisite Knowledge:* Methods by which foam prevents or controls a hazard; **principles by which foam is generated;** causes for poor foam generation and corrective measures; difference between hydrocarbon and polar solvent fuels and the concentrates that work on each; the characteristics, uses, and limitations of fire-fighting foams; **the advantages and disadvantages of using fog nozzles versus foam nozzles for foam application;** foam stream application techniques; hazards associated with foam usage; and methods to reduce or avoid hazards.

(B) *Requisite Skills:* The ability to prepare a foam concentrate supply for use, assemble foam stream components, master various foam application techniques, and approach and retreat from spills as part of a coordinated team.

NFPA 1002

2-3.7 Operate all fixed systems and equipment on the vehicle not specifically addressed elsewhere in this standard, given systems and equipment, manufacturer's specifications and instructions, and departmental policies and procedures for the systems and equipment, so that each system or piece of equipment is operated in accordance with the applicable instructions and policies.

(a) *Requisite Knowledge:* Manufacturer specifications and **operating procedures**, policies, and procedures of the jurisdiction.

(b) *Requisite Skills:* The ability to deploy, energize, and monitor the system or equipment and to recognize and correct system problems.

3-1.1 Perform the specified routine tests, inspections, and servicing functions specified in the following list in addition to those contained in the list in 2-2.1, given a fire department pumper and its manufacturer's specifications, so that the operational status of the pumper is verified.

- Water tank and other extinguishing agent levels (if applicable)
- Pumping systems
- **Foam systems**

(a) *Requisite Knowledge:* Manufacturer specifications and requirements, policies, and procedures of the jurisdiction.

(b) *Requisite Skills:* The ability to use hand tools, recognize system problems, and correct any deficiency noted according to policies and procedures.

3-2.3 Produce a foam fire stream, given foam-producing equipment, so that properly proportioned foam is provided.

(a) *Requisite Knowledge:* **Proportioning rates and concentrations, equipment assembly procedures, foam systems limitations,** and manufacturer specifications.

(b) *Requisite Skills:* The ability to operate foam proportioning equipment and connect foam stream equipment.

6-1.1 Perform the specified routine tests, inspections, and servicing functions specified in the following list, in addition to those contained in 2-2.1, given a wildland fire apparatus and its manufacturer's specifications, so that the operational status is verified.

- Water tank and/or other extinguishing agent levels (if applicable)
- Pumping systems
- **Foam systems**

(a) *Requisite Knowledge:* Manufacturer specifications and requirements, policies, and procedures of the jurisdiction.

(b) *Requisite Skills:* The ability to use hand tools, recognize system problems, and correct any deficiency noted according to policies and procedures.

NFPA 1021

2-6.1 Develop a preincident plan, given an assigned facility and preplanning policies, procedures, and forms, so that all required elements are identified and the appropriate forms are completed and processed in accordance with policies and procedures.

(a) *Prerequisite Knowledge:* Elements of a preincident plan, basic building construction, **basic fire protection systems and features**, basic water supply, basic fuel loading, and fire growth and development.

(b) *Prerequisite Skills:* The ability to write reports, to communicate verbally, and to evaluate skills.

NFPA 1081

5.3.1 Attack an incipient stage fire, given a hose line flowing up to 473 L/min (125 gpm), appropriate equipment, and a fire situation, so that the fire is approached safely, exposures are protected, the spread of fire is stopped, agent application is effective, the fire is extinguished, and the area of origin and fire cause evidence are preserved.

(A) Requisite Knowledge. **Types of handlines used for attacking incipient fires**, precautions to be followed when advancing hose lines to a fire, observable results that a fire stream has been properly applied, dangerous building conditions created by fire, principles of exposure protection, and dangers such as exposure to products of combustion resulting from fire condition.

(B) Requisite Skills. The ability to recognize inherent hazards related to the material's configuration; operate handlines; prevent water hammers when shutting down nozzles; open, close, and adjust nozzle flow; advance charged and uncharged hose; extend hose lines; operate hose lines; evaluate and modify water application for maximum penetration; assess patterns for origin determination; and evaluate for complete extinguishment.

5.3.2 Activate a fixed fire protection system, given a fixed fire protection system, a procedure, and an assignment, so that the steps are followed and the system operates.

(A) Requisite Knowledge. **Types of extinguishing agents, hazards associated with system operation, how the system operates**, sequence of operation, system overrides and manual intervention procedures, and shutdown procedures to prevent damage to the operated system or to those systems associated with the operated system.

(B) Requisite Skills. The ability to operate fixed fire protection systems via electrical or mechanical means.

5.3.3 Utilize master stream appliances, given an assignment, an extinguishing agent, and a master stream device, so that the agent is applied to the fire as assigned.

(A) Requisite Knowledge. **Safe operation of master stream appliances, uses for master stream appliances, tactics using fixed master stream appliances**, and property conservation.

(B) Requisite Skills. The ability to put into service a fixed master stream appliance, and to evaluate and forecast a fire's growth and development.

6.2.3 Operating as a member of a team, attack an exterior fire, given a water source, an attack line, personal protective equipment, tools, and an assignment, so that team integrity is maintained, the attack line is properly deployed for advancement, access is gained into the fire area, appropriate application practices are used, the fire is approached safely, attack techniques facilitate suppression given the level of the fire, hidden fires are located and controlled, the correct body posture is maintained, hazards are avoided or managed, and the fire is brought under control.

(A) Requisite Knowledge. Principles of fire streams; **types, design, operation, nozzle pressure effects, and flow capabilities of nozzles**; precautions to be followed when advancing hose lines to a fire; observable results that a fire stream has been correctly applied; dangerous conditions created by fire; principles of exposure protection; potential long-term consequences of exposure to products of combustion; physical states of matter in which fuels are found; and the application of each size and type of attack line, the role of the backup team in fire attack situations, attack and control techniques, and exposing hidden fires.

(B) Requisite Skills. The ability to prevent water hammers when shutting down nozzles; open, close, and adjust nozzle flow and patterns; apply water using direct, indirect, and combination attacks; advance charged and uncharged 38 mm (1½ in.) diameter or larger hose lines; extend hose lines; replace burst hose sections; operate charged hose lines of 38 mm (1½ in.) diameter or larger; couple and uncouple various handline connections; carry hose; attack fires; and locate and suppress hidden fires.

6.3.3 Utilize master stream appliances, given an assignment, an extinguishing agent, and a master stream device and supply hose, so that the appliance is set up correctly and the agent is applied as assigned.

(A) Requisite Knowledge. Correct operation of master stream appliances, **uses for master stream appliances, tactics using master stream appliances, selection of the master stream appliance for different fire situations,** the effect of master stream appliances on search and rescue, ventilation procedures, and property conservation.

(B) Requisite Skills. The ability to correctly put in service a master stream appliance and evaluate and forecast a fire's growth and development.

7.3.3 Utilize master stream appliances, given an assignment, an extinguishing agent, a master stream device, and supply hose, so that the appliance is set up correctly and the agent is applied as assigned.

(A) Requisite Knowledge. **Correct operation of master stream appliances; uses for master stream appliances; tactics using master stream appliances; selection of the master stream appliance for different fire situations;** the effect of master stream appliances on search and rescue, ventilation procedures, and property conservation.

(B) Requisite Skills. The ability to correctly put in service a master stream appliance and to evaluate and forecast a fire's growth and development.

7.3.4 Operating as a member of a team, extinguish an ignitable liquid fire, given an assignment, an attack line, personal protective equipment, a foam proportioning device, a nozzle, foam concentrates, and a water supply, so that the correct type of foam concentrate is selected for the given fuel and conditions, a correctly proportioned foam stream is applied to the surface of the fuel to create and maintain a foam blanket, fire is extinguished, re-ignition is prevented, and team protection is maintained.

(A) Requisite Knowledge. Methods by which foam prevents or controls a hazard; **principles by which foam is generated**; causes for poor foam generation and corrective measures; difference between hydrocarbon and polar solvent fuels and the concentrates that work on each; the characteristics, uses, and limitations of fire-fighting foams; **the advantages and disadvantages of using fog nozzles versus foam nozzles for foam application; foam stream application techniques**; hazards associated with foam usage; and methods to reduce or avoid hazards.

(B) Requisite Skills. The ability to prepare a foam concentrate supply for use, assemble foam stream components, master various foam application techniques, and approach and retreat from fires/spills as part of a coordinated team.

NFPA 1003

3-3.2 Extinguish an aircraft fuel spill fire, given PrPPE, an assignment, an ARFF vehicle hand line flowing a minimum of 95 gpm (359 L/min) of AFFF extinguishing agent, and a fire sized to the AFFF gpm flow rate divided by 0.13 (gpm/0.13 = fire square footage) (L/min/0.492 = 0.304 m^2), so that the agent is applied using the proper techniques and the fire is extinguished in 90 seconds.

(a) Requisite Knowledge: The fire behavior of aircraft fuels in pools, physical properties and characteristics of aircraft fuel, **agent application rates and densities**.

(b) Requisite Skills: Operate fire streams and apply agent.

3-3.3 Extinguish an aircraft fuel spill fire, given PrPPE, an ARFF vehicle turret, and a fire sized to the AFFF flow rate of 0.13 gpm (0.492 L/min) divided by the square feet of fire area, so that the agent is applied using the proper technique and the fire is extinguished in 90 seconds.

(a) Requisite Knowledge: Operation of ARFF vehicle agent delivery systems, the fire behavior of aircraft fuels in pools, physical properties and characteristics of aircraft fuel, **agent application rates and densities**.

(b) Requisite Skills: Apply fire-fighting agents and streams using ARFF vehicle turrets.

3-3.4 Extinguish a three-dimensional aircraft fuel fire, given PrPPE, an assignment, and ARFF vehicle hand line(s) using primary and secondary agents, so that a dual agent attack is used, the agent is applied using the proper technique, the fire is extinguished, and the fuel source is secured.

(a) Requisite Knowledge: The fire behavior of aircraft fuels in three-dimensional and atomized states, physical properties and characteristics of aircraft fuel, **agent application rates and densities**, and methods of controlling fuel sources.

(b) Requisite Skills: Operate fire streams and apply agents, secure fuel sources.

NFPA 1051

5.1.1 The Wildland Fire Fighter I shall meet the job performance requirements defined in Sections 5.1 through 5.5.

(A) Requisite Knowledge. Fireline safety, use, and limitations of personal protective equipment, agency policy on fire shelter use, basic wildland fire behavior, **fire suppression techniques**, basic wildland fire tactics, the fire fighter's role within the local incident management system, and first aid.

(B) Requisite Skills. Basic verbal communications and the use of required personal protective equipment.

5.5.6 Describe the methods to reduce the threat of fire exposure to improved properties given a wildland or urban/interface fire, suppression tools, and equipment so that improvements are protected.

(A) Requisite Knowledge. Wildland fire behavior, wildland fuel removal, structure protection methods, and **equipment and personnel capabilities**.

(B) Requisite Skills. The application of requisite knowledge to protect structures.

6.5.2 Select fireline construction methods, given a wildland fire and line construction standards, so that the technique used is compatible with the conditions and meets agency standards.

(A) Requisite Knowledge. Resource capabilities and limitations, fireline construction methods, and agency standards.

(B) Requisite Skills. None specified.

9.5.6 Structure Protection Plan. Develop and monitor a structure protection plan, given incident intelligence, current and predicted fire behavior, community data, and available resources, so that various structures and other improvements that are or may be threatened during a wildland/urban interface incident are protected and the plan is modified as needed.

(A) Requisite Knowledge. Knowledge of the availability and capability of fire apparatus and equipment that can be involved in an incident, the elements of a structure protection plan, incident objectives, and the weather forecast.

(B) Requisite Skills. The ability to develop and implement a structure protection plan and to constantly evaluate the wildland fire situation and change and modify the structure protection plan accordingly.

NFPA 472

5.4.4 Performing Defensive Control Actions. Given a plan of action for a hazardous materials incident within their capabilities, the first responder at the operational level shall demonstrate defensive control actions set out in the plan and shall meet the following related requirements:

(1) **Using the type of fire-fighting foam or vapor suppressing agent and foam equipment furnished by the authority having jurisdiction, demonstrate the effective application of the fire-fighting foam(s) or vapor suppressing agent(s) on a spill or fire involving hazardous materials.**

(2) Identify the characteristics and applicability of the following foams:
   (a) Protein
   (b) Fluoroprotein
   (c) Special purpose
      i. Polar solvent alcohol-resistant concentrates
      ii. Hazardous materials concentrates
   (d) Aqueous film-forming foam (AFFF)
   (e) High expansion

(3) Given the required tools and equipment, demonstrate how to perform the following defensive control activities:
   (a) Absorption
   (b) Damming
   (c) Diking
   (d) Dilution
   (e) Diversion
   (f) Retention
   (g) Vapor dispersion
   (h) Vapor suppression

(4) Identify the location and describe the use of the mechanical, hydraulic, and air emergency remote shutoff devices as found on cargo tanks.

(5) Describe the objectives and dangers of search and rescue missions at hazardous materials incidents.

(6) Describe methods for controlling the spread of contamination to limit impacts of radioactive materials.

Chapter 3
Foam Proportioning and Delivery Equipment

Fire-fighting foam concentrates can be delivered through fixed, semifixed, and mobile or portable systems. This chapter explains the various components of foam delivery systems: Foam proportioning devices, portable foam application devices, medium- and high-expansion foam generating devices, and high-energy foam generating systems. Each of the foam delivery systems is described in Chapter 4, Foam Delivery Systems. The four primary components of a foam delivery system are as follows:

- Water supply under pressure
- Proportioner
- Delivery device
- Foam concentrate source

The water under pressure may come through a pump from a water tank, an impounded water supply, or a municipal or private water supply system. The proportioner mixes the water with the predetermined quantity of foam concentrate based on one of the following two principles: induction or injection. The delivery device is usually a solid stream, fixed-flow, or automatic-stream fog nozzle attached to a hoseline, master stream appliance, or monitor. The nozzle may be either a standard handline nozzle or one specifically designed to spread foam agents. The foam concentrate source may be individual containers of foam concentrate or large-capacity tanks mounted on vehicles or in a fixed location (**Figure 3.1**). The exceptions to these foam delivery systems are the batch and premix systems. In the case of these two systems, a proportioner is not needed or used.

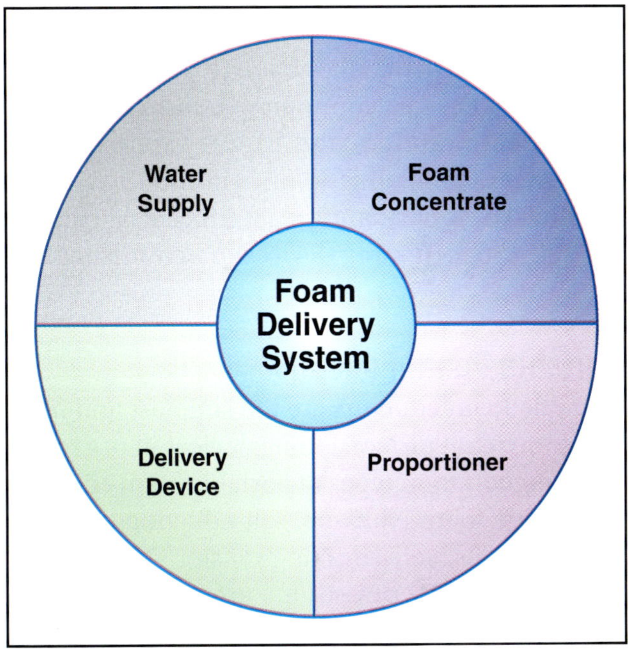

Figure 3.1 The four components that make up a foam system: water supply, proportioner, delivery device, and foam concentrate.

Foam Proportioning Devices

[NFPA 472: 5.4.4; NFPA 1001: 6.3.1; NFPA 1002: 2-3.7; NFPA 1003: 3-3.2, 3-3.4; NFPA 1021: 2-6.1; NFPA 1081: 5.3.1, 7.3.4; NFPA 1051: 5.5.1, 5.5.6, 6.5.2, 9.5.6]

The process of foam proportioning sounds simple: Add the proper amount of foam concentrate into the water stream through induction or injection and produce an effective foam solution (see bullets on next page). Unfortunately, this process is not as easy as it sounds. The correct proportioning of foam concentrate into a fire stream requires equipment that must operate within strict design specifications. Failure to operate even the best foam proportioning equipment as designed can result in poor-quality foam or no foam at all.

In general, foam proportioning devices operate by one of two basic principles:

- ***Induction*** — The pressure of the water stream flowing through an orifice creates a Venturi action that inducts (drafts) foam concentrate into the water stream.
- ***Injection*** — Pressurized proportioning devices inject foam concentrate into the water stream at a desired ratio and at a higher pressure than that of the water.

This section details the various types of foam proportioning devices commonly found in portable, apparatus-mounted, and fixed-system applications. Also included is information on two other methods of proportioning foam concentrate that do not use proportioning devices: batch mixing and premixing.

Portable Foam Eductors

Portable foam eductors are the simplest and most common foam proportioning devices in use today. The four common types of portable foam eductors are in-line foam eductors, self-educting handline foam nozzles, self-educting master stream foam nozzles, and portable around-the-pump proportioners.

In-Line Foam Eductors

In the fire service, the primary foam proportioning device is the *in-line foam eductor:* a pipe with a smoothbore nozzle inside **(Figure 3.2)**. In-line foam eductors (also known as *inductors, line proportioners,* or *ratio controllers*) are very reliable. In-line foam eductors are almost foolproof when it comes to proportioning accuracy — as long as inlet pressure and back pressure rules are not violated. In fact, experts in the foam industry and the fire service consider the simple in-line foam eductor to be among the most efficient pieces of foam proportioning equipment. The following paragraphs describe various elements of in-line foam eductors.

Rated flow. The in-line foam eductor's metering valve is designed to proportion the water and foam concentrate at the eductor's rated flow, which is a function of how much pressure is on the small nozzle inside the pipe. Most fire service in-line foam eductors develop their rated flow at about 200 psi (1 379 kPa). A 95-gpm-rated (360 L/min) eductor, which is the most popular model, has a ½ inch (13 mm) nozzle inside that flows about 95 gpm at 200 psi (360 L/min at 1 379 kPa) **(Figure 3.3)**. At 150 psi (1 034 kPa), it flows 82 gpm (310 L/min), and at 250 psi (1 724 kPa), it flows 106 gpm (401 L/min). One-and-a-half-inch (38 mm) (outlet size) eductors have flow ranges anywhere from 50 to 125 gpm (189 L/min to 473 L/min) and are usually identified with a permanent stamp on the concentrate metering valve or somewhere on the eductor body. If no flow rate is indicated, tests have to be performed to determine the proper calibration of the eductor.

Venturi process. High-speed water (120 mph [161 kmph]) flowing through the eductor passes a vent hole (Venturi) in the eductor that terminates at the

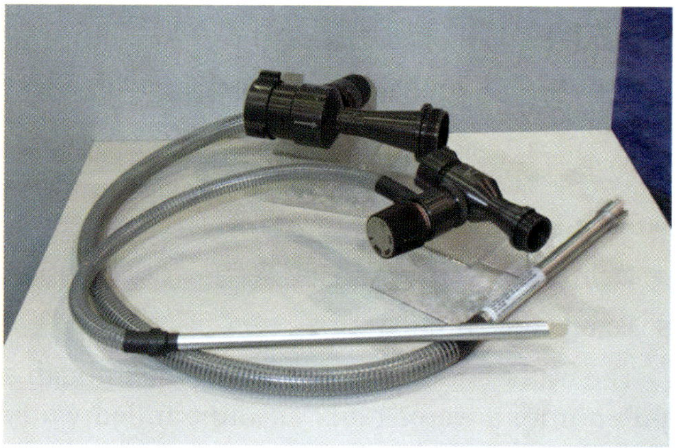

Figure 3.2 In-line foam eductors are available in a variety of styles.

Figure 3.3 Markings on this in-line foam eductor indicate that it is rated for 95 gpm (360 L/min) and can be set at ¼-, ½-, 1-, 3-, or 6-percent foam concentrate.

end of the pickup tube. A high-speed liquid that passes such a vent creates a vacuum that allows foam concentrate to travel up the pickup tube, through the metering valve, past a one-way check valve, and into the water stream. Finished foam is produced when the foam solution leaves the discharge nozzle where air is entrained with the foam solution.

Concentrate lift restrictions. In-line foam eductors should not be higher than 6 feet (1.8 m) from the foam concentrate container from which it is drawing (**Figure 3.4**). Lifting foam concentrate more than 6 feet (1.8 m) may cause proportioning inaccuracy or a total shutdown of the eduction process. However, pickup tubes can be longer than 6 feet (1.8 m). If the pickup tube needs to be longer, 10 feet (3 m) can be added. The manufacturer probably provides a short tube because of concentrate lift restrictions. If longer tubing is required, increase the inside diameter of the tube.

Filter. A filter basket is sometimes used at the inlet of the foam pickup tube to prevent impurities, sediment, or thick concentrations from clogging the eductor or nozzle (**Figure 3.5**). Usually a fine mesh screen is set in a recess to prevent the tube from sticking to the bottom of the foam concentrate container. When using synthetic foam concentrates (Class A, high-expansion, AFFF, or AR-AFFF), it is advisable to remove the filter screen because 6 gpm (23 L/min) flow may be too much for the filter to handle. This situation is particularly true when AR-AFFF foam concentrates are employed.

Check valve. A check valve located at the eductor's foam concentrate inlet closes when the nozzle is turned off or a kink in the hoseline restricts the flow of foam solution, causing pressure to build up in the hoseline. The check valve prevents water from entering the foam concentrate container or apparatus foam tank through the pickup tube. If the check valve freezes open, water can contaminate (weaken) the foam concentrate. It is vitally important to flush and inspect the check valve following each use to ensure proper operation in the future.

Concentrate ratio. Foam concentrates are manufactured and labeled according to the intended percentage of its use with water. The word *percentage* in the foam world really means *ratio*. Three percent means three parts concentrate to ninety-seven parts water (3:97). Six percent means ninety-four parts water to six parts foam concentrate (6:94), and 1 percent means ninety-nine parts water to one part foam concentrate (1:99).

Proportioning accuracy. If a ratio of 3 percent is set, and the 95-gpm (360 L/min) eductor is at 200 psi (1 379 kPa), about 3 gpm (11 L/min) of

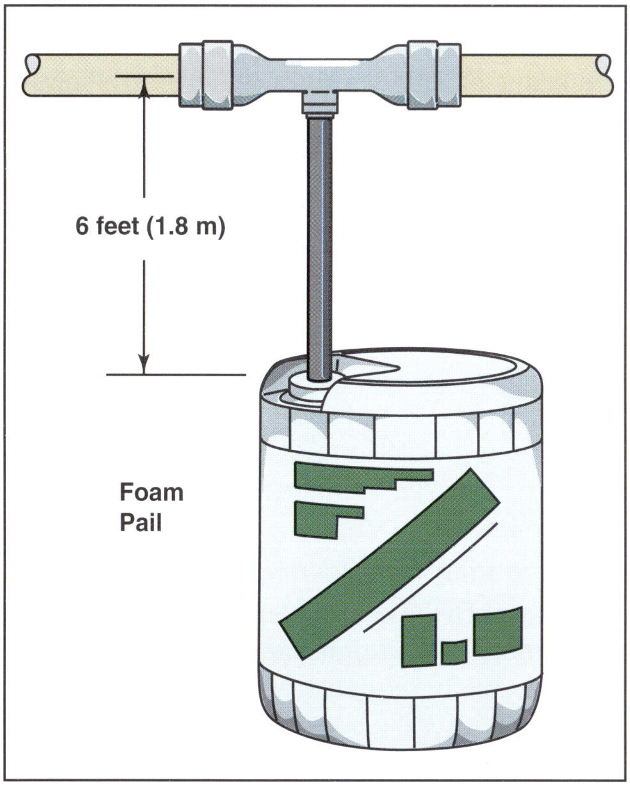

Figure 3.4 The in-line eductor should be no more than 6 feet (1.8 m) above the foam concentrate container.

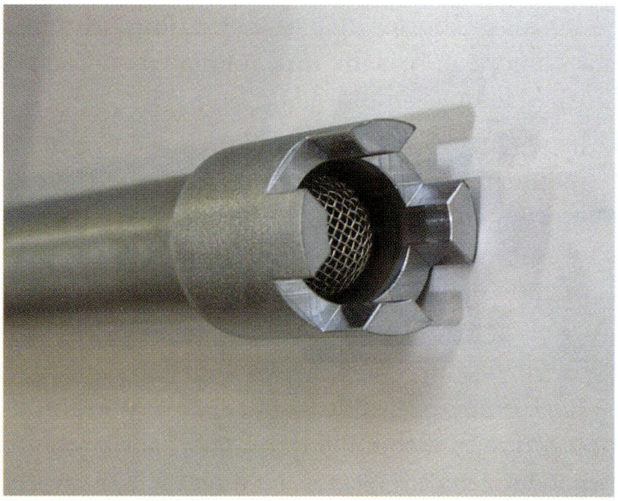

Figure 3.5 Foam pickup tubes are equipped with filters to prevent impurities or sediment from entering and clogging the eductor orifice.

concentrate will be pulled into the eductor; at 6 percent, slightly less than 6 gpm (23 L/min) will flow. At 1 percent, 99 parts water and 1 part concentrate are mixed. If less than 200 psi (1 379 kPa) is at the eductor's inlet, less than rated flow can be expected and the foam solution will be somewhat "rich" (containing a higher proportion of foam concentrate than intended). If more than 200 psi (1 379 kPa) is at the inlet, the foam solution proportions are slightly "lean" (containing a lower proportion of foam concentrate than intended). NFPA 11, *Standard for Low-, Medium- and High-Expansion Foam* (2002), allows Class B foam concentrate to proportion as much as 1-percent rich. However, there can be no lean Class B foam solutions. If 3-percent foam concentrate is proportioned at 2.5 percent, there may not be enough foam concentrate in the solution to create proper finished foam.

Proportioning accuracy can mean life or death when dealing with Class B liquid fuel fires. Foam solutions that are too rich may result in the rapid expenditure of the concentrate before finished foam can completely cover a flammable liquid or extinguish a fire. Foam solution that is too lean will not create an effective foam blanket over a liquid fuel, which may allow flammable vapors to escape and reignite. Either situation can result in injuries to personnel working near a spill or fire.

In Class A situations, proportioning inaccuracy simply means that water doesn't soak into the burning material as effectively as it would at the proper ratio. The foam blanket drainage may also be longer or shorter than expected. Proportioning inaccuracy rarely is life threatening when Class A finished foam is employed.

If an eductor is not designed to provide ½- or ¼-percent proportioning options on the eductor's foam concentrate metering valve, dilute the foam concentrate, one to one (1:1) with water for ½ percent and 2:1 for ¼ percent. Then set the meter on the in-line foam eductor at 1 percent and proceed with the operation of the foam system.

Back pressure. In-line foam eductors don't function properly if there is too much back pressure (what your thumb feels when it restricts the water flow from a garden hose). Back pressure is caused by restrictions at the discharge nozzle, kinks in hoselines, the effects of elevation on the nozzle or hoseline, and friction loss in the hoseline **(Figure 3.6)**. Most U.S.-made fire department type in-line foam eductors can tolerate approximately 130 psi (896 kPa) of back pressure (about 65 percent of its 200 psi [1 379 kPa] inlet pressure). Adding 200 feet (60 m) of 1¾-inch (45 mm) hose between the nozzle and a 95-gpm (360 L/min) eductor adds 30 psi (207 kPa) of back pressure. Add a 95-gpm (360 L/min) nozzle that is rated at 100 psi (689 kPa) and the eductor can tolerate no more back pressure. A kink in the discharge hose, a slightly closed nozzle bail, or the nozzle being 10 or more feet (3 m) higher than the eductor creates too much back pressure, causing the proportioning system to shut down. For this reason, in-line foam eductors don't work well when supplying ladder pipes or standpipe systems. Every foot (0.3 m) of elevation adds about a half-pound (0.23 kg) or 0.434 psi (3 kPa) of back pressure.

Nozzles. In-line foam eductors work well with all types of nozzles, including automatic nozzles. One hundred twenty-five gpm (473 L/min) and larger in-line foam eductors tend to be problematic in terms of back pressure when more than 50 feet (15 m) of 1½-inch (38 mm) hose or more than 100 feet (30 m) of 1¾-inch (45 mm) hose is used. See Appendix D, Hydraulic Calculations Chart.

When an onboard foam system uses a bypass foam eductor with two 1½-inch (38 mm) elbows in the discharge plumbing, it is possible that the system will not properly operate a 100-psi (689 kPa) nozzle. The reason for the problem is that 1¾-inch (45 mm) hose friction loss at 125 gpm (473 L/min) is about 23 psi per 100 feet (159 kPa per 30 m). If a 100-psi (689 kPa) nozzle is used, the total hose distance between the nozzle and eductor can only be two lengths (100 feet [30 m]) because another 50 feet (15 m) would add about 12 psi (83 kPa) more back pressure, which exceeds the available pressure (130 psi [896 kPa]) for hose and nozzle at the eductor's outlet.

Transit time. The time that it takes the foam solution to reach the nozzle from the eductor is called *transit time.* Transit time varies depending on the capacity of the eductor and the length of the hoseline. With 200 feet (60 m) of 1¾-inch (45 mm) hose, it takes about 20 seconds for the nozzle operator to see foam solution flowing through the nozzle. Transit time also delays the effects of run-

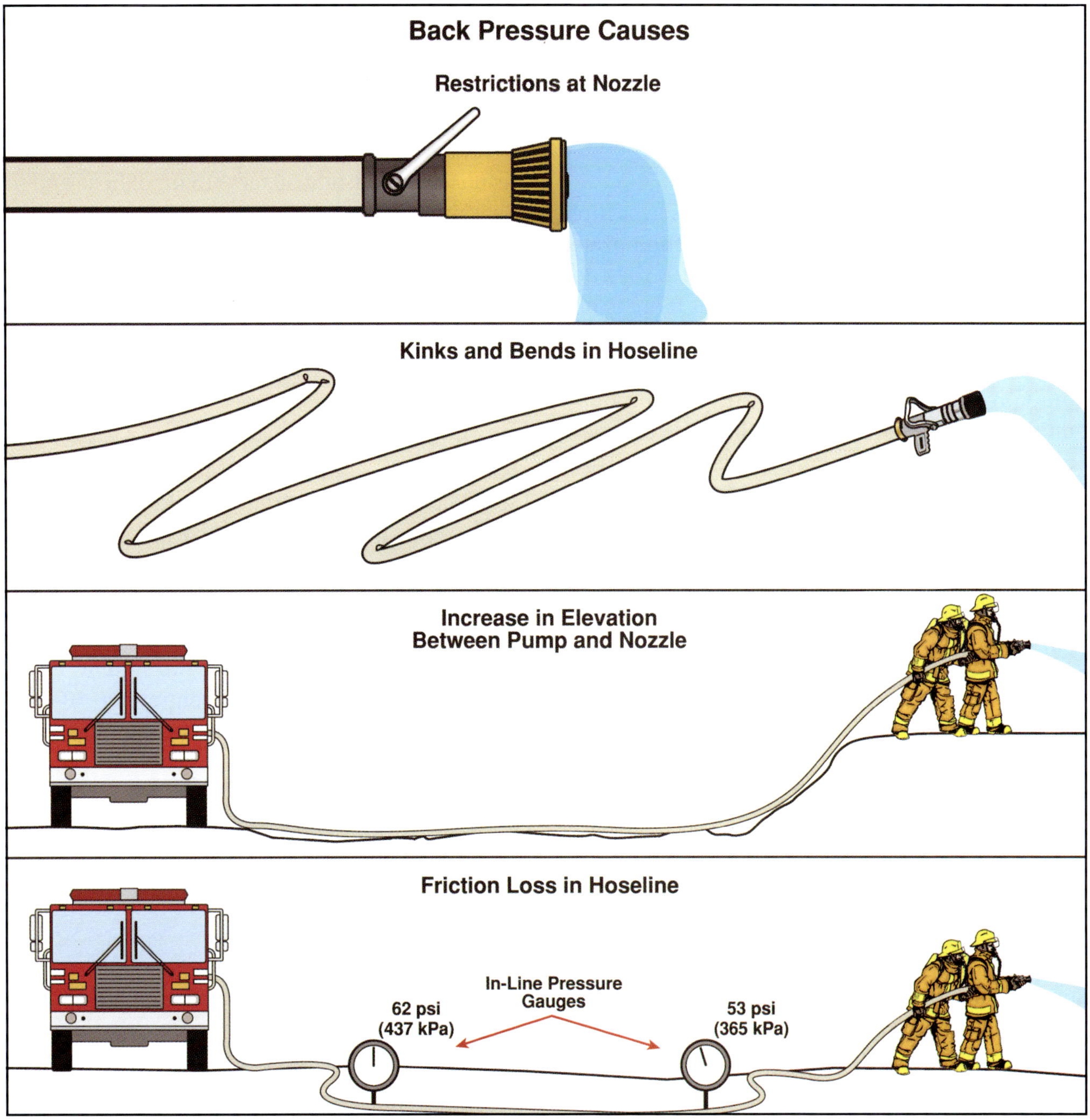

Figure 3.6 Several causes of pressure loss or gain (back pressure) are created by changes between nozzle and pump: nozzle restrictions, hoseline kinks/bends, elevation increase between pump and nozzle, and hoseline friction loss.

ning out of foam concentrate or having too much back pressure. It can take as long as 30 to 45 seconds for the solution to get to the nozzle from a 60-gpm (227 L/min) in-line foam eductor. The larger the hose, the longer it takes. The same situation is true for any onboard foam proportioning system.

In the case of large-size in-line foam eductors, if the hose is charged before the pickup tube is placed in the foam concentrate container, it takes much longer for the foam solution to reach the nozzle because plain water in the hose has to be replaced with foam solution. On the other hand, if the pickup tube is placed in the foam concentrate container before charging a dry line, the transit time is not an issue because solution rather than plain water fills the line. Immediately after opening the discharge (at 100 psi [689 kPa] or higher), foam concentrate begins moving up the tube and into the eductor. This situation happens because the eductor feels no back pressure. The eductor continues to pull foam concentrate up the tube until the foam solution (3 percent [3:97 ratio] foam concentrate) has reached the closed nozzle shutoff. When the nozzle is opened, the operator gets foam solution immediately.

It is essential to effective foam operations that fire and emergency responders operating the nozzle understand the concept of transit time. If they do not, they may prematurely report that there is no foam solution at the nozzle. This negative report causes the pump operator to lower or raise the pump pressure and open or constrict the proportioning meter on the eductor. While these changes are taking effect, the solution that was almost to the nozzle in the first place arrives and flows during the time it took the pump operator to change the pump pressure or change the percentage setting. Then the amount of foam solution increases or decreases giving the false impression that the eductor is malfunctioning. If personnel are trained to expect an average transit time between 20 and 30 seconds, operations will be both effective and efficient.

Eductor testing. In the event that an in-line foam eductor fails to operate, remove it from service. Test it using the following steps:

Step 1: Connect the subject eductor to a pump outlet, using a reducer if necessary.

Step 2: Connect 50 feet (15 m) of hose with no nozzle attached to the eductor.

Step 3: Place the pickup tube in a pail containing 4 gallons (15 L) of water.

Step 4: Set the proportioning meter for 3 percent.

Step 5: Secure the open hose butt so it doesn't flop around on the ground.

Step 6: Open the discharge (all the way) and throttle up until 200 psi (1 379 kPa) is established.

If testing a 95-gpm (360 L/min) eductor that is operating correctly, it will drain more than 3 gallons (11 L) but less than 4 gallons (15 L) of water from the bucket in about 60 seconds. A 60-gpm (227 L/min) in-line foam eductor drains almost 2 gallons (8 L) in 60 seconds, and a 125-gpm (473 L/min) eductor drains slightly more than 3.5 gallons (13 L) in 60 seconds. Rich test results are usually obtained when testing with water in place of foam concentrate. Foam equipment manufacturers actually use water to create foam equivalent tables when testing foam proportioning systems.

Add a nozzle and repeat the steps listed. If the results are good, add 50 feet (15 m) of hose and repeat the test. If the results are still good, continue adding hose in 50-foot (15 m) sections (creating additional back pressure) until the eductor stops suctioning water (**Figure 3.7**).

Remember that the nozzle flow rate must be the same or higher than the eductor's flow rate. A 60-gpm (227 L/min) nozzle on a 95-gpm (360 L/min) eductor shuts down the system because of excessive back pressure. A well-maintained automatic nozzle self-adjusts to the flow that the eductor gives it.

Care and maintenance. Like all fire-fighting tools and equipment, in-line foam eductors must have proper care and maintenance. Following each use, eductors must be flushed to remove any foam concentrate that may dry and clog the system (**Figure 3.8**). Dried Class A foam concentrate or AR-AFFF concentrate causes suction plumbing clogs or oc-

Figure 3.7 Water can be used in place of foam to test the operation of an in-line foam eductor.

Figure 3.8 Following each use, thoroughly flush foam eductors, nozzles, and hoselines to prevent corrosion of metal parts.

clusions that either prevent foam concentrate from flowing or limit foam concentrate flow. When the foam operations are complete, substitute a 5-gallon (19 L) pail of freshwater for foam concentrate. Run the system until the water pail is empty. Be certain that kinks in foam concentrate pickup tubes do not cause cracking or take on a permanent form that restricts the flow of concentrate through the pickup tube. A cracked pickup tube or loose check-valve fitting allows air rather than foam concentrate to enter.

AFFF and Class A foams tend to strip factory lubricants from the interiors of structural fog nozzles. If nozzle controls are difficult to operate, lubricate the moving parts with an approved lubricant.

Some pickup tubes have a filter at the inlet. Thicker AR-AFFF foams may have problems passing through eductor filters that have not been Underwriters Laboratories Inc. (UL) tested and approved. The remedy is to unscrew the filter retainer or filter basket and use the tube without the filter. Pickup tube ends (otherwise known as *wands*) should be cut at about a 30-degree angle when the filter basket is removed. This procedure prevents the tube from sticking to the bottom of the foam concentrate container, which may stop or restrict foam concentrate flow. If the filter is used, remove, inspect, and thoroughly clean it before replacing it.

Self-Educting Handline Foam Nozzles

The self-educting handline foam nozzle operates on the same basic principle as the in-line foam eductor. However, this eductor is built into the nozzle rather than into the hoseline **(Figure 3.9)**. Finished foam

Figure 3.9 Portable self-educting foam nozzles (such as this unit intended for wildland fire fighting) must have the foam concentrate located near the nozzle.

is produced when the foam solution is discharged from the nozzle and air is entrained with the foam solution. Nozzle use requires the foam concentrate to be available where the nozzle is operated. If the foam nozzle is moved, the foam concentrate also needs to be moved. The logistical problems of relocation are magnified by the gallons (liters) of concentrate required. Use of a self-educting handline foam nozzle eductor may compromise firefighter safety: Firefighters cannot always move quickly, and they must leave their concentrate containers behind if they are required to retreat for any reason. However, the self-educting handline foam nozzle has several advantages:

- It is easy to use.
- It is inexpensive.
- It works with pressures lower than those required by in-line foam eductors.
- A wide variety of flow rate versions are available.

Self-Educting Master Stream Foam Nozzles

Self-educting master stream foam nozzles are very similar to the self-educting handline nozzles mentioned in the previous section. These nozzles are available with flow capabilities up to 12,000 gpm (45 425 L/min) **(Figure 3.10)**. The self-educting master stream foam nozzle uses a modified Venturi design to draw foam concentrate into the water stream.

A jet ratio controller (JRC) may be used to supply foam concentrate to a self-educting master stream foam nozzle **(Figure 3.11)**. The jet ratio controller is

Figure 3.10 Self-educting master stream nozzles with flow capacities up to 12,000 gpm (45 425 L/min) are mounted on top of apparatus.

Figure 3.11 Jet ratio controllers are designed to supply large volumes of foam concentrate long distances from the master stream nozzle.

a type of in-line foam eductor that allows the foam concentrate supply to be as far as 3,000 feet (914 m) away from the self-educting master stream foam nozzle (**Figure 3.12**). This distance allows fire and emergency responders who are involved in operating fire pumps and maintaining the foam concentrate supply to be a safe distance from a fire. The JRC also allows an elevation change of up to 50 feet (15 m).

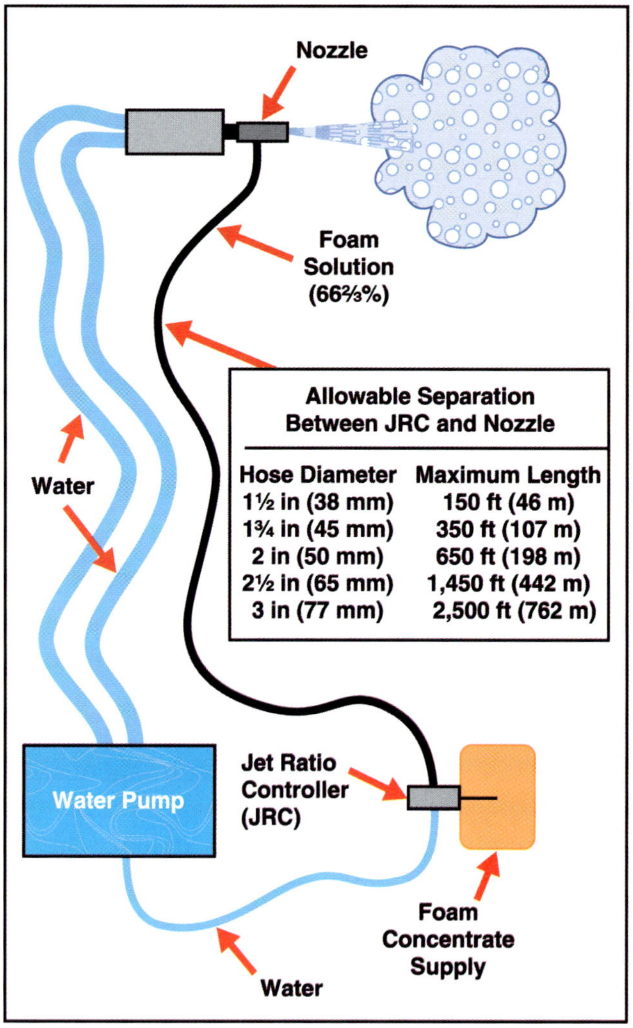

Figure 3.12 Example of the placement of a jet ratio controller in a hoseline.

The flow of water to the JRC represents about 2½ percent of the total flow in the system. As water flows through the JRC (as with a standard in-line foam eductor), a Venturi effect is created that draws concentrate through the pickup tube and into the hoseline. The difference is that the JRC proportions the concentrate at a 66⅔-percent solution. This rich solution is then pumped to a self-educting master stream foam nozzle where it is further proportioned with the water supplied by the fire pump down to a discharge proportion of 3 percent. To achieve a proper proportion, it is important that the JRC and nozzle have the same delivery rate in gpm (L/min).

JRCs are not limited to portable applications. They may be used in fixed-system applications as well. More information on fixed foam systems is contained in Chapter 4, Foam Delivery Systems.

Portable Around-the-Pump Proportioners

Portable around-the-pump (ATP) foam proportioning kits (sometimes called *portable crash trucks*) are now available (**Figure 3.13**). These kits attach to the auxiliary suction valve of any apparatus with a fire pump. A short hose that goes from the pump's discharge back into the pump inlet via the ATP unit drives them. The portable around-the-pump proportioner can turn any pumper into a foam truck. These kits are very reliable and require little or no maintenance. They offer solution flows from 50 to 4,500 gpm (189 L/min to 17 034 L/min) at a 1-percent ratio, 50 to 1,500 gpm (189 L/min to 5 678 L/min) at 3-percent ratios, and 750 gpm (3 839 L/min) at 6-percent ratios. Like a portable in-line foam eductor, this device uses no electronic interface, flow meter, or foam concentrate pump. Miniature versions are available for use with Class A foam or 1-percent aqueous film forming foam (AFFF) concentrate.

Around-the-pump proportioners are foam eductors that operate on the suction or supply side of the pump rather than the discharge side of the pump. Since they rely on suction-side plumbing and are designed to draft foam concentrate into the pump's inlet, more than atmospheric pressure (14.7 psi) on pump suction will shut down the eductor. These devices are very easy to operate when using the apparatus water tank to supply the pump. They are more difficult to operate when using hydrant pressure. The pump operator must maintain incoming supply pressure at 10 psi (69 kPa) or less to prevent the eductor from shutting down.

Figure 3.13 Around-the-pump proportioners can be used to make any type of fire pump into a foam pumper.

Apparatus-Mounted and Fixed-System Proportioners

[NFPA 1002: 2-3.7, 3-1.1, 3-2.3, 6-1.1; NFPA 1003: 3-3.3]

Foam proportioning devices are commonly mounted on structural, industrial, wildland, and aircraft rescue and fire fighting apparatus as well as fireboats (**Figure 3.14**). The majority of the foam proportioning devices described in this section can be used with both Class A and Class B foam concentrates. Because there is little difference in apparatus-mounted and fixed-system proportioners, they are both discussed in this section. Maintenance of apparatus-mounted foam eductor systems is also described. More detailed information on the overall design of fixed systems is contained in Chapter 4, Foam Delivery Systems. The types of foam proportioning devices discussed in this section include the following:

- Installed in-line eductors
- Around-the-pump proportioners
- Bypass-type balanced-pressure proportioners
- Variable-flow variable-rate direct-injection devices
- Variable-flow demand-type balanced-pressure proportioners
- Variable-flow variable-orifice balanced-pressure proportioners
- Bladder-tank balanced-pressure proportioners
- Fixed-flow direct-injection proportioners

Figure 3.14 Onboard foam systems have specific control panels mounted on the pump panel that indicate the proportioner setting, quantity of foam in the tank, and controls for operating and flushing the system.

The devices described in this section are called low-energy foam systems. A *low-energy* foam device imparts pressure on the foam solution solely by the use of a fire pump. These devices introduce air into the solution when it travels through the proportioning device or is discharged from the nozzle. *High-energy* foam devices introduce compressed air into the foam solution before it is discharged into the hoseline. The high-energy foam generating systems are described at the end of this chapter.

Installed In-Line Eductor Systems

Installed in-line eductors use the same principles of operation as do portable in-line eductors. The only difference is that these eductors are permanently attached to the apparatus pumping system (**Figure 3.15**). The same precautions regarding hose lengths, matching nozzle and eductor flows, and inlet pressures listed for portable in-line eductors also apply to installed in-line eductors. Foam concentrate may be supplied to these devices from either pickup tubes (using 5 gallon [19 L] pails) or from foam concentrate tanks installed on the apparatus.

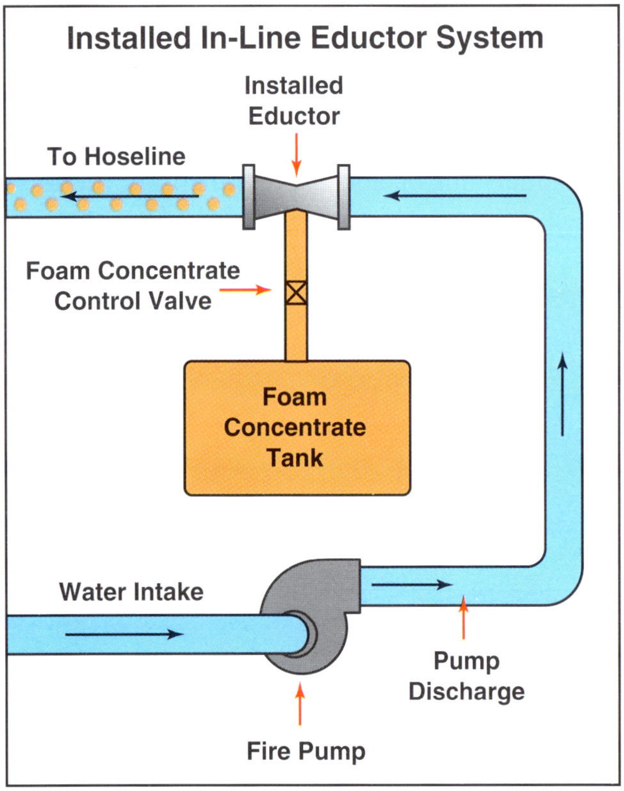

Figure 3.15 Diagram of the operation of an installed in-line foam eductor system on an apparatus.

In many cases, a special version of this device (a bypass eductor) is used. In the bypass (water-only) mode, a valve directs the water through a second chamber of the eductor that has no orifice or restrictions (**Figure 3.16**). This mode is commonly used when foam is not desired, and the discharge is functioning as a normal plain-water attack line. It should be noted that the eductor limits the discharge (gpm [L/min]) capacity of the attack line depending on the capacity of the eductor.

When finished foam is desired, the diverter valve is moved to the foam position; this action directs the water flow through the eductor/orifice chamber (**Figure 3.17**). An adjacent metering valve is usually present to accommodate various types of foam concentrates.

Installed in-line eductors are most commonly used to proportion Class B foam concentrates. Take care to adhere to the operating rules for using these devices. Because Class A foam concentrates are normally used at very low concentrations (0.1 to 1 percent), installed in-line eductors are probably not the first choice for proportioning them.

Around-the-Pump Proportioners

An around-the-pump proportioner is the most common type of built-in proportioner installed in mobile fire apparatus today. This proportioner is also installed on some fixed systems. It is especially

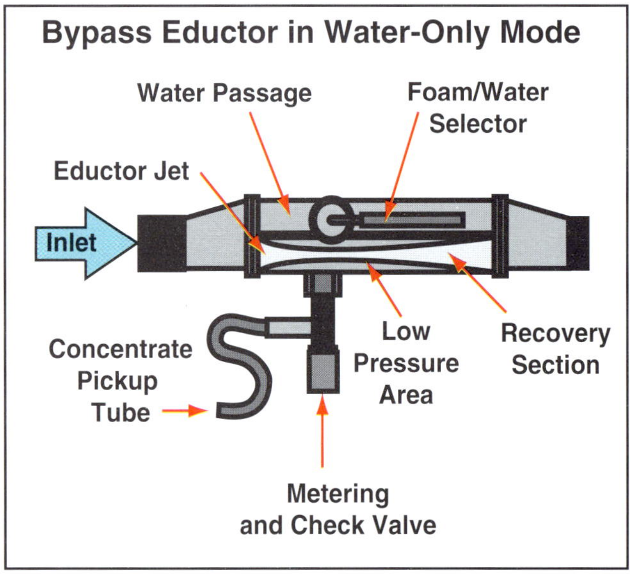

Figure 3.16 When the foam eductor system is in the bypass mode, it discharges plain water rather than foam.

Figure 3.17 The apparatus-mounted foam system is activated by operating the foam diverter lever into one of three positions: Tank A, Tank B, and Flush.

useful when there is low water pressure or when a power supply is not available for a separate foam concentrate pump.

The around-the-pump proportioning system consists of a small return (bypass) waterline connected from the discharge side of the fire pump back to the intake side of the pump (**Figure 3.18**). An inline eductor is positioned on this bypass line. A valve positioned on the bypass line, just off the pump discharge piping, controls the flow of water through the bypass line. When the valve is open, a small amount of water (10 to 40 gpm [39 L/min to 151 L/min]) discharged from the pump is directed through the bypass piping. As this water passes through the eductor, the resulting Venturi effect draws foam concentrate out of the foam concentrate tank and into the bypass piping. The resulting foam solution is then supplied back to the intake side of the pump where it is then discharged to the delivery devices.

Around-the-pump proportioning systems are available with a wide range of flow options. Many ATP systems will allow flow from as little as 60 gpm (227 L/min) to as high as 2,000 gpm (7 571 L/min).

A major disadvantage of the around-the-pump proportioner is that the pump cannot make use of incoming pressure. If the inlet water supply is any greater than 10 psi (69 kPa), the foam concentrate will not enter the pump intake. Versions of these systems are capable of handling intake pressures up to 40 psi (276 kPa). ATP systems must be dedicated

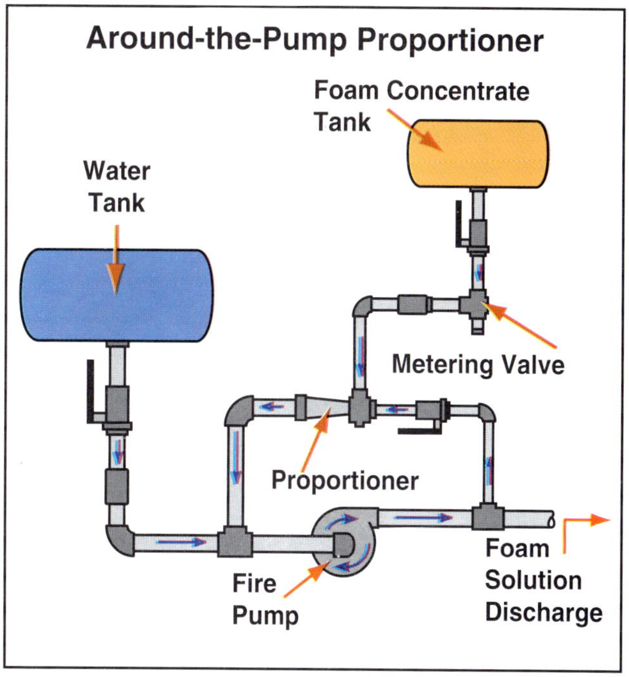

Figure 3.18 Around-the-pump proportioners are commonly found on structural fire apparatus.

solely to foam operation. An around-the-pump proportioner does not allow plain water and finished foam to be discharged from the pump at the same time.

Bypass-Type Balanced-Pressure Proportioners

The bypass-type balanced-pressure proportioner is one of the most accurate methods of foam proportioning. It is most commonly used in fixed systems and large-scale mobile apparatus applications such as airport crash vehicles and refinery fire-fighting apparatus. The primary advantage of the bypass-type balanced-pressure proportioner is its ability to monitor the demand for foam concentrate and accurately adjust the amount of concentrate supplied. Bypass-type balanced-pressure proportioners have the ability to discharge finished foam from some outlets and plain water from others at the same time. Thus, a single apparatus can supply both foam attack lines and protective cooling waterlines simultaneously.

Systems equipped with a bypass-type balanced-pressure proportioner have a foam concentrate line connected to each fire pump discharge outlet (**Figure 3.19**). A foam concentrate pump separate from the main fire pump supplies this line. The pump draws the foam concentrate from a supply source

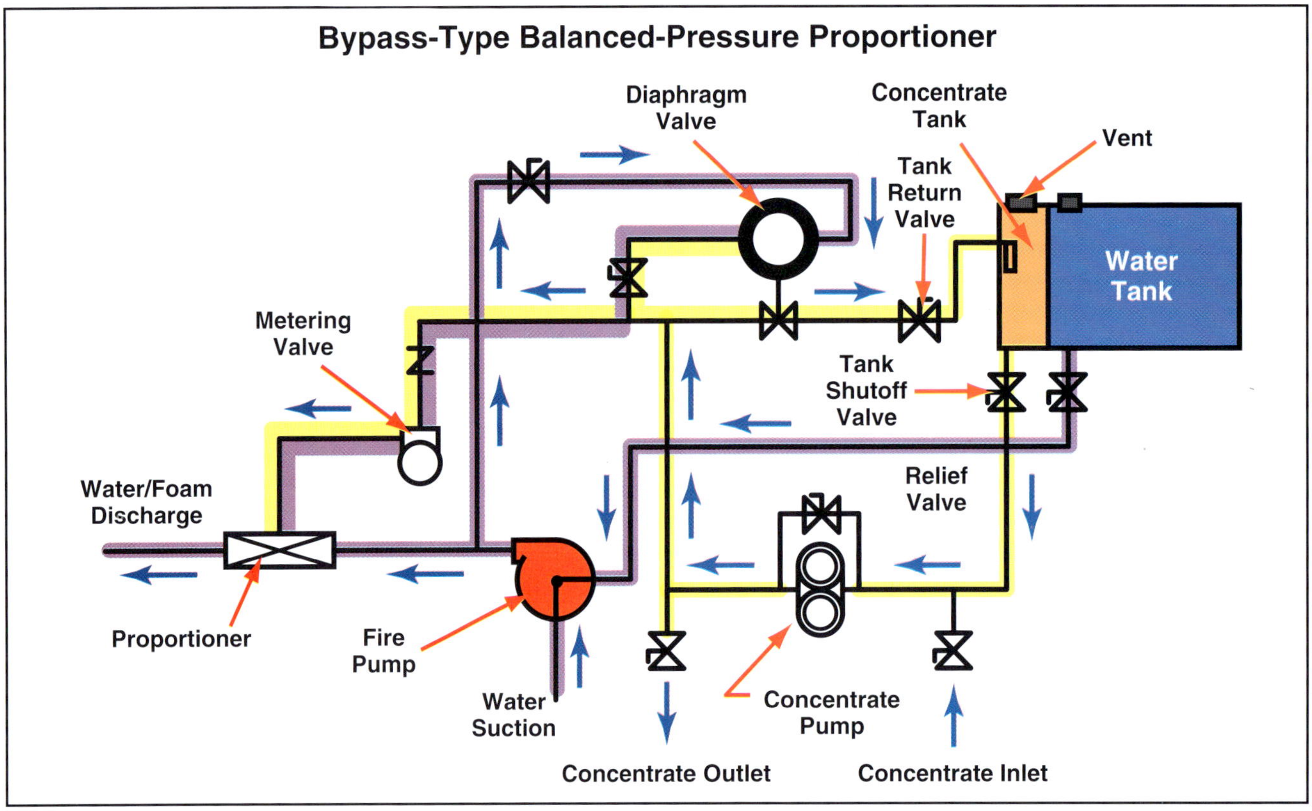

Figure 3.19 Example of a bypass-type balanced-pressure foam proportioning system.

(fixed tank and/or auxiliary supply). This pump is designed to supply foam concentrate to the outlet at the same pressure at which the fire pump is supplying water to that discharge. A hydraulic pressure control valve jointly monitors the pump discharge and the foam concentrate pressure from the foam concentrate pump to ensure that concentrate pressure and water pressure are balanced.

The orifice of the foam concentrate line is adjustable at the point where it connects to the discharge line. If 3-percent foam concentrate is desired, the foam concentrate discharge orifice is set to 3 percent of the total size of the water discharge outlet. If 6-percent foam concentrate is desired, the foam concentrate discharge orifice is set to 6 percent of the total size of the water discharge outlet. Because the foam concentrate and water are supplied at the same pressure and the sizes of the discharges are proportional, the foam solution is proportioned correctly.

When foam concentrate demand is low, such as when the system is at idle mode, a great deal of foam concentrate is circulated between the foam pump and the foam storage tank. When demand for foam concentrate is high, very little foam concentrate is bypassed back to the foam storage tank. Returning foam concentrate back to the storage tank can be a problem where AR-AFFF concentrates are used. AR-AFFF concentrates will froth (bubble) as a result of constant discharge back to the storage tank. AR-AFFF concentrates that do not defroth (settle down) result in leaner foam solution each time the foam concentrate is circulated back to the tank. Bypass-type balanced-pressure foam systems are better suited to regular AFFF and Class A foam concentrates than to AR-AFFF concentrates.

Variable-Flow Variable-Rate Direct-Injection Systems

This type of proportioner operates from power supplied from the apparatus electrical system. Large-volume systems may use a combination of electric and hydraulic power. Monitoring the water flow and controlling the speed of a positive-displacement foam concentrate pump controls the foam concentrate injection, thus injecting concentrate at the desired ratio. Because the water flow governs the foam concentrate injection, water pressure is not a factor. Full flow through the fire pump discharges is possible because there are no flow-restricting passages in the proportioning system.

Variable-flow variable-rate direct-injection systems provide foam concentrate rates from 0.1 to 3 percent, although greater concentrate flows may be possible in some applications. The control unit has a digital display that shows the current water or foam solution flow rate, the total amount of water or solution flowed to this time, the current foam concentrate flow rate, and the amount of foam concentrate used to this time. These systems can be used with all Class A foam concentrates and many Class B concentrates. These systems may not be used with alcohol-resistant foam concentrates due to the high viscosity of the concentrate. These systems are supplied from atmospheric pressure foam concentrate tanks on the apparatus.

Several advantages to variable-flow variable-rate direct-injection systems exist. One advantage is their ability to proportion at any flow rate or pressure within the design limits of the system. Another advantage is that the system automatically adjusts to changes in water flow when nozzles are either opened or closed. Also, nozzles may be either above or below the pump without affecting the foam proportioning process. And finally, this system may be used with high-energy foam generating systems (discussed later in this chapter). The disadvantage of these systems is that the foam concentrate injection point must be within the piping before any manifolds or distribution to multiple fire pump discharges.

Variable-Flow Demand-Type Balanced-Pressure Proportioners

The variable-flow demand-type balanced-pressure proportioning system (also called a *pumped/demand system*) is a versatile system used in both fixed systems and mobile apparatus applications. In this system, a variable-speed mechanism that is either hydraulically or electrically controlled drives a foam concentrate pump. The foam concentrate pump supplies foam concentrate to a Venturi-type proportioning device built into the waterline (**Figure 3.20**). When activated, the foam concentrate

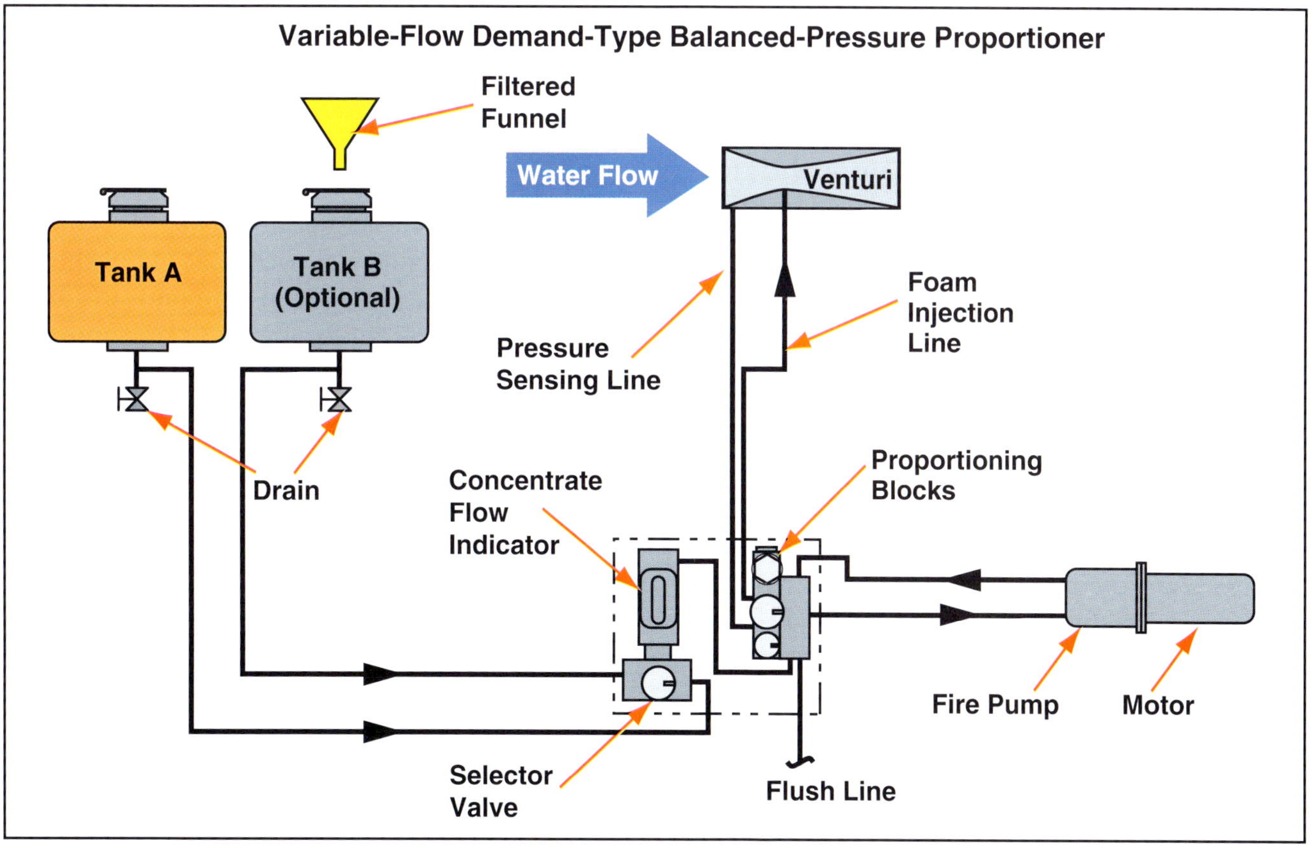

Figure 3.20 The foam concentrate pump supplies foam concentrate to a Venturi-type proportioning device that is built into the water line. *Courtesy of KK Products, Inc.*

pump output is automatically monitored so that the flow of foam concentrate is commensurate with the flow of water to produce an effective foam solution.

Several advantages are found with variable-flow demand-type balanced-pressure proportioning systems. First, the foam concentrate flow and pressure match system demand. Second, there is no recirculation back to the foam concentrate tank. The system is maintained in a ready-to-pump condition and requires no flushing after use. Last, water and/or finished foam can be discharged simultaneously from any combination of outlets up to rated capacity. A limitation of these systems is that the fire pump discharges have ratio controllers (which reduce the discharge area), thus pressure drops across the discharge are higher than those seen on standard pumpers.

Variable-Flow Variable-Orifice Balanced-Pressure Proportioners

This type of balanced-pressure proportioning system uses a flow-sensitive piston in the water supply line to regulate a variable orifice in the foam concentrate line to alter the quantity of foam concentrate. The metering orifice (which is supplied by a foam concentrate pump) increases/decreases its diameter depending on the system demand for foam solution. These systems are particularly useful for fixed-system applications that require large flow rates. They are not practical in smaller applications.

Bladder-Tank Balanced-Pressure Proportioners

This type of system is found on both fire apparatus and fixed foam systems. It consists of one or more foam concentrate tanks that connect both to the water supply line and to the foam solution lines of the overall system. This system is designed so that a small amount of water from the supply source is pumped into the concentrate tank(s). This water volumetrically replaces the foam concentrate in the foam supply tank, forcing the foam concentrate into the system where it mixes with water (**Figure 3.21**). A bladder membrane inside the tank separates the water pumped into the tank from the foam concentrate expelled. The system allows for automatic proportioning over a wide range of flows and pressures and does not depend on an external power

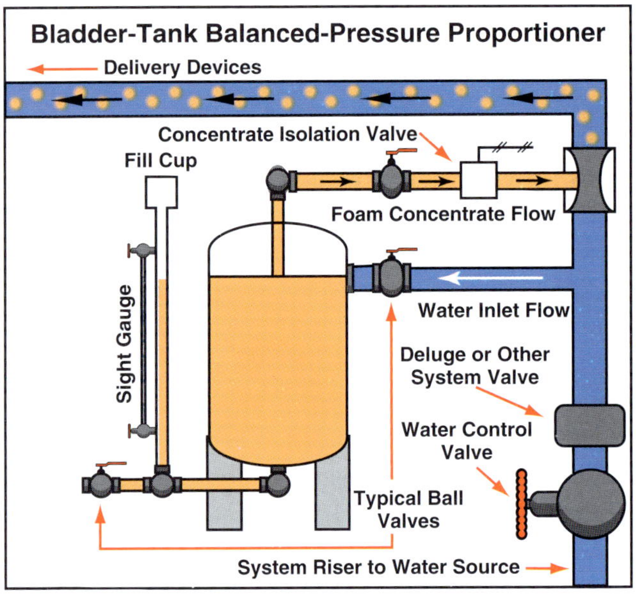

Figure 3.21 The bladder-tank balanced-pressure proportioner delivers foam solution to sprinklers, monitors, hose reels, and other delivery devices. Bladder-tank systems are limited in the duration of operation by their capacity.

source. However, the system is limited by the size of the concentrate tank. The system operating time is limited because once the concentrate is expended, the water must be drained from the tank before it can be refilled with concentrate.

Fixed-Flow Direct-Injection Proportioners

Fixed-flow, direct-injection proportioning systems (also known as *orifice-injection systems*) are commonly found on systems designed for a set flow rate of foam solution that will not vary. Separate pumps supply the water and foam concentrate to the proportioner. In order for the foam concentrate to be properly injected into the water, the foam concentrate pump runs about 10 psi (69 kPa) higher than the water pump. Two orifices are at the proportioner: one for the water stream and one for the foam concentrate. The size of the foam concentrate orifice is directly proportional to the size of the water stream orifice, based on the percent of foam concentrate used. For example, if the system is designed to be used with 6-percent foam concentrate, the foam orifice area will be 6 percent of the size of the water stream orifice. Thus, when water and foam concentrate are supplied to the proportioner at the same pressures, the proportioning rate will be correct. These types of proportioners are found only on fixed systems.

Apparatus-Mounted Foam Eductor System Maintenance

Hydraulic operating rules are the same for apparatus-mounted foam eductor systems as they are for the portable in-line foam eductors discussed earlier. What makes these systems somewhat more complex and prone to failure or misproportioning of foam solutions are complications associated with hidden valves and their linkage(s) and hoses that connect the eductor to the apparatus foam concentrate tank.

Flushing the system requires sending freshwater through the proportioning meter and eductor's foam concentrate check valve (also called a *nonreturn valve* or *backflow preventer*) **(Figure 3.22)**. Foam systems that do not flush freshwater from the foam tank outlet to the system inlet may have problems with partially blocked or completely blocked concentrate supply lines. Foam concentrate supply lines occupy one of the hottest spaces on the apparatus — the pump compartment. Foam concentrate left in the foam concentrate supply hoses can easily exceed the foam manufacturers and Underwriters Laboratories maximum concentrate storage temperature (120°F [49°C]). System difficulties or full failures can be caused by overheated foam concentrate in the tank, eductor supply hose, or pipe.

In hot climates where foam systems are idle for more than 60 days, it is recommended that the system be operated (make foam) for about 30 seconds every 2 months. This procedure keeps fresh foam concentrate in the foam system's foam concentrate supply line.

Batch Mixing

By far, the simplest means of proportioning foam is to simply pour an appropriate amount of foam concentrate into a tank of water. This method is called *batch mixing* or the *dump-in method.* To do this method, the driver/operator pours a predetermined amount of foam concentrate into the tank via the top fill opening at the time finished foam is needed **(Figure 3.23)**. The truck is then pumped normally,

Figure 3.22 Thoroughly train the apparatus driver/operator in the flushing procedures for apparatus-mounted foam eductor systems. *Courtesy of Tom Ruane.*

Figure 3.23 Batch mixing foam concentrate requires that the driver/operator pour a predetermined quantity of foam concentrate into the water in the apparatus tank.

and finished foam is discharged through any hoseline that is opened. The amount of foam concentrate needed depends on the size of the water tank and the proportion percentage for which the foam concentrate is designed.

The size of the water tank and the proportioning percentage of the foam concentrate dictate how much concentrate must be poured into the water tank. For example, when using 3-percent foam concentrate to produce 100 gallons (379 L) of foam solution, 3 gallons (11 L) of foam concentrate must be added to 97 gallons (367 L) of water. Thus, to form a proper foam solution, a 500-gallon (1 893 L) booster tank would require the addition of 15 gallons (57 L) of 3-percent foam concentrate. **Table 3.1** shows the amounts of concentrate required at various percentages for different water tank sizes.

In general, batch mixing is only used with regular AFFF (not alcohol-resistant AFFF concentrates) and Class A concentrates. The AFFF concentrate mixes readily with water and stays suspended in the solution for an extended period of time. When batch mixing AFFF, the water in the tank has to circulate for a few minutes before discharge to ensure complete mixing.

Class A foam concentrates may not retain their foaming properties if mixed in water for more than 24 hours. Further degradation may occur, depending on the product used. Another problem with this method is that lather (froth) may form when the water tank is refilled. This lather may result in either fire pump cavitation or priming difficulty. Because Class A foam concentrates are excellent cleansing agents, they may have a tendency to remove lubricants from ball valve seals. When using batch mixing, pay special attention to these seals. This method should only be used when no other proportioning system is available.

The disadvantage of this method is that all the water onboard the apparatus is converted to foam solution. This method does not allow for continuous finished foam discharge on large incidents because the stream has to be shut down while the apparatus is replenished. It is difficult to maintain the correct concentrate ratio unless the water tank is completely emptied each time.

CAUTION

Alcohol-resistant foam concentrates are generally not designed to be premixed or batch mixed because their polymer components combine rapidly with water to form a membrane that will clog the delivery system.

Premixing

The most common application for premixing is in fire-extinguishing equipment that uses stored pressure for discharge energy. These types of equipment include hand and wheeled portable fire extinguishers, skid-mounted twin-agent units, and rapid intervention vehicle-mounted systems (**Figures 3.24 a and b**). With this equipment, the foam solution is stored in a pressure vessel charged with nitrogen. The agent is discharged when the nozzle is opened. The main disadvantage of this equip-

Table 3.1
Amount of Concentrate Needed for Various Sizes of Water Tanks

Foam Concentrate Proportioning %	Water Tank Size in Gallons (Liters)				
	500 (1 893)	750 (2 839)	1,000 (3 785)	1,500 (5 678)	2,000 (7 571)
1	5 (19)	7.5 (28)	10 (38)	15 (57)	20 (76)
3	15 (57)	22.5 (85)	30 (114)	45 (170)	60 (227)
6	30 (114)	45 (170)	60 (227)	90 (341)	120 (454)

Note: Concentrate is added in gallons (liters).

Figure 3.24a Wheeled premixed foam extinguishers of the type shown may still be found at refineries and other processing facilities in North America. *Courtesy of Conoco, Inc.*

Figure 3.24b Skid-mounted premixed pump and tank units are popular with rural fire departments for use in wildland fire fighting.

ment (as well as the batch mixing method) is that the supply of agent is limited. The entire unit must be recharged to continue application.

Portable Foam Application Devices

[NFPA 1001: 6.3.1; NFPA 1081: 5.3.1, 6.2.3]

Once the foam concentrate and water have been mixed together to form a foam solution, the foam solution must then be mixed with air (aerated) and delivered to the surface of the fuel. With low-energy foam systems, the discharge of finished foam is accomplished by a foam nozzle (sometimes referred to as a *foam maker*). Low-expansion foams may be discharged through either handline nozzles or master stream devices. While standard fire-fighting nozzles can be used for applying some types of low-expansion foams, it is best to use nozzles that produce the desired result (such as fast-draining or slow-draining finished foam). The following sections highlight portable foam application devices: handline nozzles and master steam foam appliances. Foam application devices intended to be used on fixed foam systems are discussed in Chapter 4, Foam Delivery Systems.

NOTE: Foam nozzle eductors and self-educting master stream foam nozzles are considered portable foam nozzles, but they have been omitted from this section because they were discussed earlier in this chapter.

Handline Nozzles

IFSTA defines a *handline nozzle* as any nozzle that one to three firefighters can safely handle and that flows less than 350 gpm (1 325 L/min). Most handline foam nozzles flow considerably less than that amount. Nozzles can be directly connected to foam eductors. The following sections detail the handline nozzles commonly used for foam application: solid bore, fog, air-aspirating foam, and multiagent (**Figure 3.25**).

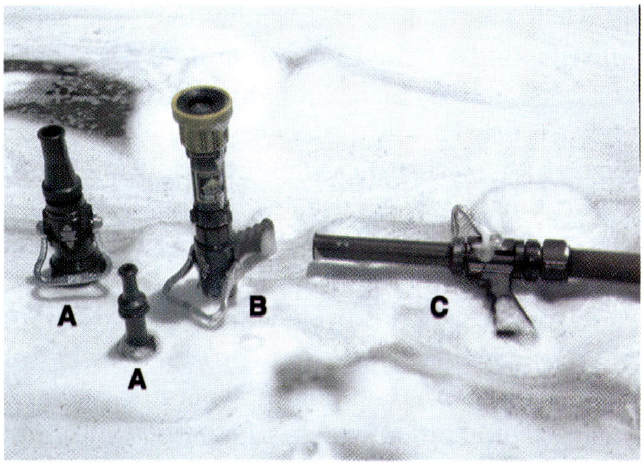

Figure 3.25 Foam handline nozzles (like the ones shown here) include (a) solid bore, (b) fog, and (c) air-aspirating types. Multiagent nozzles (not shown) are also designed for use with foam solutions. *Courtesy of Tom Ruane.*

Solid Bore Nozzles

The use of solid bore nozzles is limited to Class A, compressed-air foam system (CAFS) applications (see High-Energy Foam Generating Systems section). In these applications, the solid bore nozzle provides an effective fire stream that has maximum reach capabilities (**Figure 3.26**). Tests indicate that the reach of the CAFS fire stream can be greater than twice the reach of a low-energy fire stream. When using a solid bore nozzle in conjunction with a CAFS, disregard the standard guideline that the discharge orifice of the nozzle should be no greater than one-half the diameter of the hose. Tests show that a 1½-inch (38 mm) hoseline may be equipped with a nozzle tip up to 1¼-inch (29 mm) in diameter and still provide an effective fire stream.

Figure 3.26 The solid bore nozzle is an excellent choice for situations that demand great reach of the foam stream. *Courtesy of Tom Ruane.*

Fog Nozzles

Either fixed-flow or automatic fog nozzles can be used with foam solution to produce low-expansion, short-lasting finished foam (**Figure 3.27**). These nozzles break the foam solution into tiny droplets and use the agitation of water droplets moving through air to achieve its foaming action. Their best applications are when used with regular AFFF and Class A foam concentrates. These nozzles cannot be used with protein and fluoroprotein foam concentrates. They may be used with alcohol-resistant AFFF foam concentrates on hydrocarbon fires but should not be used on polar solvent fires because insufficient aeration occurs to handle these fires. Some nozzle manufacturers have foam aeration attachments that can be added to the end of the nozzle to increase aeration of the foam solution (**Figure 3.28**).

Fog nozzles used in structural fire fighting can generate excellent foam when fitted with clamp-on foam tubes or adjustable expansion foam nozzles (foam makers). These devices are required when applying finished foam at polar solvent incidents or when maximum vapor suppression is required during postfire operations or where spills must be secured for extended periods (**Figure 3.29**).

Figure 3.27 Fog nozzles are used to produce low-expansion, short-lasting finished foam blankets of AFFF and Class A foam. *Courtesy of Tom Ruane.*

Figure 3.28 Most nozzle manufacturers also market foam aeration attachments for their standard fog nozzles. *Courtesy of KK Products, Inc.*

Figure 3.29 Clamp-on foam tubes attach to standard structural fog nozzles for use in applying AR-AFFF foam to polar solvent fuel spills or fires.

Air-Aspirating Foam Nozzles

The conditions at each incident dictate what type of nozzle to use. The two main choices are standard fog nozzles and air-aspirating foam nozzles. Each has its advantages and disadvantages. The standard fog nozzle has greater reach than the aspirating nozzle. It also allows firefighters to adjust the stream to a fog pattern if necessary. The disadvantage of fog nozzles is that they do not fully aspirate the foam, which reduces postfire foam blanket security.

Air-aspirating nozzles often (but not always) produce fully aerated finished foam of excellent quality. However, they are usually limited to a straight-stream application, and their reach is less than that of a fog nozzle set on a straight-stream setting. On average, a foam stream from a fog nozzle in a straight-stream setting outreaches aerating nozzles by as much as 20 percent.

The air-aspirating foam nozzle inducts air into the foam solution by Venturi action **(Figure 3.30)**. These nozzles must be used with protein and fluoroprotein concentrates. They may also be used with Class A foams in wildland fire applications. These nozzles provide maximum expansion of the agent.

Figure 3.30 The air-aspirating foam nozzle uses the Venturi principle to induct air into the foam solution stream and create finished foam.

Multiagent Nozzles

Multiagent nozzles are special devices that are designed to discharge a variety of extinguishing agents separately or simultaneously. Both twin-agent and quad-agent designs are currently available.

Twin-agent nozzles are special nozzles used on fire extinguishing equipment that simultaneously discharge premixed AFFF foam solution and dry-chemical extinguishing agent **(Figure 3.31)**. The

Figure 3.31 Multiagent nozzles permit the use of dry chemical and halon substitute agents with foam solutions.

dry-chemical agent provides for a fast fire knockdown, while the finished foam produced provides cooling action and prevents reflash of the fire. The most common types of twin-agent nozzles are the side-by-side arrangements and the top-and-bottom arrangements. A separate hoseline supplies each nozzle. An individual pistol-grip control lever that allows either agent to discharge separately or together also controls each nozzle.

The quad-agent nozzle is designed to discharge water, AFFF finished foam, dry-chemical, and halon-substitute (clean agent gas or combustion inhibitor) extinguishing agents. It operates on the same principle as the twin-agent nozzle.

Master Stream Foam Appliances

[NFPA 1081: 5.3.2, 5.3.3, 6.3.3, 7.3.3]

Large-scale flammable and combustible liquid emergencies are beyond the scope of those that can be handled using handlines. Master stream foam appliances are required to deliver adequate amounts of finished foam in these emergencies. As is the case with handline nozzles, standard fixed-flow or automatic master stream fog nozzles may be used to deliver finished foam when necessary. Their performances are much the same as those described for fog nozzles in the Handline Nozzles section; therefore, they are not covered again in this section. The following sections highlight the basic types of master stream foam appliances.

Manual Foam Monitors

Manual foam monitors equipped with fixed-flow or automatic master stream fog nozzles are typically mounted in fixed locations to protect target hazards

(Figure 3.32). Manual foam monitors are commonly found at paper mills, lumberyards, petrochemical refining and storage facilities, loading docks, and aircraft hangers. Manual foam monitors may also be mounted on fire apparatus. These monitors swivel both vertically and horizontally and may be locked in either direction. If necessary, manual foam monitors may also be used to apply plain water. Manual foam monitors are available with flow capabilities ranging from 200 to 2,000 gpm (757 L/min to 7 571 L/min) and higher.

Figure 3.32 Manual foam monitors are master stream appliances that are permanently affixed to water supply pipes for the protection of storage tanks and other units at petroleum refineries. *Courtesy of Conoco, Inc.*

Automatic Oscillating Foam Monitors

Automatic oscillating foam monitors equipped with fixed-flow or automatic master stream fog nozzles are used in fixed locations where it is necessary for the foam stream to sweep back and forth across a wide area **(Figure 3.33).** The nozzle uses the energy of the foam solution flowing through it to propel itself back and forth (in the same manner as a common lawn sprinkler). Depending on the manufacturer of the nozzle, it has a range of motion 80 to 100 degrees from center in either direction.

Figure 3.33 Automatic oscillating monitors sweep over an area of coverage that ranges from 80 to 100 degrees from the center of the nozzle line. *Courtesy of Conoco, Inc.*

These nozzles may flow up to 2,000 gpm (7 571 L/min). They are also capable of flowing plain water with an effective fire stream. If desired, a release pin may be pulled to allow the nozzle to be operated manually. These type nozzles are commonly found in aircraft hangers, tank farms, and loading rack areas.

Remote-Controlled Foam Monitors

Remote-controlled foam monitors equipped with fixed-flow or automatic master stream fog nozzles are operated with a joystick-type control, toggle switches, or a combination of both. They are most commonly found on fire apparatus (particularly aircraft rescue and fire fighting [ARFF] apparatus) and fireboats **(Figure 3.34).** In these cases, the controls for the nozzle are located in the cab (apparatus), on the pump panel, or in the pilothouse (boats) **(Figure 3.35).** The remote controls allow the operator to change the direction, flow rate, and type of stream

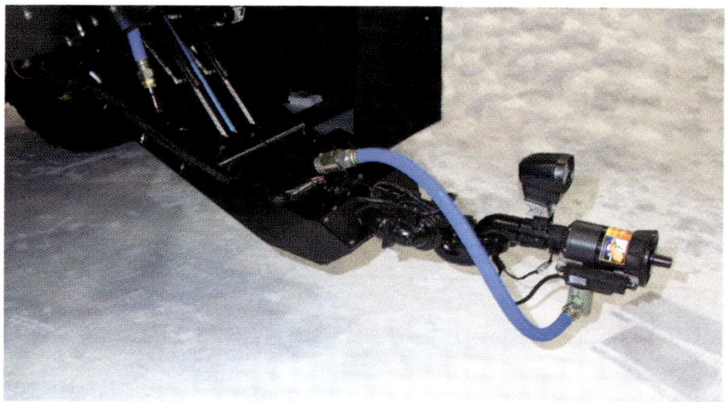

Figure 3.34 Apparatus mounted remote controlled monitors are commonly found on aircraft rescue and fire fighting vehicles.

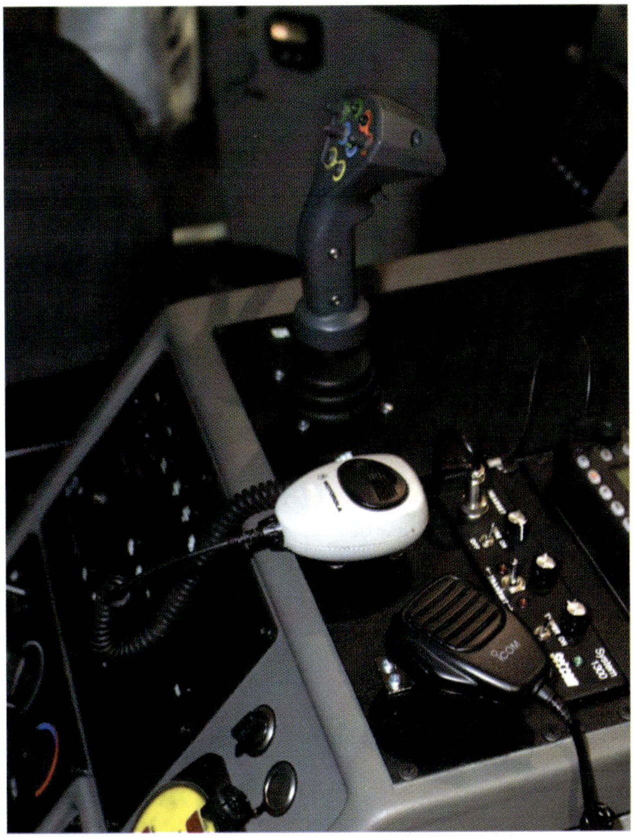

Figure 3.35 Apparatus remote monitor controls resemble modern jet aircraft operating sticks.

pattern the nozzle is flowing. The motors on these nozzles may be either electric or hydraulic or a combination of both. They may flow foam solution or water at rates of up to 3,000 gpm (11 356 L/min).

> **CAUTION**
>
> **Match flow proportioners and the nozzles used with them based on the gpm (L/min) of each device. If the proportioner and nozzle do not match, the proper foam consistency will not be obtained. Thus a 95-gpm (360 L/min) proportioner is used with a 95-gpm (360 L/min) nozzle.**

Medium- And High-Expansion Foam Generating Devices

[NFPA 1001: 6.3.1]

Medium- and high-expansion foam generators produce high-air-content, semistable finished foam. For medium-expansion foam, the air content ranges from 20 parts air to 1 part foam solution (20:1) to 200 parts air to 1 part foam solution (200:1). For high-expansion foam, the ratio is 200:1 to 1,000:1. Two basic types of medium- and high-expansion foam generators are available: the water-aspirating type and the mechanical blower.

The water-aspirating type is very similar to the other foam-producing nozzles except that it is much larger and longer (**Figure 3.36**). The back of the nozzle is open to allow airflow. The foam solution is pumped through the nozzle in a fine spray that mixes with air to form medium-expansion finished foam. The end of the nozzle has a screen or series of screens that further breaks up the foam solution and mixes it with air. These nozzles typically produce lower-air-volume finished foam than do mechanical blower generators.

Figure 3.36 Water-aspirating nozzles like this one have screens over the end that break up the foam solution and mix it with air.

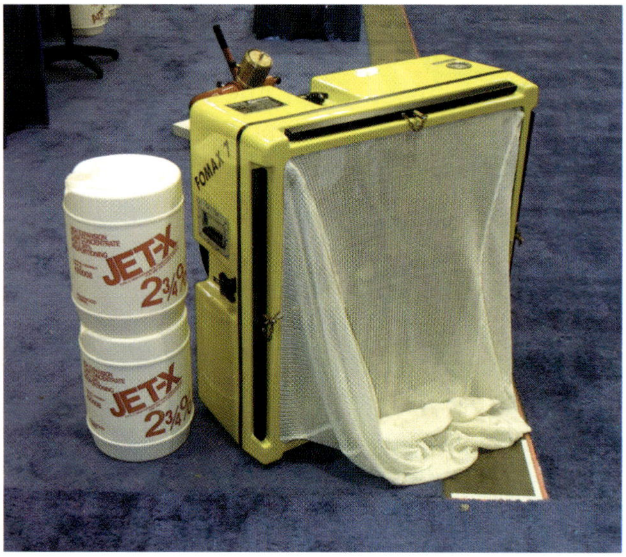

Figure 3.37 Mechanical foam blowers are commonly used for high-expansion foams in total flooding system applications.

A mechanical blower generator is similar in appearance to a smoke ejector (**Figure 3.37**). It operates on the same principle as the water-aspirating nozzle except that air is forced through the foam spray by a powered fan instead of being pulled through by water movement. This device produces finished foam with high air content and is typically associated with total-flooding applications. Its use is limited to high-expansion foam concentrates.

High-Energy Foam Generating Systems

[NFPA 1001: 6.3.1; NFPA 1002: 6-1.1]

High-energy foam generating systems differ from those previously discussed in that they introduce compressed air into the foam solution before discharge into the hoseline. The turbulence of the foam solution and compressed air going through the piping and/or hoseline creates finished foam. In addition to simply forming the foam, the addition of the compressed air also allows the foam stream to discharge a considerably greater distance than does a regular foam or water fire stream (**Figure 3.38**).

In the mid-eighties, the U.S. Bureau of Land Management (BLM) conducted research that led to the development of the high-energy Class A foam system that is now becoming common on structural and wildland fire apparatus. Instead of using air cylinders, the BLM added a rotary-air compressor to a standard fire department pumper. This system uses a standard centrifugal fire pump to supply the water. A direct-injection foam-proportioning system is attached to the discharge side of the fire pump. Once the foam concentrate and water are mixed to form a foam solution, compressed air is added to the mixture before it is discharged from the apparatus and into the hoseline. This system is commonly called a *compressed-air foam system (CAFS)* (**Figure 3.39**).

Using compressed-air foam provides the following tactical advantages:

- The effective reach of the fire stream is considerably longer than streams from low-energy systems.
- A CAFS produces uniformly sized, small air bubbles that are very durable.
- CAFS-produced finished foam adheres to the fuel surface and resists heat longer than low-energy foam.
- Hoselines containing high-energy foam solution are lighter than those containing low-energy foam solution or plain water.
- A CAFS provides a safer fire-suppression action that allows an effective attack on a fire from a greater distance than regular handlines allow.
- A CAFS provides a rapid transition between wet finished foam and dry finished foam.
- Rapid fire extinguishment can result from the use of a CAFS.

A CAFS has the following inherent limitations:

- A CAFS adds expense to a vehicle and can be a high-maintenance item.
- Hose reaction can be erratic with a CAFS if foam solution is not supplied to the hoseline in sufficient quantities.
- The compressed air accentuates the hose reaction in the event the hose ruptures.
- Additional training is required for firefighters and driver/operators who are expected to make a fire attack using a CAFS or who will operate CAFS equipment.
- Maintenance concerns include specialized training and parts inventories.

Figure 3.38 Compressed-air foam systems (CAFSs), originally used for wildland applications, are now found on structural apparatus. Uses range from structure fires to vehicle fires. *Courtesy of Tom Ruane.*

Figure 3.39 CAFS units can be purpose-built, self-contained units like this vehicle intended for quick attack and off-road use.

Most apparatus equipped with a CAFS also flow plain water if desired. In fact, most CAFS-equipped apparatus only flow finished foam through preselected discharges. Other discharges may be capable of flowing finished foam or plain water.

The fire pump on a CAFS-equipped fire vehicle is a standard centrifugal fire pump. The foam proportioning system is some type of automatic, discharge-side proportioning system (**Figure 3.40**). Foam eductors are typically not used because they are not designed to work at either the 0.1 to 1 percent eduction rates or the variable flow rates required for Class A foam concentrates. Variable-rate, automatic-flow sensing proportioners are necessary to ensure that foam concentrate is supplied to the fire stream at the proper rate.

Figure 3.40 With the addition of air-pressure gauges and foam controls, the CAFS pump panel looks different from the standard pump panel on a structural apparatus.

In general, 2 cubic feet per minute (cfm) (0.06 m^3/min) of air flow per gallon (4 L) per minute of foam solution flow produces very dry finished foam at flows up to 100 gpm (379 L/min) of foam solution. This flow produces a large amount of finished foam at a 10:1 expansion ratio. Most structural and wildland fire attacks using CAFS are done with an airflow rate of 0.5 to 1 cfm (0.015 m^3/min to 0.03 m^3/min) per gallon (4 L) of foam solution. This rate allows for adequate drainage of solution from the blanket to wet the fuel and prevent reignition. It also prevents smoldering from occurring beneath the foam blanket.

Summary

Successful foam operations depend on a thorough understanding of the types of foam concentrate and the various delivery devices available to the emergency responder. Each type of delivery device has its own characteristics and requirements. To effectively use the foam concentrate, fire and emergency services personnel must learn the characteristics of the proportioner and nozzle that they employ and practice using them during simulated fire and liquid spill situations. It is essential that the proportioner and nozzle be considered as a system and that the flow rate of in-line foam eductors and nozzles be the same. This consideration must be taken into account when the equipment is purchased and assigned to the operating units before use at an emergency incident.

Courtesy of Mike Wieder

Chapter 4

Foam Delivery Systems

Job Performance Requirements

This chapter provides information that will assist the reader in meeting the following job performance requirements from NFPA 1002, *Standard for Fire Apparatus Driver/Operator Professional Qualifications,* 1998 edition; NFPA 1021, *Standard for Fire Officer Professional Qualifications,* 1997 edition; NFPA 1081, *Standard for Industrial Fire Brigade Member Professional Qualifications,* 2001 edition; NFPA 1003, *Standard for Airport Fire Fighter Professional Qualifications,* 2000 edition; NFPA 472, *Standard for Professional Competence of Responders to Hazardous Materials Incidents,* 2002 edition; and NFPA 1051, *Standard for Wildland Fire Fighter Professional Qualifications,* 2002 edition. Colored portions of the standard are specifically addressed in this chapter.

NFPA 1002

3-1.1 Perform the specified routine tests, inspections, and servicing functions specified in the following list in addition to those contained in the list in 2-2.1, given a fire department pumper and its manufacturer's specifications, so that the operational status of the pumper is verified.

- Water tank and other extinguishing agent levels (if applicable)
- Pumping systems
- **Foam systems**

(a) *Requisite Knowledge:* Manufacturer specifications and requirements, policies, and procedures of the jurisdiction.

(b) *Requisite Skills:* The ability to use hand tools, recognize system problems, and correct any deficiency noted according to policies and procedures.

6-1.1 Perform the specified routine tests, inspections, and servicing functions specified in the following list, in addition to those contained in 2-2.1, given a wildland fire apparatus and its manufacturer's specifications, so that the operational status is verified.

- Water tank and/or other extinguishing agent levels (if applicable)
- Pumping systems
- **Foam systems**

(a) *Requisite Knowledge:* Manufacturer specifications and requirements, policies, and procedures of the jurisdiction.

(b) *Requisite Skills:* The ability to use hand tools, recognize system problems, and correct any deficiency noted according to policies and procedures.

6-2-3 Produce a foam fire stream, given foam-producing equipment, so that the proper proportion of foam is provided.

(a) *Requisite Knowledge:* Proportioning rates and concentrations, equipment assembly procedures, **foam systems limitations,** and manufacturer specifications.

(b) *Requisite Skills:* The ability to operate foam proportioning equipment and connect foam stream equipment.

7-1.1 Perform the routine tests, inspections, and servicing functions specified in the following list in addition to those contained in the list in 2-2.1, given an ARFF vehicle and the manufacturer's servicing, testing, and inspection criteria, so that the operational status of the vehicle is verified.

- **Agent dispensing systems**
- **Secondary extinguishing systems**
- Vehicle-mounted breathing air systems

(a) *Requisite Knowledge:* Manufacturer specifications and requirements, policies, and procedures of the jurisdiction.

(b) *Requisite Skills:* The ability to use hand tools, recognize system problems, and correct any deficiency noted according to policies and procedures.

7-2.1 Maneuver and position an ARFF vehicle, given an incident location and description that involves the largest aircraft that routinely uses the airport, so that the vehicle is properly positioned for safe operation at each operational position for the aircraft.

(a) *Requisite Knowledge:* Vehicle positioning for firefighting and rescue operations; **capabilities and limitations of turret devices related to reach;** and effects of topography, ground, and weather conditions on agent application, distribution rates, and density.

(b) *Requisite Skills:* The ability to determine the appropriate position for the apparatus, maneuver apparatus into proper position, and avoid obstacles to operations.

7-2.2 Produce a fire stream while the vehicle is in both forward and reverse power modulation, given a discharge rate and intended target, so that the pump is safely engaged, the turrets are deployed, the agent is delivered to the intended target at the proper rate, and the apparatus is safely moved and continuously monitored for potential problems.

(a) *Requisite Knowledge:* Principles of agent management and application, effects of terrain and wind on agent application, **turret capabilities and limitations,** tower light signals, airport markings, aircraft recognition, aircraft danger areas, theoretical critical fire area and practical critical fire area, aircraft entry and egress points, and proper apparatus placement.

(b) *Requisite Skills:* The ability to provide power to the pump, determine the appropriate position for the apparatus, maneuver apparatus into proper position, avoid obstacles to operations, apply agent, and determine the length of time an extinguishing agent will be available.

7-2.3 Produce a fire stream, given a rate of discharge and water supplied from the sources specified in the following list, so that the pump is safely engaged, the turrets are deployed, the agent is delivered to the intended target at the proper rate, and the apparatus is continuously monitored for potential problems.

- The internal tank
- Pressurized source
- Static source

(a) *Requisite Knowledge:* Principles of agent management and application, effects of terrain and wind on agent application, **turret capabilities and limitations,** tower light signals, airport markings, aircraft recognition, aircraft danger areas, theoretical critical fire area and practical critical fire area, aircraft entry and egress points, and proper apparatus placement.

(b) *Requisite Skills:* The ability to provide power to the pump, determine the appropriate position for the apparatus, maneuver apparatus into proper position, avoid obstacles to operations, apply agent, and determine the length of time an extinguishing agent will be available.

8-1.1 Perform routine tests, inspections, and servicing functions specified in the following list, in addition to those specified in the list in 2-2.1, given a fire department mobile water supply apparatus, so that the operational readiness of the mobile water supply apparatus is verified.

- Water tank and other extinguishing agent levels (if applicable)
- Pumping system (if applicable)
- Rapid dump system (if applicable)
- **Foam system** (if applicable)

(a) *Requisite Knowledge:* Manufacturer specifications and requirements, policies, and procedures of the jurisdiction.

(b) *Requisite Skills:* The ability to use hand tools, recognize system problems, and correct any deficiency noted according to policies and procedures.

NFPA 1021

2-6.1 Develop a preincident plan, given an assigned facility and preplanning policies, procedures, and forms, so that all required elements are identified and the appropriate forms are completed and processed in accordance with policies and procedures.

(a) *Prerequisite Knowledge:* Elements of a preincident plan, basic building construction, **basic fire protection systems and features,** basic water supply, basic fuel loading, and fire growth and development.

(b) *Prerequisite Skills:* The ability to write reports, to communicate verbally, and to evaluate skills.

NFPA 1081

5.3.2 Activate a fixed fire protection system, given a fixed fire protection system, a procedure, and an assignment, so that the steps are followed and the system operates.

(A) Requisite Knowledge. Types of extinguishing agents, hazards associated with system operation, **how the system operates, sequence of operation,** system overrides and manual intervention procedures, and shutdown procedures to prevent damage to the operated system or to those systems associated with the operated system.

(B) Requisite Skills. The ability to operate fixed fire protection systems via electrical or mechanical means.

6.1.2.1 Utilize a pre-incident plan, given pre-incident plans and an assignment, so that the industrial fire brigade member implements the responses detailed by the plan.

(A) Requisite Knowledge. The sources of water supply for fire protection or other fire-extinguishing agents, site-specific hazards, **the fundamentals of fire suppression and detection systems including specialized agents,** and common symbols used in diagramming construction features, utilities, hazards, and fire protection systems.

(B) Requisite Skills. The ability to identify the components of the pre-fire plan such as fire suppression and detection systems, structural features, site-specific hazards, and response considerations.

6.3.8 Activate a fixed fire protection system, given personal protective equipment, a fixed fire protection system, a procedure, and an assignment, so that the correct steps are followed and the system operates.

(A) Requisite Knowledge. Different types of extinguishing agents, hazards associated with system operation, **how the system operates,** sequence of operation, system overrides and manual intervention procedures, and shutdown procedures to prevent damage to the operated system or to those systems associated with the operated system.

(B) Requisite Skills. The ability to operate fixed fire suppression systems via electrical or mechanical means and to properly shut down fixed fire suppression systems.

7.1.2.3 Utilize a pre-incident plan, given pre-incident plans and an assignment, so that the industrial fire brigade member implements the pre-incident plan.

(A) Requisite Knowledge. The sources of water supply for fire protection or other fire-extinguishing agents, site-specific hazards, **the fundamentals of fire suppression and detection systems including specialized agents,** and common symbols used in diagramming construction features, utilities, hazards, and fire protection systems.

(B) Requisite Skills. The ability to identify the components of the pre-incident plan such as fire suppression and detection systems, structural features, site-specific hazards, and response considerations.

7.3.2 Activate a fixed fire protection system, given personal protective equipment, a fixed fire protection system, a procedure, and an assignment, so that the procedures are followed and the system operates.

(A) Requisite Knowledge. Different types of extinguishing agents on site, **manual fire suppression activities within areas covered by fixed fire suppression systems,** hazards associated with system operation, **how the system operates,** sequence of operation, system overrides and manual intervention procedures, and shutdown procedures to prevent damage to the operated system or to those systems associated with the operated system.

(B) Requisite Skills. The ability to operate fixed fire suppression systems via electrical or mechanical means and to shut down fixed fire suppression systems.

NFPA 1003

3-3.3 Extinguish an aircraft fuel spill fire, given PrPPE, an ARFF vehicle turret, and a fire sized to the AFFF flow rate of 0.13 gpm (0.492 L/min) divided by the square feet of fire area, so that the agent is applied using the proper technique and the fire is extinguished in 90 seconds.

(a) Requisite Knowledge: **Operation of ARFF vehicle agent delivery systems,** the fire behavior of aircraft fuels in pools, physical properties and characteristics of aircraft fuel, agent application rates and densities.

(b) Requisite Skills: Apply fire-fighting agents and streams using ARFF vehicle turrets.

NFPA 472

6.2.4.2 The hazardous materials technician shall identify the impact of the following fire and safety features on the behavior of the products during an incident at a bulk storage facility and explain their significance in the risk assessment process:

(1) **Fire protection systems**

(2) Monitoring and detection systems

(3) Product spillage and control (impoundment and diking)

(4) Tank spacing

(5) Tank venting and flaring systems

Chapter 4
Foam Delivery Systems

In this chapter, various methods to assemble the four primary components of foam delivery systems (water supply under pressure, proportioner, delivery devices, and foam concentrate source) described in Chapter 3, Foam Proportioning and Delivery Equipment, are discussed. Foam fire protection systems are placed in locations where a significant probability for a flammable or combustible liquid fire exists. These systems may be in fixed locations or mounted on mobile platforms. As with other types of fixed fire-suppression systems (such as automatic sprinklers or halon systems), the purpose of a fixed foam system is to control a fire in its early stages. Early intervention reduces or eliminates the need to employ manual methods of fire fighting. This chapter highlights the more common types of foam systems in use and the target hazards they protect. It also highlights mobile and portable foam systems found on fire apparatus.

Foam Systems

[NFPA 1002: 6-2.3; NFPA 1021: 2-6.1; NFPA 1081: 5.3.2, 6.1.2.1, 6.3.8, 7.1.2.3, 7.3.2; NFPA 472: 6.2.4.2]

All foam systems can be classified into one of the following broad categories:

- Fixed
- Semifixed
- Mobile (foam fire apparatus)
- Portable

The following sections contain a basic description of each type of system. The remainder of the chapter shows how each system is applied to protect real-life target hazards.

Fixed

A fixed foam extinguishing system is a complete installation piped from a central foam proportioning station. Fixed foam systems apply finished foam through fixed delivery outlets directly to the target hazard **(Figure 4.1)**. If a pump is required, it is usually permanently installed. Fixed systems may be either the total flooding type or local application type. Both types use either low-, medium-, or high-expansion foam concentrate. Either some type of products-of-combustion detection system or a manual means activates most fixed systems. This fixed system normally has a lengthy or unlimited water supply and may have large foam concentrate supplies as well.

Figure 4.1 Fixed foam systems protect a variety of hazards at refineries including storage tanks, piping, processing units, and pumping stations. *Courtesy of Conoco, Inc.*

Semifixed

A semifixed system is actually a combination of one or more components of a foam delivery system that is fixed, mobile, or portable (according to the NFPA definition). Semifixed foam systems are designed to make the application of finished foam to a target hazard much easier than by strictly manual means. Any combination of the four components used to deliver finished foam is considered a semifixed system (**Figure 4.2**).

Mobile

Mobile foam systems are mounted either on a fire apparatus or trailer. These units are driven or towed to the scene of a spill or fire. Once on the scene, firefighters can use onboard foam and/or water to attack the incident. The most common types of mobile systems are found on industrial, airport fire apparatus, and many municipal fire apparatus (**Figure 4.3**). These apparatus are discussed in more detail later in this chapter.

Portable

Portable systems are those in which all of the equipment necessary to produce the finished foam may be transported by hand. An example of a portable device is a handheld foam fire extinguisher (**Figure 4.4**).

Figure 4.2 Semifixed systems provide hose connections for mobile and portable foam systems plus extinguishing agents to protect a hazard. *Courtesy of Conoco, Inc.*

Figure 4.3 One of the more common types of mobile foam systems is the airport rescue and fire fighting vehicle (like the one shown here).

Figure 4.4 Portable foam systems may consist of hand extinguishers or wheeled units. The unit shown is a multiagent system that includes premixed foam and dry chemical extinguishing agents.

Foam-Water Sprinkler Systems

[NFPA 1021: 2-6.1; NFPA 1081: 5.3.2, 6.1.2.1, 6.3.8, 7.3.2; NFPA 472: 6.2.4.2]

Most automatic sprinkler systems deliver plain water to a fire area when the need arises. A foam-water sprinkler system with aspirating heads provides an excellent foam blanket and consistent expansion because of its pre-engineered characteristics. In some situations, a water sprinkler head can apply aqueous film forming foam (AFFF), alcohol-resistant aqueous film forming foam (AR-AFFF), or film forming fluoroprotein (FFFP) foam. The sections that follow highlight some of the more important elements of these systems. The following NFPA standards are applicable to foam-water automatic sprinkler systems:

- NFPA 11, *Standard for Low-, Medium, and High-Expansion Foam* (2002)
- NFPA 13, *Standard for the Installation of Sprinkler Systems* (2002)
- NFPA 16, *Standard for the Installation of Foam-Water Sprinkler and Foam-Water Spray Systems* (1999)
- NFPA 409, *Standard on Aircraft Hangars* (2001)

Sprinkler System Types

In general, the same types of sprinkler systems used for plain water are also used for foam-water sprinkler systems. The addition of foam proportioning and concentrate storage equipment sets them apart from regular systems. The four types of foam sprinkler systems are as follows: wet pipe, dry pipe, deluge, and preaction. Each of these systems may be equipped with either standard sprinklers or aspirating-type foam-water sprinklers. Finished foam is produced as the water-foam solution exits the piping system through the sprinkler head. A detailed description of these sprinkler heads is provided later in this chapter.

Wet Pipe

In a wet-pipe sprinkler system, either water or foam solution is in the piping at all times. This system is the most reliable, the fastest acting, and the simplest responding type of sprinkler system. When a closed sprinkler's heat-sensitive element fuses, water is immediately applied to the area. This flow of water in the system trips a sprinkler valve that in turn sends an alarm signal to the appropriate location **(Figure 4.5)**. As the water in the piping discharges onto the fire, the new water entering the system through the sprinkler valve mixes with foam concentrate to form foam solution. When all the water that was originally in the piping discharges, foam solution begins to flow from the open sprinkler(s), creating finished foam. Wet-pipe systems cannot be used where freezing conditions are likely to damage piping. Standard water sprinkler systems can usually be converted to foam-water systems when AFFF, AR-AFFF, or Class A foam concentrates are used. AFFF is the most common foam concentrate used in foam-water sprinkler systems. When using AFFF concentrate, the piping to the sprinklers can sometimes be primed with foam solution instead of plain water to enable immediate effective discharge. A test discharge connection is located downstream from the foam proportioner. This connection permits testing of the foam solution for proper concentration. The test connection should be of sufficient size to meet the minimum flow rate for the particular proportioner.

Dry Pipe

In a dry-pipe system, the piping above the dry-pipe sprinkler valve on the system riser contains pressurized air rather than foam solution **(Figure 4.6)**.

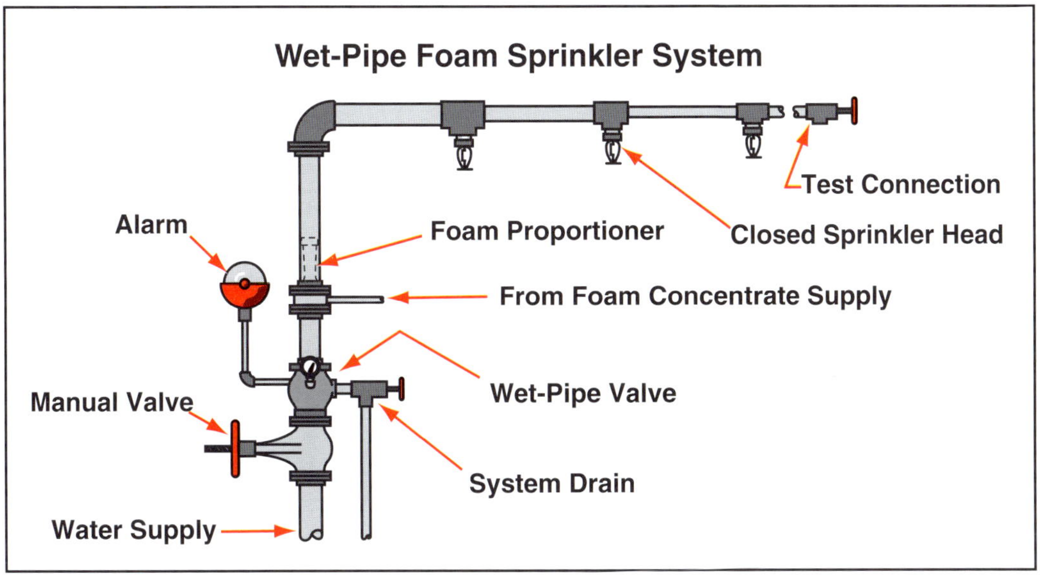

Figure 4.5 Components of a wet-pipe foam sprinkler system.

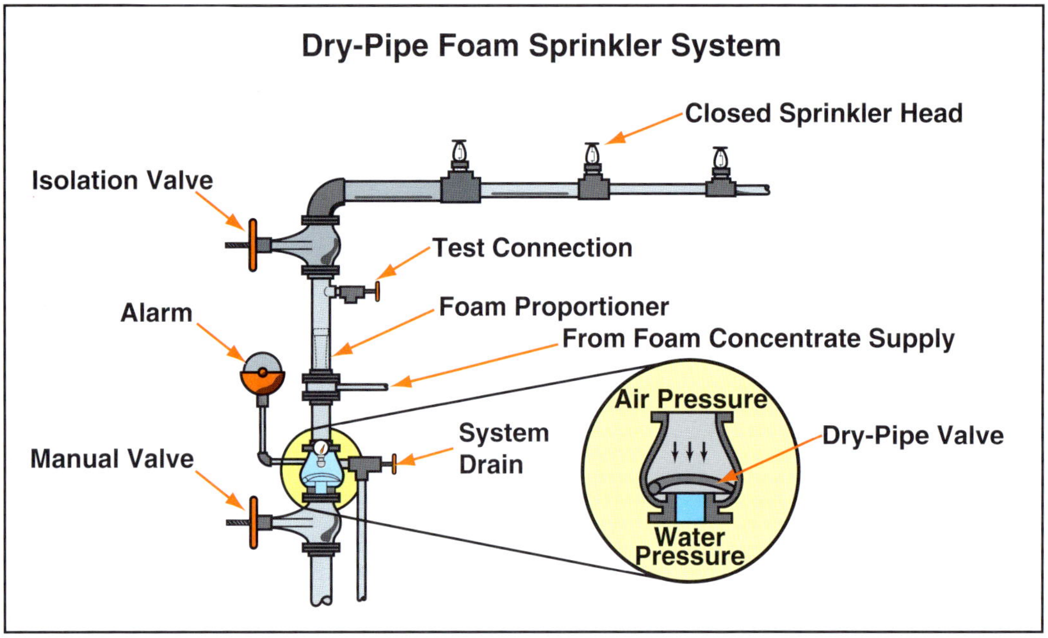

Figure 4.6 Components of a dry-pipe foam sprinkler system.

When a closed sprinkler fuses, the air escapes from the piping. When sufficient air pressure is lost in the piping, the dry-pipe valve opens and allows foam solution into the system. Foam solution then flows into the piping grid and finished foam discharges from the fused sprinkler(s). The dry-pipe valve is designed so that a moderate air pressure prevents a higher foam solution pressure from opening the valve. This system takes longer to apply a finished foam blanket to the fire area than it does for a wet-pipe system.

This system provides protection in areas subject to freezing temperatures. Piping must be drained either of water or foam solution following operation to prevent freezing. Typically, pendent sprinklers are used because they promote better drainage than do upright heads. Approved dry-pendant sprinklers are available if they must be used. As with a wet-pipe system, a dry-pipe system has a test connection built into it. Standard dry-pipe systems can usually be converted to AFFF foam-water systems.

Deluge

A deluge system is the most common foam-water sprinkler system. In the deluge system, all sprinklers are always open, and no foam solution is in the piping. Most deluge systems have a detection system that senses fire and opens the foam solution control valve. Heat and/or flame detection systems are the most common types of actuators used to trigger deluge valves. Other systems are only manually actuated. Once the valve is open, foam solution flows through the piping system and finished foam discharges from all sprinklers simultaneously. This system is commonly employed in locations where an immediate, large flow of extinguishing agent is required to control a fire. Dip tank/quenching operations, parking garages, flammable liquid loading racks, woodworking shops, and airplane hangars are examples of typical occupancies that might use these systems (**Figure 4.7**).

Preaction

A preaction foam-water sprinkler system is similar to a dry-pipe system. A products-of-combustion detection system opens the water-control valve on the riser, allowing water to enter the sprinkler piping. When the sprinkler head or heads activate due to the increased temperature in the fire area, water flows from the head or heads. The foam concentrate control valve that injects foam into the water stream may activate at the same time the water-control valve operates or when a second detection system signals it to operate (**Figure 4.8**). The foam concentrate control valve is also the proportioner for the system, injecting the correct ratio of foam concentrate for the required delivery. Once the foam concentrate tank is empty, the sprinkler system continues to provide a water spray from the open heads.

Figure 4.7 The controls for a deluge foam-water sprinkler system in a loading rack.

System Design Considerations

A number of important considerations are involved in the design of a foam-water sprinkler system. These include the size and type of hazard protected, the spacing of sprinklers, and the flow capabilities of the system. With the exception of deluge systems, sprinkler systems have only a portion of the sprinklers flowing at the same time. This portion of the total number of sprinklers is called the *design* or *demand area.* The design area is based on numerous considerations: The most important are the occupancy type and level of hazard present. Directions on determining the design area are found in the various NFPA standards associated with sprinkler and foam-water sprinkler design.

In addition to determining how many of the sprinklers will flow, fire and emergency responders must also determine how much foam-water solution each sprinkler needs to flow in order to produce an adequate finished foam blanket to control the fire. The amount, type, and configuration of the fuel require case-by-case determinations of the flow required from each sprinkler. The application rate (commonly referred to as *density* or *design density*) is usually expressed in gallons per minute per square foot (gpm/ft^2) or liters per minute per square

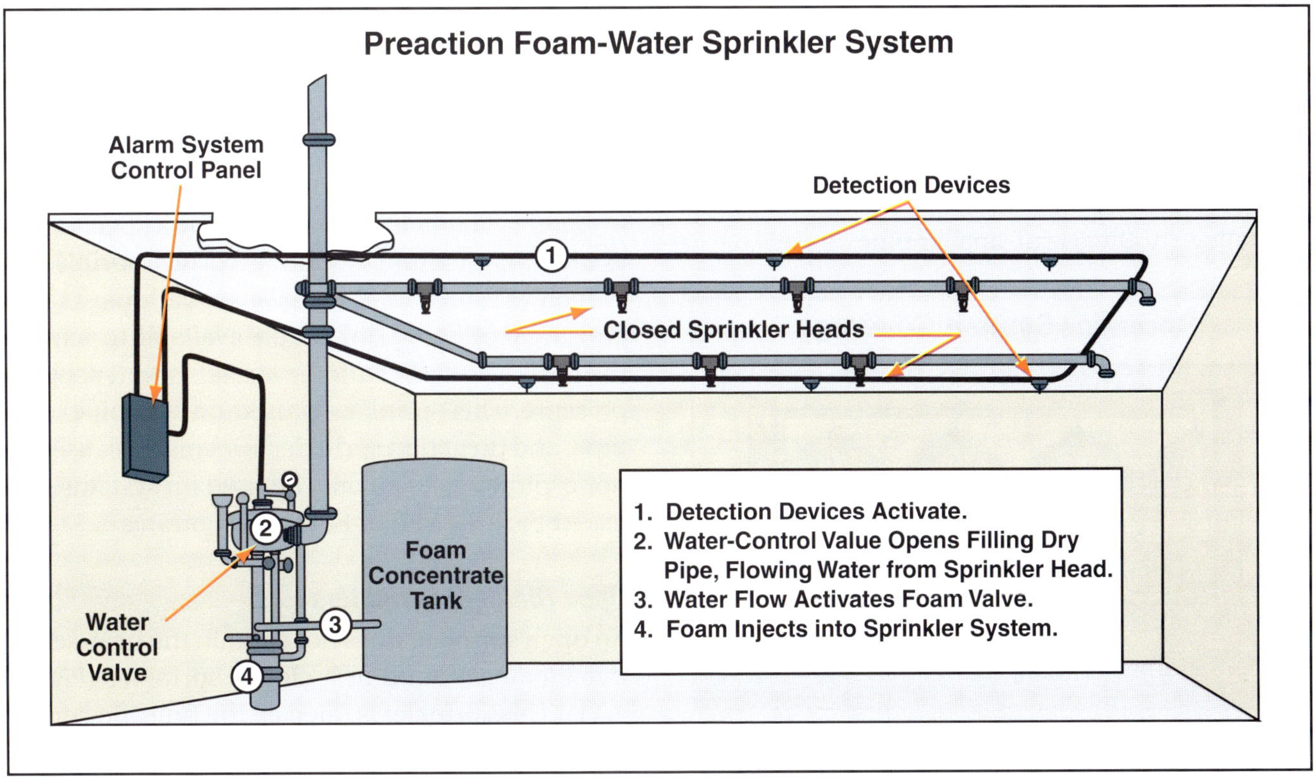

Figure 4.8 An illustration of a preaction foam-water sprinkler system and how it operates.

meter (L/min/m²) of area. Characteristics of some flammable products may require higher densities and special concentrates. Again, directions on determining the design density are found in the NFPA standards associated with sprinkler and foam-water sprinkler design, Underwriters Laboratories Inc. (UL) or Factory Mutual (FM) listings, and sprinkler system manufacturers' recommendations. Although the hazard protected has a direct bearing on the duration of a foam-water solution discharge needed to generate an adequate finished foam blanket, all the standards specify that a 10-minute minimum supply of foam concentrate must be available for discharge into the system.

The sprinkler provides for a relatively uniform distribution of water or finished foam over a given area. The solution strikes a deflector in the sprinkler head and breaks into a distribution pattern. Proper spacing of sprinklers results in overlapping patterns to ensure uniform coverage and effective fire control **(Figure 4.9).** As with determining the design area size, the spacing of sprinklers is determined by the occupancy type and level of hazard present. The NFPA standards provide direction for spacing, given various parameters.

The size of piping to the sprinkler heads may be determined by either pipe schedule methods (charts) or hydraulic calculations to ensure an adequate supply of water and consistent distribution from the sprinklers. Most modern systems, particularly complicated foam-water systems, are hydraulically calculated. These systems are more economical than pipe schedule systems because they allow for the use of smaller pipe.

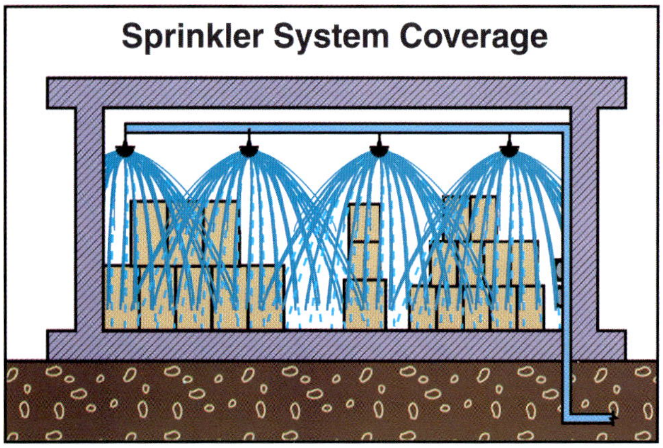

Figure 4.9 Sprinkler systems are designed to provide complete coverage of a protected area through overlapping spray patterns.

Sprinkler Types

Once a foam concentrate has been added to water (to form a foam solution), there must be a place in the system where air is added to the solution to produce expanded foam. A discharge device is used to add air to the solution. In this case, the discharge device is either an aspirated or nonaspirated sprinkler.

In an aspirated discharge device, foam solution passes through an orifice, moves past air inlets into a mixing or expansion area, and then discharges. In a nonaspirated device, foam solution passes through an orifice and stream deflector to produce droplets of solution. These droplets combine with air between the device outlet and the fuel surface to produce finished foam. Nonaspirated devices are only used with film forming foam solutions that require less energy to expand than protein-based or nonfluorochemical synthetic foams.

Three types of sprinklers may be used on foam-water sprinkler systems: closed (nonaspirating), open (deluge, nonaspirating), and foam-water (aspirating). A directional foam spray nozzle is used on some foam sprinkler systems. Each type of sprinkler/nozzle and its individual application is discussed in the following sections.

Closed Sprinklers

The closed sprinkler is the most common type and is available for upright, pendant, or sidewall installations **(Figures 4.10 a–c).** The sprinkler is held closed by a heat-sensitive element (a fusible link or frangible glass bulb) and is rated for a specific operating temperature. A standard orifice sprinkler is ½ inch (13 mm) in size; however, various smaller and larger orifice sprinklers are available to achieve the desired density with the available water supply pressure. This sprinkler is used on wet-pipe, dry-pipe, and preaction sprinkler systems. Because it is nonaspirating, it can only be used on systems that use AFFF, AR-AFFF, or FFFP concentrates.

Open (Deluge) Sprinklers

An open sprinkler does not contain the heat-sensitive element found in a closed sprinkler **(Figure 4.11).** This sprinkler is open all the time. Because it is nonaspirating, it can only be used on systems that use AFFF, AR-AFFF, or FFFP concentrates.

Figure 4.10a Upright sprinkler head installation.

Figure 4.10b Pendent sprinkler head installation.

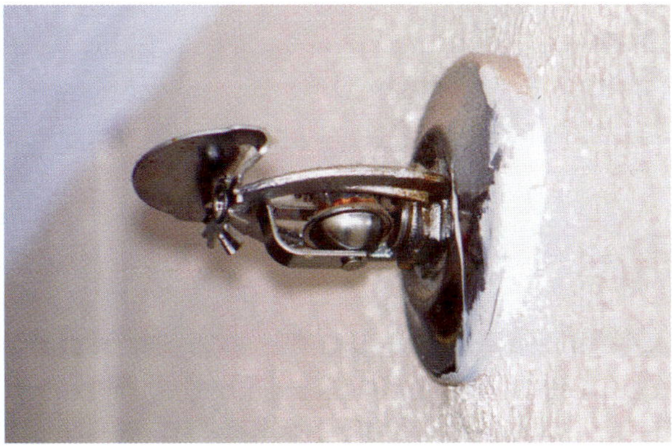

Figure 4.10c Side-wall sprinkler head installation.

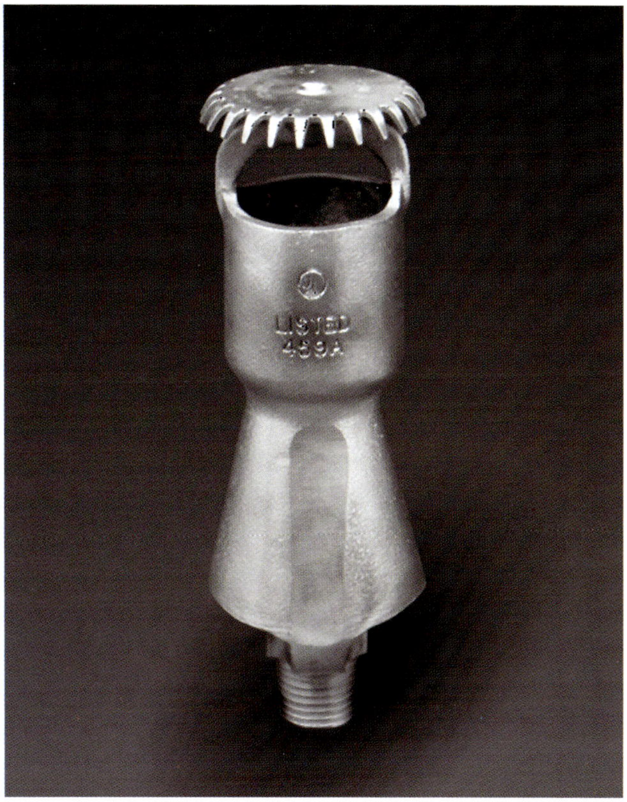

Figure 4.11 An open deluge sprinkler head. This is the most common type of sprinkler head used in foam sprinkler systems. *Courtesy of National Foam, Inc.*

Foam-Water Sprinklers

A foam-water sprinkler is an open-type device that is found on deluge foam-water systems (**Figure 4.12**). A foam-water sprinkler resembles an aspirating nozzle in that it uses the Venturi principle to mix air into the foam solution before discharge. Foam-water sprinklers come in upright and pendant designs. The deflectors on this sprinkler

Figure 4.12 Foam-water sprinklers are used to protect electric transformers like the one shown. *Courtesy of National Foam, Inc.*

must be adapted to meet the specific installation requirements. Patterns of coverage are similar to conventional sprinkler heads.

Directional Foam Spray Nozzles

Although not technically considered a sprinkler, a directional foam spray nozzle is used on some foam sprinkler systems. It is usually similar to a standard foam nozzle but is mounted in a fixed location. This type of nozzle is frequently used in truck loading rack systems to provide additional finished foam to the truck undercarriage. It is also commonly used in aircraft hangars to provide protection beneath planes (**Figure 4.13**). Other types of nozzles (both air aspirating and nonaspirating) are available to meet the needed coverage for specific hazard applications.

Figure 4.13 Example of an under-wing foam spray nozzle.

Outdoor Storage Tank Protection Systems

[NFPA 1021: 2-6.1; NFPA 1081: 5.3.2, 6.1.2.1, 6.3.8, 7.3.2; NFPA 472: 6.2.4.2]

The only practical way to adequately protect outdoor petroleum storage tanks is with foam systems. Because the design of petroleum storage tanks differs widely, each requires foam systems tailored to its particular characteristics. The following sections describe fixed cone roof tanks, external floating roof tanks, and internal floating roof tanks that firefighters may encounter and the types of foam system used to protect each.

In the fire and emergency services, the following three accepted delivery methods for applying finished foam are available:

- ***Type I discharge outlet*** — Device that conducts and delivers finished foam onto the burning surface of a liquid without submerging it or agitating the surface; no longer manufactured and considered obsolete; however, may still be used in some isolated facilities and areas (**Figure 4.14**).
- ***Type II discharge outlet*** — Device that delivers finished foam onto the burning liquid and partially submerges it and produces restricted agitation of the surface (**Figure 4.15**).
- ***Type III discharge method*** — Includes master streams and handlines that deliver finished foam in a manner that causes it to fall directly onto the surface of the burning liquid and does so in a way that causes general agitation (**Figure 4.16**); includes use of monitors and handlines (see details in Chapter 7, Class B Foam Fire Fighting: Class B Liquids at Fixed Sites).

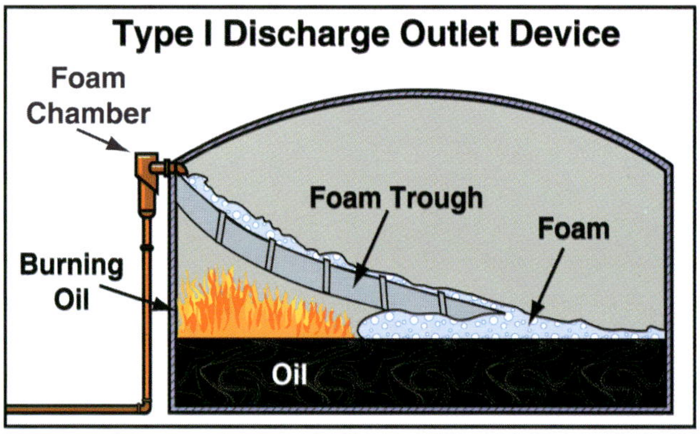

Figure 4.14 Type I foam discharge outlet: Foam discharges from the foam chamber at the top of the tank and flows down the trough onto the surface of the tank contents. This type of device is obsolete but may be encountered in old storage tanks.

Figure 4.15 Type II foam discharge outlet device: The foam chamber is located on the reverse side of the wall in this cut-away training tank. The foam pours down onto the surface of the fuel extinguishing the fire.

Figure 4.16 Type III discharge method: In this example, firefighters advance a foam handline in the center while two water fog handlines provide overhead exposure protection for them.

Fixed Cone Roof Tanks

An ordinary fixed cone roof tank (sometimes referred to as a *dome roof tank*) stores flammable, combustible, and corrosive liquids. It has a cone-shaped, pointed roof **(Figure 4.17).** Tanks over 50 feet (15 m) in diameter are designed with a weak roof-to-shell seam that breaks when/if the container becomes overpressurized. A disadvantage of this type of tank is that when it is partially full, the remaining portion of the tank contains a potentially dangerous vapor space that can be explosive if the area is exposed to an ignition source. These tanks and recommended foam fire fighting methods are described in more detail in Chapter 7, Class B Foam Fire Fighting: Class B Liquids at Fixed Sites. Four acceptable methods of protecting cone roof tanks are available: surface application (foam chambers), subsurface injection, semisubsurface injection, and foam monitors and handlines. The Type II fixed systems applications are discussed in the sections that follow, while the Type III monitors and handlines are discussed in Chapter 7, Class B Foam Fire Fighting: Class B Liquids at Fixed Sites.

Figure 4.17 Fixed cone roof storage tank. *Courtesy of Conoco, Inc.*

Surface Application

Fixed discharge outlets (commonly referred to as *foam chambers*) apply the finished foam to the surface of the burning fuel. The finished foam is applied by one or more foam chambers installed on the shell of the tank just below the roof joint **(Figure 4.18).** If two or more foam chambers are used on one tank, they must be equally spaced around the perimeter

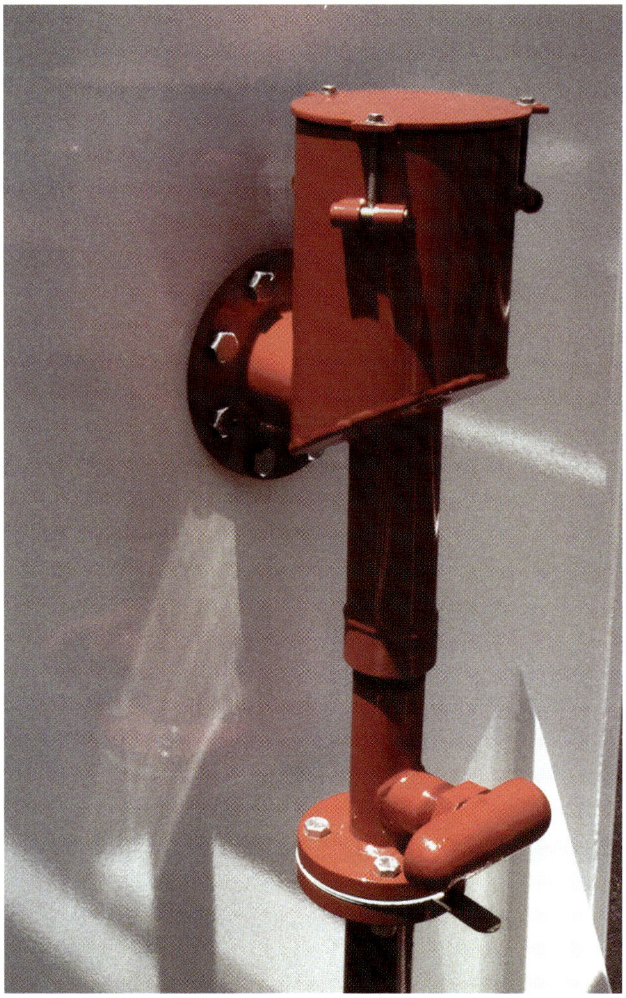

Figure 4.18 Foam chambers, like the one shown, are mounted on the top exterior of cone roofed storage tanks.

Table 4.1
Number of Fixed Foam Discharge Outlets for Fixed-Roof Tanks Containing Hydrocarbons or Flammable and Combustible Liquids Requiring Alcohol-Resistant Foams

Tank Diameter (or Equivalent Area) Feet	Meters	Minimum Number of Discharge Outlets
Up to 80	Up to 24	1
Over 80 to 120	Over 24 to 36	2
Over 120 to 140	Over 36 to 42	3
Over 140 to 160	Over 42 to 48	4
Over 160 to 180	Over 48 to 54	5
Over 180 to 200	Over 54 to 60	6

Reprinted with permission from NFPA 11, *Standard for Low-, Medium-, and High-Expansion Foam*, Copyright © 2002, National Fire Protection Association, Quincy, MA 02269. This reprinted material is not the complete and official position of the National Fire Protection Association on the referenced subject, which is represented only by the standard in its entirety.

of the tank. **Table 4.1** shows the minimum number of foam chambers required for a tank based on its size. A foam solution pipe is extended from the proportioning source outside the dike wall to the foam chambers. A deflector is located inside the tank shell to deflect the discharge against the shell (**Figure 4.19**).

The foam chamber contains an orifice plate (sized for the required flow and inlet pressure), air inlets, an expansion area, and a discharge deflector to direct the gentle application of the expanded foam down the inside of the tank. This device also contains a vapor seal that prevents the entrance of vapors into the foam chamber and the supply pipe. Foam chambers are typically supplied by a fixed or semifixed system arrangement. The piping to the foam chambers is supplied by a fixed fire pump and foam proportioning system or mobile apparatus. The minimum fixed system application rate of foam for hydrocarbon fires (as specified by NFPA 11) is 0.10 gpm/foot^2 (4.1 L/min/m^2). The duration of application varies from 20 to 55 minutes, depending on the exact type of discharge devices and the flash point of the fuel. The application rate and discharge time for tanks containing polar solvent fuels vary widely. Consult the manufacturer of the foam concentrate used in the system for the requirements for each specific case.

Surface applications using foam chambers have several advantages. The first is that each system is engineered specifically for a particular application. Another advantage is that not as much foam con-

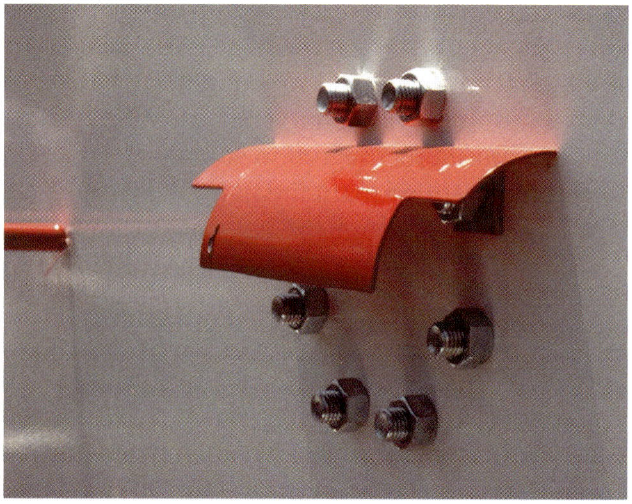

Figure 4.19 Inside the storage tank, a deflector directs the foam against the interior wall.

centrate is lost or wasted when compared to extinguishing the fire using monitors or nozzles. Surface application may be used on both hydrocarbon and polar solvent fuels and with a variety of foam concentrates. The primary disadvantage with these systems is that they can be damaged by an initial explosion or fire. This situation requires manual fire-fighting techniques to control the incident.

Subsurface Injection

One way to avoid the potential problem of having surface application equipment damaged beyond use before foam application is to inject finished foam below the surface of the fuel and away from the exact location of a fire/explosion. This method is called *subsurface injection* **(Figures 4.20 a and b).** Subsurface injections are permitted by NFPA 11 for use in vertical fixed-roof atmospheric storage tanks that contain liquid hydrocarbons. In this type of system, finished foam is injected into the base of the tank and allowed to float to the top of the fuel where it forms a blanket over the surface of the fuel. Finished foam that is subsurface injected expands less than finished foam that is applied through surface application or manual application equipment. A 4:1 expansion ratio is common for subsurface injection.

The finished foam may be discharged into the tank using independent foam delivery lines or through the tank's product fill line. If independent foam lines are used, they must be spaced equally around the edge of the tank. **Table 4.2** shows the number of discharges required based on the size of the tank and amount of fuel in the tank. It is

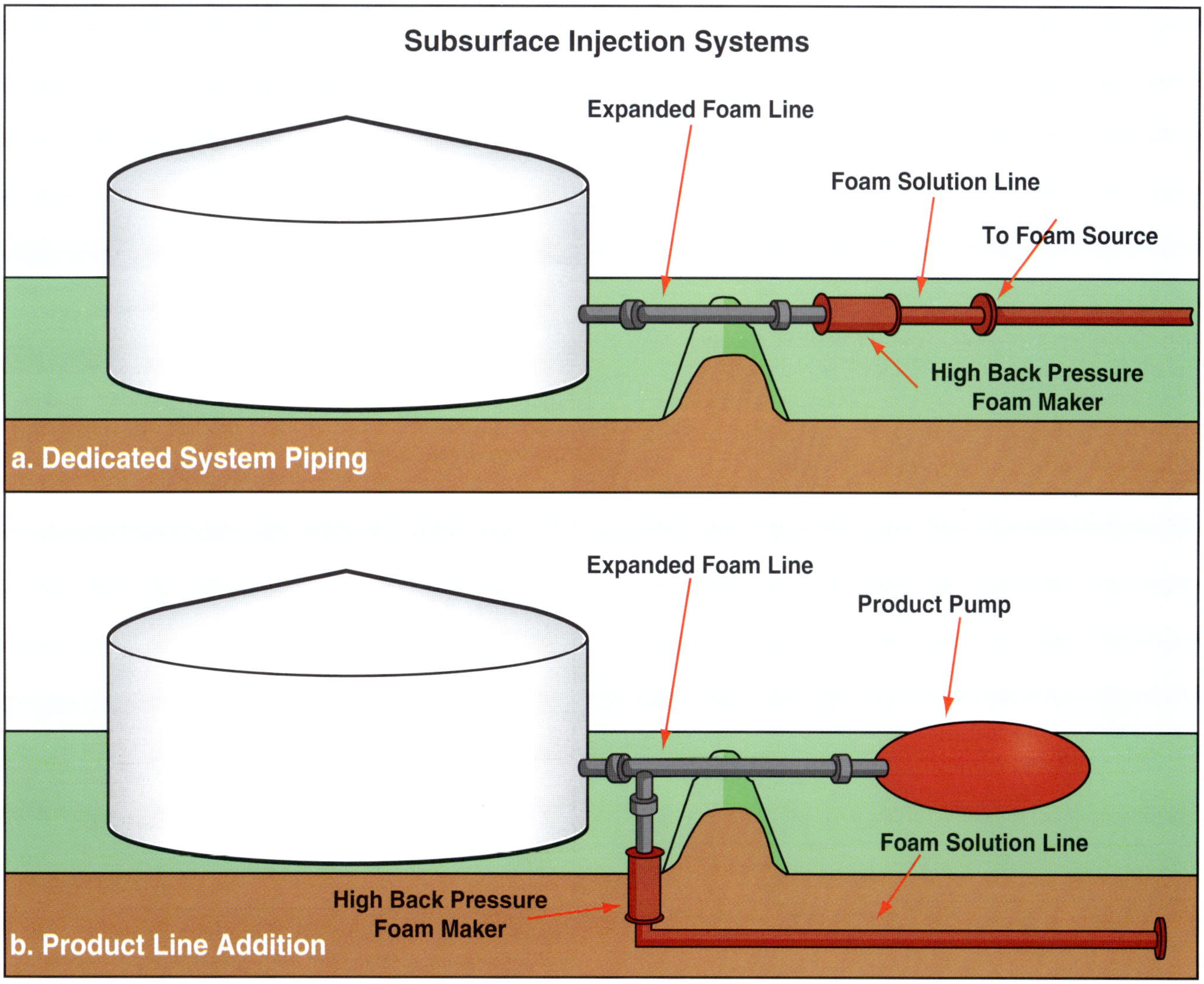

Figures 4.20 a and b Two types of subsurface injection systems. *(a)* Dedicated system piping: Foam is injected through a dry pipe that is dedicated to the foam system. *(b)* Product line addition: Foam is injected into the tank product line.

important that finished foam discharges above the layer of water commonly found resting in the bottom of the tank. The water collects in the bottom of the tank as a result of condensation and leaks. Attempting to pump finished foam through the layer of water results in the foam's destruction.

Table 4.2
Minimum Number of Subsurface Foam Discharge Outlets for Fixed-Roof Tanks Containing Hydrocarbons

Tank Diameter		Minimum Number of Discharge Outlets	
Feet	Meters	Flash Point Below 100° (37.8°C)	Flash Point 100°F (37.8°C) or Higher
Up to 80	Up to 24	1	1
Over 80 to 120	Over 24 to 36	2	1
Over 120 to 140	Over 36 to 42	3	2
Over 140 to 160	Over 42 to 48	4	2
Over 160 to 180	Over 48 to 54	5	2
Over 180 to 200	Over 54 to 60	6	3
Over 200	Over 60	6 plus 1 outlet for each additional 5,000 ft² (465 m²)	3 plus 1 outlet for each additional 7,500 ft² (697 m²)

Notes:

(1) Liquids with flash points below 73°F (22.8°C), combined with boiling points below 100°F (37.8°C), require special consideration.

(2) Table 4.2 is based on extrapolation of fire test data on 25-ft (7.5-m), 93-ft (27.9-m), and 115-ft. (34.5-m) diameter tanks containing gasoline, crude oil, and hexane, respectively.

(3) The most viscous fuel that has been extinguished by subsurface injection where stored at ambient conditions [60°F (15.6°C)] had a viscosity of 2,000 ssu (440 centistokes) and a pour point of 15°F (-9.4°C). Subsurface injection of foam generally is not recommended for fuels that have a viscosity greater than 2,000 ssu (440 centistokes) at their minimum anticipated storage temperature.

(4) In addition to the control provided by the smothering effect of the foam and the cooling effect of the water in the foam that reaches the surface, fire control and extinguishment can be enhanced further by the rolling of cool product to the surface.

Most subsurface injection systems are semifixed systems. The piping that supplies the discharge(s) runs from the discharge(s) to an intake manifold outside the dike area. An independent fire pump (usually a pumper) connected to either a fire hydrant or static water supply source pumps the foam solution into the system **(Figure 4.21).** Some subsurface injection systems are completely fixed systems supplied by a dedicated fire pump and foam concentrate supply.

Because of the amount of piping involved in these systems and the back pressure of the fuel in the storage tank, a high back pressure foam maker is required. The high back pressure foam maker (or forcing foam maker) is an in-line aspirator used to deliver finished foam under pressure **(Figure 4.22).** High back pressure aspirators supply air directly to

Figure 4.21 Mobile fire apparatus provide the foam solution for some types of subsurface foam injection systems. *Courtesy of Paul Valentine.*

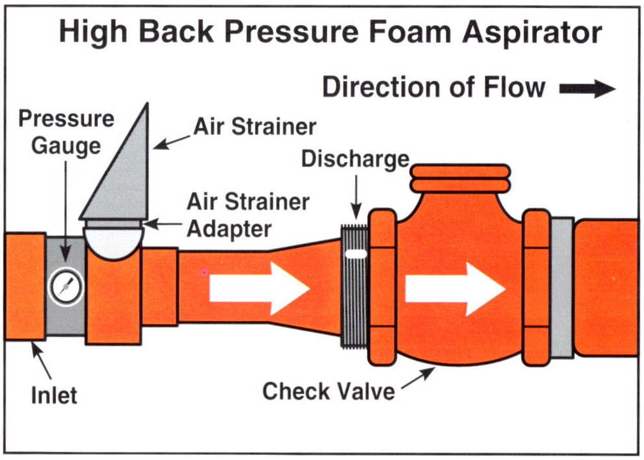

Figure 4.22 Because of the high pressure created by the fuel in the storage tank, high back pressure foam aspirators may be installed in the subsurface foam injection line. Air pressure is added to the foam to help it overcome the resistance of the fuel.

the foam solution through Venturi action. This action typically produces low-air content but homogeneous and stable finished foam.

For both the subsurface injection system and the surface application method, NFPA 11 recommends the application rate of 0.10 gpm/foot^2 (4.1 L/min/m^2) for hydrocarbon fires. The duration of application varies from 30 to 55 minutes, depending on the flash point of the fuel. Fluoroprotein, AFFF, and FFFP concentrates are most commonly used for subsurface injection. Regular protein foam concentrates cannot be used for subsurface injection. Protein finished foam becomes saturated with the flammable liquid and burns after rising to the surface. Subsurface injection systems cannot protect tanks containing polar solvent fuels. The fuel would destroy the finished foam before it ever reached the surface.

NOTE: Subsurface foam systems cannot be used with hydrocarbon products (such as Bunker C oil and asphalt) that have viscosities above 2,000 SSU/SUS (440 centistokes) at 60°F (15°C) or with any fuels heated above 200°F (93°C). SSU/SUS (Saybolt Universal Seconds) and centistokes are measurements of viscosity used in the petroleum industry and defined by the American Society for Testing and Materials (ASTM) standards.

Two primary advantages of subsurface injection systems are as follows:

- Finished foam is efficiently delivered to the surface of the fuel without being affected by wind or thermal updrafts.
- The chance of subsurface foam equipment being damaged by the initial explosion or fire is substantially less than that of fixed, surface-application equipment.

Semisubsurface Injection

The equipment used for semisubsurface injection systems is basically the same as that used for regular subsurface injection systems. The primary difference is at the actual delivery point of the foam: Regular subsurface injection systems inject finished foam through fixed orifices in the bottom of the tank, while semisubsurface injection systems discharge finished foam through a flexible hose that rises from the bottom of the tank through the fuel to the surface.

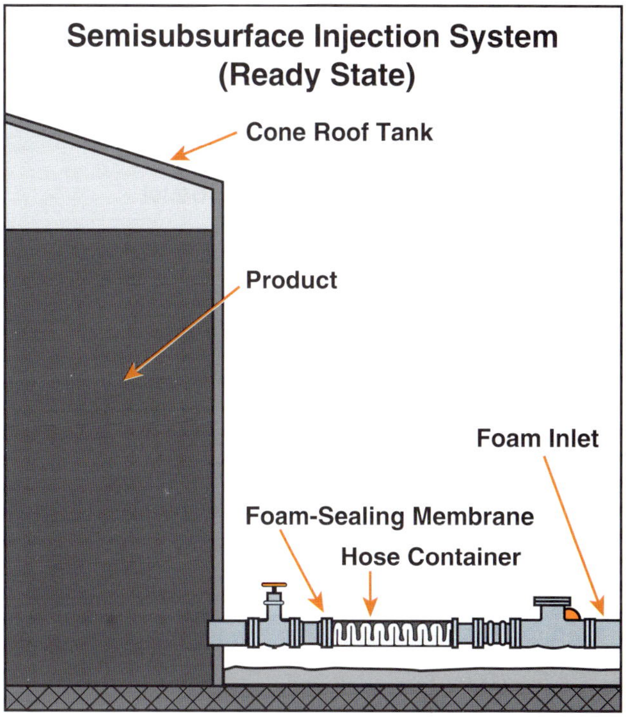

Figure 4.23 A semisubsurface foam injection system before use. A membrane at the tank end of the injection hose prevents foam from leaking into the tank and prevents tank contents from damaging the hose.

Under nonfire conditions, the flexible hose is contained within a housing located at the base of the tank **(Figure 4.23)**. When the foam system is placed into service, the hose is released from the housing. The buoyancy of the foam solution inside the hose causes it to float to the surface of the fuel. Once deployed, the hose discharges finished foam directly to the surface of the burning fuel **(Figure 4.24, page 114)**.

The application rates and discharge times listed previously for subsurface injection systems also apply to semisubsurface injection systems. The same advantages also apply, with one notable addition: Because semisubsurface injection systems actually apply finished foam directly to the surface of the fuel, they may be used in tanks that contain polar solvent fuels. Obviously, alcohol-resistant (polar solvent) foam concentrates must be used in these situations. The flow rates and application times vary depending on the type of fuel and the foam concentrate manufacturer's instructions.

External Floating Roof Tanks

The external floating roof tank (sometimes referred to as the *open-top floating roof tank*) is similar to the cone roof tank except that it has no fixed roof. A

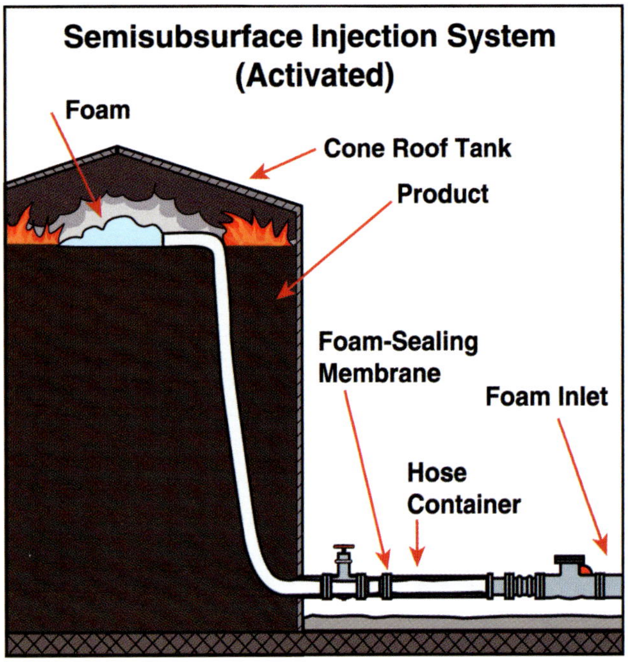

Figure 4.24 When the semisubsurface foam injection system activates, the membrane ruptures, and the hose releases into the tank. The hose floats to the surface of the fuel, releasing the foam.

Figure 4.25 The exterior floating roof tank may have catwalks around the perimeter of the top edge to permit access to the tank roof and seal area. This unit is used for flammable liquids fire-fighting training.

pontoon-type roof or double-deck roof floats directly on the flammable liquid surface (**Figure 4.25**). The space between the roof and the tank shell is equipped with a mechanical shoe or tube seal. These tanks are excellent for the storage of flammable fuels such as gasoline that release large quantities of vapors. By having the roof float on the surface of the liquid, no vapor space poses an explosion hazard.

The only place where the surface of the flammable liquid may be exposed is at the point where the shell and roof come together. Fixed foam protection systems are usually designed to extinguish a fire in this "seal" area. Full-surface fires are only possible if the entire floating roof sinks or tilts. If this situation occurs, it is necessary to launch a full-scale attack using monitors and/or mobile foam apparatus.

Two basic seal designs are commonly used on external floating roof tanks: the mechanical shoe seal and the tube seal (**Figure 4.26**). The mechanical shoe seal (also called the *pantograph seal*) consists of a fabric seal that is anchored to the top of the roof and rides on the inside of the tank wall. The actual mechanical shoe (or *pantograph*) is attached below the fabric seal to keep the roof

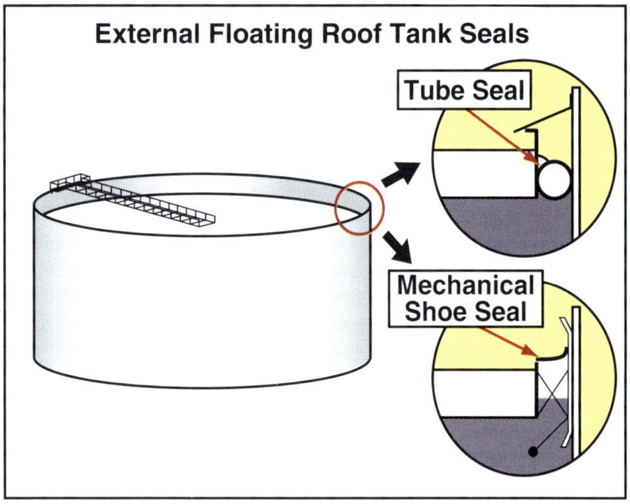

Figure 4.26 Mechanical shoe and tube seals found on external floating roof tanks.

properly aligned within the tank. The tube seal is constructed of urethane foam that is contained within an envelope. The seal is connected to the edge of the roof around the entire circumference of the tank. A secondary weather shield is usually installed above the main seal.

Three basic types of foam systems are used to protect external floating roof tanks: foam chambers on the tank shell, foam discharges on the floating roof, and handline systems. Handline systems are discussed in Chapter 7, Class B Foam Fire Fighting: Class B Liquids at Fixed Sites, while the others are discussed in the sections that follow.

Tank Shell Foam Chambers

This system involves the discharge of finished foam into the seal area from foam chambers mounted on steel plates above the top rim of the tank shell. A finished foam dam is required to retain the foam over the seal or weather shield **(Figure 4.27).** This dam is normally 12 or 24 inches (300 or 600 mm) in height. The dam must be located at least 1 foot (0.3 m) but no more than 2 feet (0.6 m) from the wall of the tank. When a secondary seal is installed, the finished foam dam should extend at least 2 inches (50 mm) above the top of the secondary seal.

NFPA 11 contains the requirements for foam application rate, duration of discharge, and spacing of the foam chambers. In general, an application rate of 0.3 gpm/foot^2 (12.2 L/min/m^2) and a discharge duration of 20 minutes are specified for all tanks, regardless of whether they contain hydrocarbon or polar solvent fuels. The foam chambers must be 40 feet (12.2 m) apart on tanks with 12-inch (300 mm) dams and 80 feet (24.4 m) apart on tanks with 24-inch (600 mm) dams. Regular fluoroprotein, AFFF, and FFFP concentrates or their alcohol-resistant counterparts may be used in these systems.

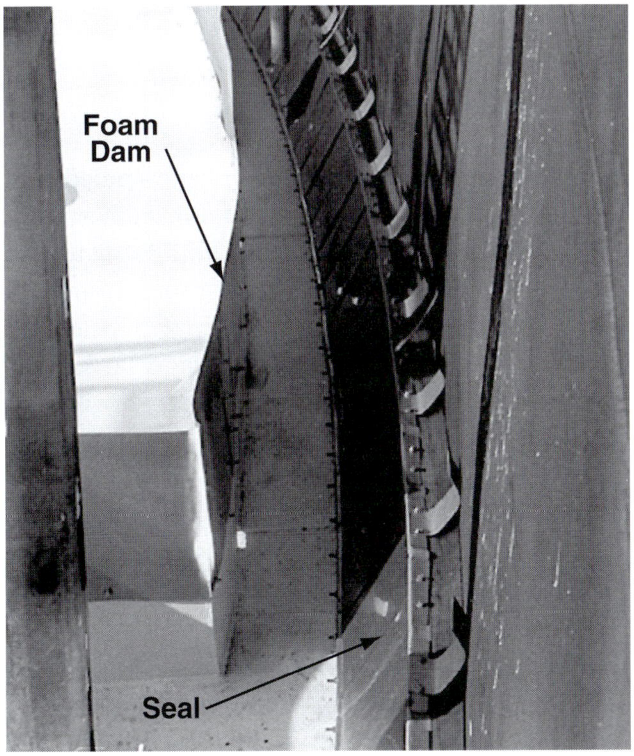

Figure 4.27 A foam dam keeps foam in the seal area where a fire can most likely be contained.

Floating Roof Foam Discharges

The second method for protecting external floating roofs is to have the foam discharges located on the roof itself. These systems fall into one of two categories: those that provide finished foam to the top of the seal and those that supply foam beneath the seal.

Systems that provide top-of-seal protection are basically the same as foam chambers located on the tank shell described previously. The only real difference is that the foam discharges and the piping that supply them are located on the roof **(Figure 4.28).** A feed line that either runs up the side of the tank and down the stairway or through the inside of the tank to the underside of the floating roof supplies the piping. A flexible hose is used near the top of the system to allow for movement of the roof. A circle of piping then follows the edge of the tank and connects to the foam makers. Tanks that are protected by top-of-seal protection must have finished foam dams of the same specifications previously listed for systems with foam chambers mounted on the walls.

Systems that provide below-the-seal protection have basically the same design as top-of-seal systems. The primary difference is that the foam discharge orifices actually penetrate the seal. This difference allows finished foam to be applied directly to the surface of the fuel as opposed to the top of the seal. These systems may be of the fixed or semifixed variety.

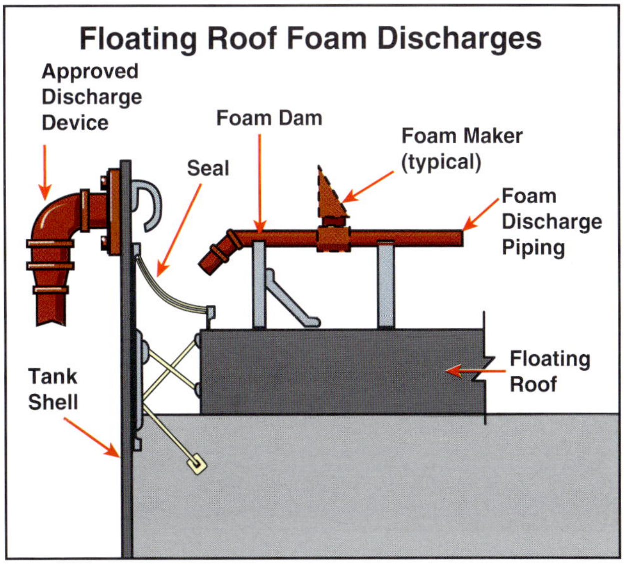

Figure 4.28 In some cases, fixed foam discharges are directly on the tank's floating roof.

Internal Floating Roof Tanks

The internal floating roof tank (sometimes referred to as a *covered floating roof tank*) is a combination of the fixed cone roof tank and the external floating roof tank. It has a fixed cone roof and a pan or deck-type float inside that rides directly on the product surface. Open vents are provided around the shell between the fixed roof and the floating roof **(Figures 4.29 a and b).**

NFPA 11 states that tanks with either steel double-deck roofs or pontoon floating roofs may be protected by a system designed to extinguish a seal fire. Any of the systems described in the external floating roof tanks section (with the exception of handline systems) may be used on these types of tanks. All other types of internal floating roof tanks must be protected against a full surface area fire. Any of the surface application systems described in the Fixed Cone Roof Tank section of this chapter may be used. According to NFPA 11, subsurface and semisubsurface injection systems cannot be used

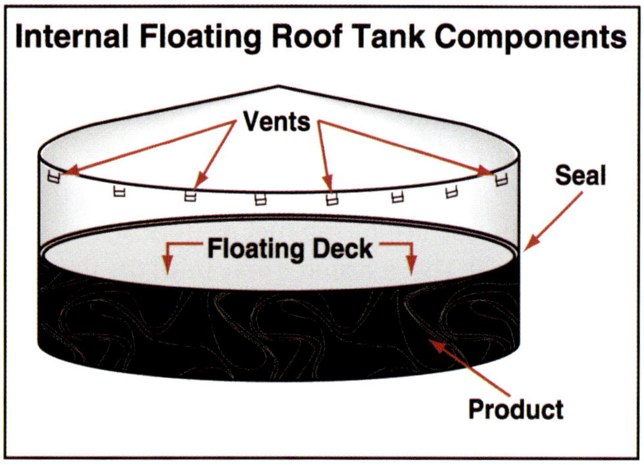

Figure 4.29a Components of an internal floating roof tank.

Figure 4.29b Vent opening can be seen located at intervals around the upper edge of this internal floating roof tank. *Courtesy of Paul Valentine.*

for protecting internal floating roof tanks. However, the use of these systems is becoming commonplace in today's industry. The application rates and discharge duration times described for external floating roof tanks and fixed cone roof tanks apply for internal floating roof tanks, depending on which type of system is used.

Diked/Nondiked Area Protection Systems

[NFPA 1021: 2-6.1; NFPA 1081: 5.3.2, 6.1.2.1, 6.3.8, 7.3.2; NFPA 472: 6.2.4.2]

The areas surrounding flammable and combustible liquid processing and storage facilities may be either diked or nondiked, depending on the age of the facility, its location, or the laws or standards that are applied by the local jurisdiction. Diked areas provide a passive protection for spills by confining the spill to the immediate vicinity of the leak. Active protection includes foam distribution systems that are used to apply finished foam onto the spill within the diked area. Nondiked areas do not necessarily have these fixed foam distribution systems and may be found both inside the production and storage facility and outside of it. The use of foam concentrates in both situations is included in the following sections.

Diked Areas

Diked areas are areas that are bounded by either natural or manmade barriers to keep spilled fuel within these boundaries **(Figure 4.30).** It is assumed that any flammable/combustible liquid spill within the dike will achieve depths greater than 1 inch (25 mm). Dikes are commonly found in two locations: around individual large fuel storage tanks and groups of small fuel tanks (particularly horizontal tanks). Three common types of systems are used to protect diked areas: fixed low-level foam discharge outlets, foam monitors, and foam-water sprinkler systems. The following sections detail each system as it applies to protecting diked areas.

Fixed Low-Level Foam Discharge Outlets

The fixed foam maker method consists of installing piping around the outside wall of the dike and connecting a series of equally spaced foam discharge outlets to the piping to discharge finished foam into

Figure 4.30 Diked areas surrounding storage tanks not only confine spilled fuel but also the water and foam used to extinguish a tank fire. *Courtesy of Paul Valentine.*

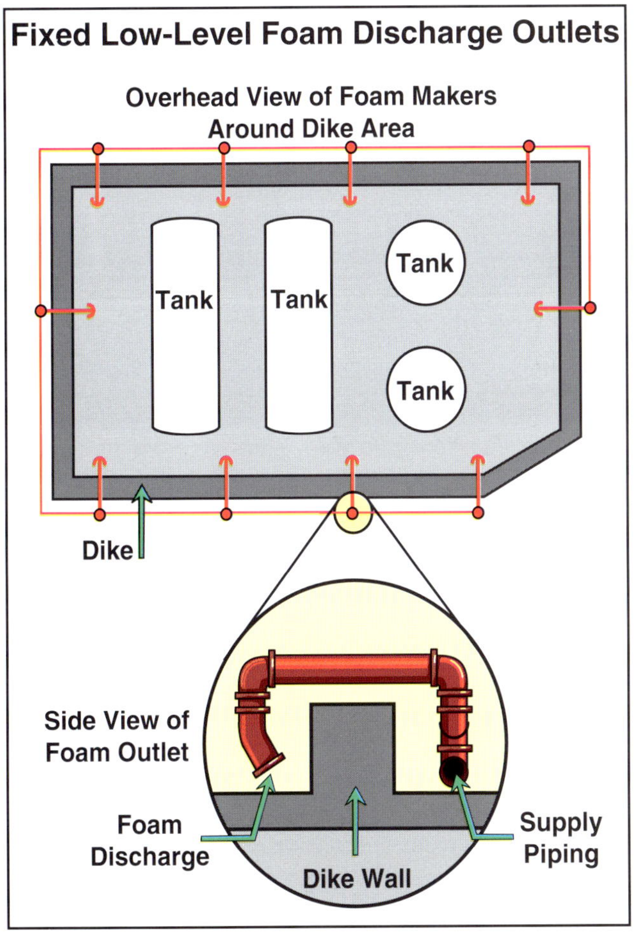

Figure 4.31 Storage tanks may be protected by fixed low-level foam outlets that can provide a foam blanket within the diked area.

the dike (**Figure 4.31**). These discharge outlets are similar to those described for external floating roof tanks. These systems may be fixed or semifixed. On hydrocarbon products only, the recommended rate of application for this type of system is 0.1 gpm/foot^2 (4.1 L/min/m^2) of dike area. The duration of discharge needs to be at least 30 minutes for flammable liquid hydrocarbons and 20 minutes for combustible liquid hydrocarbons. Polar solvent liquids require higher rates. Consult the foam manufacturer for further information. Discharge outlets that flow less than 60 gpm (240 L/min) must be spaced no more than 30 feet (9 m) apart. Discharge outlets that flow greater than 60 gpm (240 L/min) may be spaced up to 60 feet (18 m) apart.

Foam Monitors

Foam monitors can be used as an alternative to the fixed foam maker method. Monitors that are properly sized and spaced to discharge finished foam into the dike area may be mounted outside the dike. Position and aim the monitors so that the streams are directed against the dike wall, tank surfaces, or other objects. This procedure prevents the streams from plunging directly into the surface of the burning fuel.

Because this method of foam application is less precise than the low-level discharge method, NFPA requires greater flow rates. When using the monitor method, the required flow rate is 0.16 gpm/foot^2 (6.5 L/min/m^2). The discharge duration is the same as for the low-level discharge method.

Foam-Water Sprinkler Systems

Foam-water sprinkler systems (previously described in this chapter) may be used to protect diked spill areas. Their design is in accordance with NFPA 16. If the diked area contains large obstructions, supplementary low-level discharge outlets may be required to provide uniform coverage.

Nondiked Areas

Nondiked areas are any locations where flammable or combustible liquids might be spilled but not contained within a system of predesigned barriers. In reality, this location could be anywhere outside the diked areas of a fuel storage facility, including off-site locations. In almost every case, attacks on these incidents are made using mobile foam apparatus and/or fixed monitors with foam stations.

NFPA 11 recommends the following guidelines for foam application in these circumstances:

Hydrocarbon Fires:

- Flow protein and fluoroprotein finished foam at 0.16 gpm/foot^2 (6.5 L/min/m^2) for at least 15 minutes.
- Flow AFFF and FFFP finished foam at 0.10 gpm/foot^2 (4.1 L/min/m^2) for at least 15 minutes.

Polar Solvent Fires:

- Consult the foam concentrate manufacturer's recommendations for the particular fuel involved in the incident.

Loading Rack Protection Systems

[NFPA 1021: 2-6.1; NFPA 1081: 5.3.2, 6.1.2.1, 6.3.8, 7.3.2; NFPA 472: 6.2.4.2]

Loading racks are fixed facilities where either truck tank cars or railroad tank cars are bulk loaded with flammable and combustible liquids (**Figure 4.32**). Fuel is transferred to the vehicles from either pipelines or storage tanks. Depending on the design of the system, the fuel may be either transferred under pressure or by gravity. These facilities provide the following different possible fire scenarios:

- Ground spill fires
- Gravity-fed, three-dimensional flowing fuel fires
- Pressure-fed, three-dimensional flowing fuel fires

The loading rack itself is typically an open, canopy-type structure. A myriad of piping, valves, and controls are located within the canopy. Depending on the types of vehicles filled at the rack, the piping may be designed to either fill tanks from the top or the bottom. The entire assembly is usually surrounded by a low-profile dike or curb designed to contain any spill that may occur (**Figure 4.33**). Some loading racks are designed with special drainage systems to divert spilled fuel to a holding area.

Two types of systems are used to protect loading rack facilities: deluge-type foam-water sprinkler systems and fixed foam monitors. Both of these systems are very similar in design to those described earlier in the chapter for tank protection. The only major difference in foam-water sprinkler systems used for these applications is the addition of ground-spray nozzles designed to rapidly deploy finished foam to the underside of the railcar, truck, or piping. Their design criteria is consistent with NFPA 16.

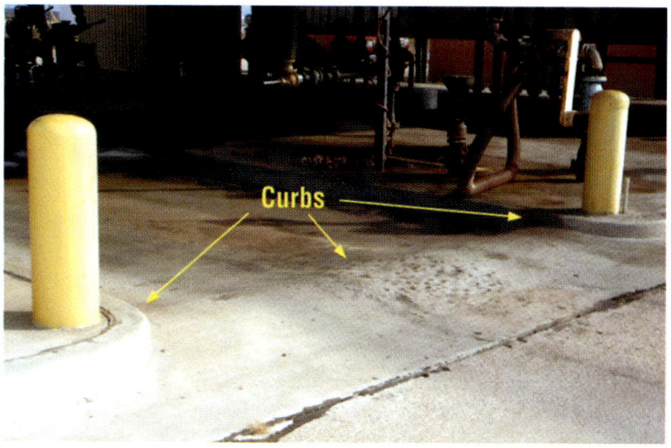

Figure 4.33 Loading racks are generally protected by curbs or dikes that can confine spilled liquids. *Courtesy of Conoco, Inc.*

Figure 4.32 Transport and railcar loading racks may be protected by fixed foam systems.

Table 4.3
Minimum Application Rates and Discharge Times for Loading Racks Protected by Foam Monitor Nozzle Systems

Foam Type	Minimum Application Rate		Minimum Discharge Time (min)	Product Being Loaded
	gpm/ft²	(L/min)/m²		
Protein and Fluoroprotein	0.16	(6.5)	15	Hydrocarbons
AFFF, FFFP, and alcohol-resistant AFFF or FFFP	0.10*	(4.1)	15	Hydrocarbons
Alcohol-resistant foams	Consult manufacturer for listings on specific products		15	Flammable and combustible liquids requiring alcohol-resistant foam

* If a fuel depth of more than 1 in. (25.4 mm) can accumulate within the protected area, the application rate shall be increased 0.16 gpm/ft² [to (6.5 L/min)/m²].

Fixed monitor systems must provide coverage to all portions of the loading area, including the undersides of the vehicles in the area. **Table 4.3** shows the minimum application rates and discharge durations for fixed monitor systems that protect loading racks.

Aircraft Hangar Protection Systems

[NFPA 1021: 2-6.1]

Aircraft hangars are large structures that contain aircraft for the purposes of storage, shelter from weather, maintenance, repairs, and/or alterations. NFPA 409 classifies aircraft hangars into three groups. The characteristics of each of these groups and the protection systems required for each are discussed in the sections that follow.

Group I Hangars

NFPA 409 defines a Group I hangar as one that meets any of the following conditions:

- An aircraft access door height in excess of 28 feet (8.5 m)
- A single fire area in excess of 40,000 square feet (3 716 m²)
- One that houses an aircraft whose tail height exceeds 28 feet (8.5 m) **(Figure 4.34, p. 120)**
- One that houses strategically important military aircraft as determined by the U.S. Department of Defense

Group I hangars must be equipped with an automatic, deluge-type, foam-water sprinkler system. The maximum protection area under any one deluge system is 15,000 square feet (1 394 m²). The general design of the system must be consistent with those discussed earlier in this chapter. If protein, fluoroprotein, or AFFF concentrates are used in conjunction with air-aspirating sprinklers, the minimum application rate must be at least 0.20 gpm/foot² (8.1 L/min/m²) of finished foam. If AFFF concentrate is used in conjunction with nonair-aspirating sprinklers, the minimum application rate must be 0.16 gpm/foot² (6.5 L/min/m²) of finished foam. The system must have a foam concentrate supply sufficient to last for 7 to 10 minutes, depending on the design of the system.

Hangars that house aircraft that have wing areas in excess of 3,000 square feet (279 m²) must have a supplementary protection system installed in addition to the deluge foam-water system. The

Figure 4.34 Group I aircraft hangar.

purpose of this system is to apply finished foam directly to the floor area beneath the aircraft. The most common types of systems used in these applications are low-expansion foam systems using oscillating foam monitors **(Figure 4.35).** These systems require a flow rate of 0.16 gpm/foot^2 (6.5 L/min/m^2) for systems using protein and fluoroprotein foam and 0.10 gpm/foot^2 (4.1 L/min/m^2) for systems using AFFF. The system must be capable of flowing finished foam for at least 10 minutes.

High-expansion foam systems may also be used for supplementary protection. These systems must be capable of covering the protected area to a depth of 3 feet (1 m) within one minute. These systems must be in compliance with NFPA 11A, *Standard for Medium- and High-Expansion Foam Systems* (1999). These systems are discussed in more detail later in this chapter.

Group I hangars are also required to have foam-water handlines installed in areas where aircraft are located. The hoselines must be at least 1½ inches (38 mm) in diameter and capable of flowing 60 gpm (240 L/min) of finished foam **(Figure 4.36).** Spacing and design of the hoses should follow NFPA 11 and NFPA 14, *Standard for the Installation of Standpipe, Private Hydrants, and Hose Systems* (2000). The system must have a foam concentrate supply capable of flowing foam solution for 20 minutes. Other areas of the structure (those not actually containing the aircraft) must be equipped with water hoselines.

Group II Hangars

Group II hangars have an aircraft access door that is less than 28 feet (8.5 m) tall and a single fire area that is less than 40,000 square feet (3 716 m^2) in size **(Figure 4.37).** There are allowable fire areas, based

Figure 4.35 A directional spray foam nozzle is commonly used in aircraft maintenance facilities to provide protection to areas underneath aircraft wings.

Figure 4.36 Group I hangers are also equipped with preconnected foam-water handlines that are connected to the foam system.

on the building construction type (**Table 4.4**). NFPA 409 allows for the choice of one of the following three options for the protection of Group II hangars:

- System of the same design and capabilities as described for Group I hangars
- Automatic sprinkler system and an automatic low-level, low-expansion foam system
- Automatic sprinkler system and an automatic high-expansion foam system

Automatic sprinkler systems used in Group II hangars must comply with NFPA 13 with a few additional specified requirements. Unlike the deluge-type foam systems required in Group I hangars, these sprinkler systems can be the closed-sprinkler type. The maximum area that can be protected by any single system is 25,000 square feet (2 323 m^2). The system must be capable of discharging at least 0.17 gpm/$foot^2$ (6.9 L/min/m^2) over any 5,000 square-foot (465 m^2) area in the system, including the most remote part. The temperature rating for closed sprinklers is between 325 and 375°F (162°C and 190°C).

Figure 4.37 Group II aircraft hanger.

Table 4.4
Fire Areas for Group II Aircraft Hangers

Type of Construction	Single Fire Area Equal to or Greater Than But Not Larger Than: ft^2	m^2
Type I (443 and 332)	30,001 ≥ 40,000	2,787 ≥ 3,716
Type II (222)	20,001 ≥ 40,000	1,858 ≥ 3,716
Type II (111), Type III (211), and Type IV (2HH)	15,001 ≥ 40,000	1,394 ≥ 3,716
Type II (000)	12,001 ≥ 40,000	1,115 ≥ 3,716
Type III (200)	12,001 ≥ 40,000	1,115 ≥ 3,716
Type V (111)	8,001 ≥ 40,000	743 ≥ 3,716
Type V (000)	5,001 ≥ 40,000	465 ≥ 3,716

Note: Types of aircraft hangar construction and the subtypes shown in parenthesis are found in NFPA 220, *Standard on Types of Building Construction* (1999).

Low-expansion foam systems in Group II hangars are the fixed-monitor type. The monitors should be aimed at the aircraft parking area. Each nozzle must have a manual shutoff valve. The minimum application rate for protein and fluoroprotein systems is 0.16 gpm/$foot^2$ (6.5 L/min/m^2). The minimum application rate for AFFF systems is 0.10 gpm/$foot^2$ (4.1 L/min/m^2). Heat detection devices automatically activate the system, and the system is capable of flowing finished foam for at least 10 minutes. High-expansion foam systems for Group II hangars are basically the same as those used in Group I hangars.

Group III Hangars

Group III hangars have aircraft access doors less than 28 feet (8.5 m) in height and a single fire area that is less then those given for the various types of building construction in **Table 4.5.** NFPA 409 does

Table 4.5
Maximum Fire Areas for Group III Hangars

Type of Construction	Maximum Single Fire Area: ft^2	m^2
Type I (443) and (332)	30,000	2,787
Type II (222)	20,000	1,858
Type II (111), Type III (211), and Type IV (2HH)	15,000	1,394
Type II (000)	12,000	1,115
Type III (200)	12,000	1,115
Type V (111)	8,000	743
Type V (000)	5,000	465

Note: Types of aircraft hangar construction, and the subtypes shown in parenthesis, are found in NFPA 220 *Standard on Types of Building Construction* (1999).

not require the installation of fixed fire protection systems in Group III hangars. The standard does require the installation of portable fire extinguishers in accordance with NFPA 10, *Standard for Portable Fire Extinguishers* (1998).

Medium- and High-Expansion Foam Systems

[NFPA 1021: 2-6.1; NFPA 1081: 5.3.2, 6.1.2.1, 6.3.8, 7.3.2]

Medium- and high-expansion foam systems are similar in basic design to low-expansion foam systems. All of the following same elements are required:

- Reliable water supply
- Foam concentrate supply
- Fire pump
- Foam concentrate proportioner
- Foam discharge devices

The primary difference with medium- and high-expansion foam systems is the finished product: the finished foam itself. Whereas low-expansion finished foam is applied to form a one-dimensional blanket that covers a liquid fuel, medium- and high-expansion finished foams are intended to fill a space and provide three-dimensional coverage. While medium- and high-expansion foams can be used outdoors with limited success, they are primarily intended for indoor use in confined spaces. Their excellent flow characteristics allow them to travel throughout an area to quickly fill the space. With Class A fires, the smothering effect of high-expansion finished foam controls the fire. To a lesser extent, the water in the foam also aids extinguishment by cooling the fuel. When used on some special hazards such as liquefied natural gas (LNG) spills, a layer of ice may form on the fuel surface to aid in the prevention of ignition.

Three basic types of medium- and high-expansion foam systems are available: total flooding, local application, and portable. The following sections discuss the relative merits and designs of each system.

Total Flooding

A total flooding system is used in locations where permanent structures or enclosures are around the hazard areas. These structures or enclosures allow the finished foam to build and maintain until the fire is controlled or completely extinguished. A total flooding system is effective on Class A and Class B fuels and is commonly found in storage vaults, warehouses (particularly those storing rolled paper), basements, and aircraft hangars **(Figure 4.38)**.

Three basic types of fires that total flooding systems are successful in extinguishing are as follows:

- Fires on the surface of Class A and Class B liquid fuels
- Smoldering fires deep-seated within Class A materials
- Some three-dimensional flammable liquid fires

In some special circumstances — such as high-piled, rolled-paper storage — medium- and high-expansion foam systems are used together with automatic sprinkler systems to provide a level of

Figure 4.38 High-expansion foam systems (see ceiling in photo) are commonly used to protect rolled-paper storage areas.

protection greater than either system could provide by itself. Sprinkler systems that are used in conjunction with foam systems must conform to NFPA 13.

A total flooding medium- and high-expansion foam system is a deluge system activated by product-of-combustion (heat, smoke, or flame) detectors. The actuation system must have a backup power supply to ensure operation in the event the main power supply fails. Once the actuator detects fire conditions, a signal is sent to the deluge valve that trips the foam system into operation. Once water is flowing, the foam proportioning system is activated to send the proper amount of foam concentrate into the water stream. In particular, bladder-tank balanced-pressure proportioners are commonly used with medium- and high-expansion foam systems. The foam solution is then supplied to the fixed discharge devices. Two types of discharge devices are used on these systems: mechanical blower and water-aspirating. These devices are discussed in detail in Chapter 3, Foam Proportioning and Delivery Equipment.

In addition to starting the flow of finished foam, the actuator system must also trigger the closing of all doors, windows, and other openings in the protected area. This procedure prevents finished foam from escaping the area. In addition, the protected area must be equipped with air vents that are above the level to which the finished foam rises. This procedure allows air to expel from the area as finished foam discharges into it. If the air were not vented, the buildup in pressure (caused by compression) as the finished foam fills the room may create pressure above the top of the foam that prevents it from reaching the desired levels.

Total flooding systems must be designed to provide a level of finished foam in the protected area that is at least 1.1 times the height of the highest hazard in the area. Under no circumstances should the level of finished foam be less than 2 feet (0.6 m) above the highest hazard **(Figure 4.39)**. The amount of time that the system must supply finished foam to this level depends on the type of hazard and the type of building construction **(Table 4.6, p. 124)**. This level of finished foam must be maintained in the protected area for at least 30 minutes in sprinklered areas and 60 minutes in unsprinklered areas.

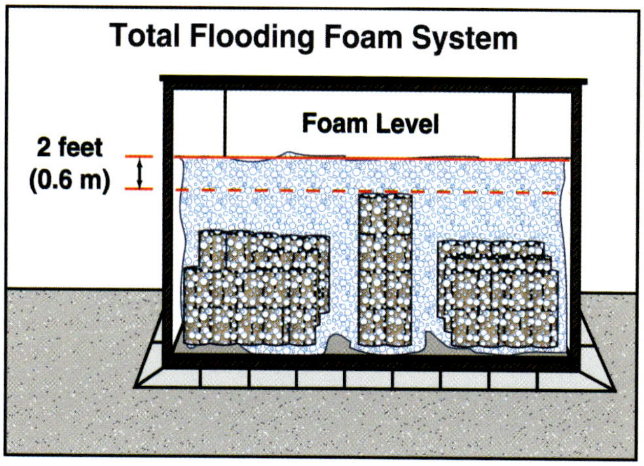

Figure 4.39 The level of the high-expansion foam should be at least 2 feet (0.6 m) above the highest portion of the hazard.

The system must meet one of the following two requirements (whichever of the two is less); however, the system must always be capable of flowing at full capacity for at least 15 minutes:

- Equip with a foam concentrate supply that allows continuous operation of the system for at least 25 minutes at full capacity.
- Generate at least four times the amount of finished foam required to reach the desired level in the protected area.

Local Application

Local application systems are effective on Class A and Class B liquid fires in areas that are not totally enclosed. These areas include curbed areas, diked areas, trenches, and open tanks. The basic components and operation of these systems are the same as those described for total flooding systems. The primary differences are in the performance requirements. Local application systems are required to cover the protected area to a depth of 2 feet (0.6 m) within 2 minutes of system actuation. The foam concentrate supply must support continuous operation of the system for at least 12 minutes.

Portable

Portable medium- and high-expansion foam equipment is primarily used by industrial and municipal firefighters who respond to incidents that cannot be controlled using standard fire-fighting tactics. Basement fires are a common situation where many fire departments find the use of medium- or high-

Table 4.6
Maximum Submergence Time for High-Expansion Foam Measured from Start of Foam Discharge[a] (Minutes)

Hazard	Light or Unprotected Steel Construction		Heavy or Protected or Fire-Resistive Construction	
	Sprinklered	Not Sprinklered	Sprinklered	Not Sprinklered
Flammable liquids [flash points below 100°F (38°C)] having a vapor pressure not exceeding 40 psia (276 kPa)[a]	3	2	5	3
Combustible liquids [flash points of 100°F (38°C) and above][b]	4	3	5	3
Low-density combustibles (i.e., foam rubber, foam plastics, rolled tissue, or crepe paper)	4	3[c]	6	4[c]
High-density combustibles (i.e., rolled paper kraft or coated banded)	7	5[c]	8	6[c]
High-density combustibles (i.e., rolled paper kraft or coated unbanded)	5	4[c]	6	5[c]
Rubber tires	7	5[c]	8	6[c]
Combustibles in cartons, bags, or fiber drums	7	5[c]	8	6[c]

Notes:

(a) This submergence time is based on the maximum of 30 seconds delay between fire detection and start of foam discharge. Any delays in excess of 30 seconds shall be deducted from the submergence times given.

(b) Polar solvents are not included in this table. Flammable liquids having boiling points less than 38°C (100°F) might require higher application rates. See NFPA 30, *Flammable and Combustible Liquids Code.* Where use of high-expansion foam is contemplated on these materials, the foam equipment supplier shall substantiate suitability for the intended use.

(c) These submergence times might not be directly applicable to high-piled storage above 15 ft (4.6 m) or where fire spread through combustible contents is very rapid.

expansion foam useful. Most commonly, in-line eductors are used to supply foam solution via hoselines to handheld or otherwise portable foam generators. Placements of the foam generator and appropriate flow rates are dictated by the requirements of the incident. When using portable medium- and high-expansion foam equipment to attack structural fires, vertical ventilation of structures must be performed before finished foam application. This procedure allows fire gases and heat displaced by the finished foam to exit the structure. It also allows finished foam to fill the area more completely by either reducing or eliminating the air-compression effect that occurs when trying to pump finished foam into a closed room.

Mobile Foam Apparatus

[NFPA 1002: 3-1.1, 6-1.1, 7-1.1, 7-2.1, 7-2.2, 7-2.3, 8-1.1; NFPA 1003: 3-3.3]

Mobile foam apparatus are used to combat large-scale flammable and combustible liquid fires in areas not protected by fixed systems. Mobile foam apparatus are found at airports, at petrochemical and other industrial locations, and in municipal fire departments. All of these apparatus use various combinations of the foam proportioning systems, discharge devices, and foam concentrates discussed to this point. The manufacturer of the apparatus provides operating instructions specific to each particular piece of apparatus, based on the exact design of that rig. The purpose of the following sections is to review the various types of foam vehicles that are available and their basic uses and capabilities.

Aircraft Rescue and Fire Fighting Apparatus

Aircraft rescue and fire fighting (ARFF) apparatus are used to provide immediate suppression of flammable liquid fires and suppression of spill vapors on airport properties. In rare instances, ARFF vehicles may respond off airport property to assist municipal firefighters at large-scale flammable liquid incidents (**Figure 4.40**). Requirements for ARFF apparatus are contained in NFPA 414, *Standard for Aircraft Rescue and Fire Fighting Vehicles* (2001). NFPA 414 divides ARFF apparatus into three general classifications:

- Major fire fighting vehicles
- Rapid intervention vehicles (RIVs)
- Combined-agent vehicles

Although the information is this section is based on NFPA 414, other sources of information and requirements pertaining to ARFF apparatus are available. Additional requirements for airports in the United States are found in Federal Aviation Administration (FAA) regulations Title 14 (Aeronautics and Space) *CFR* Part 139, *Certification and Operations: Land Airports Serving Certain Air Carriers.* Airports outside the United States follow the International Civil Aviation Organization (ICAO) Annex 14, *International Standards and Recommended Practices, Aerodromes,* to determine ARFF vehicle requirements. The length of the largest passenger aircraft currently being used at an airport determines the number of required ARFF vehicles and amount of foam capability. The following three sources of information are also helpful:

- IFSTA **Aircraft Rescue and Fire Fighting** book
- FAA, AC 150/5220-10, *Guide Specifications for Water/Foam Type Aircraft Fire and Rescue Trucks*
- FAA, AC 150/5220-14A, *Airport Fire and Rescue Vehicle Specification Guide*

Figure 4.40 Flammable liquid spill fires on streets and highways may require the use of aircraft rescue and fire fighting apparatus away from the airport.
Courtesy of Ron Jeffers.

Major Fire Fighting Vehicles

Major fire fighting vehicles are the largest of all ARFF fire apparatus (**Figure 4.41, p. 126**). They provide the bulk of the major extinguishing capabilities of the apparatus responding to an airport emergency. Despite their immense size, these vehicles are operated both on and off paved surfaces (whichever the incident requires).

The FAA divides aircraft fire-fighting vehicles into four classes based on the water tank size of the vehicle (**Table 4.7, p.126**). Each class has specific operating capability requirements for things such as acceleration speed, turret discharge rates, and fire stream reach. Although the acceleration speeds differ depending on the class of the vehicle, all classes of major fire fighting vehicles must be able to obtain speeds in excess of 65 mph (100 kmph).

Rapid Intervention Vehicles

A rapid intervention vehicle is another type of ARFF apparatus; it is considerably smaller and faster than major fire fighting vehicles (**Figure 4.42, p. 126**). The role of the RIV is to get to the scene of an emergency as fast as possible (usually well ahead of the larger vehicles) and begin a fast attack on the incident. The theory is that if a fire incident can be attacked quickly by an RIV, it can be extinguished or at least under control by the time the larger vehicles get

Table 4.7
Federal Aviation Administration ARFF Vehicle Classes and Capabilities

Class	Minimum Water Tank Capacity In Gallons (Liters)	Minimum Rated Turret Flow In Gallons (Liters)	Straight Stream Range In Feet (Meters)	Dispersed Stream Width In Feet (Meters)	Dispersed Stream Length In Feet (Meters)	Maximum Acceleration Time 0–50 mph (0–80 kmph) In Seconds
1	1,000 (4 000)	500 (2 000)	160 (49)	35 (10)	60 (18)	25
2	1,500 (6 000)	750 (3 000)	190 (58)	35 (10)	65 (20)	30
3	2,500 (10 000)	1,250 (5 000	230 (70)	35 (10)	70 (21)	40
4	3,000+ (12 000+)	1,500 (6 000)	250 (76)	35 (10)	75 (23)	45

Figure 4.41 An aircraft rescue and fire-fighting vehicle can apply foam from the top turret, bumper, and under-chassis nozzles while the vehicle is in motion. *Courtesy of Christchurch International Airport, New Zealand.*

Figure 4.42 Rapid intervention vehicle used at airports. *Courtesy of Joel Woods.*

there. RIVs are required to accelerate from 0 to 50 mph (0 kmph to 80 kmph) in 20 seconds and must be capable of top speeds in excess of 65 mph (100 kmph).

Combined-Agent Vehicles

Combined-agent vehicles are similar to rapid intervention vehicles and are also used in ARFF operations. Their primary differences are that they are smaller and carry less extinguishing agent **(Figure 4.43). Table 4.8** shows the amount of extinguishing agent and required acceleration time for each of the three classes of combined-agent vehicles. Class 1 vehicles are not required to have turrets. Class 2 and Class 3 vehicles are required to have turrets that allow for simultaneous discharge of foam and dry chemical.

Figure 4.43 Combined-agent vehicle capable of applying both premixed foam and dry-chemical or halon substitutes.

Industrial Foam Pumpers

Many industrial facilities that contain large quantities of flammable and combustible liquids are equipped with large-capacity foam pumpers **(Figure 4.44).** These foam pumpers are staffed by trained members of the site fire brigade in the event

Table 4.8
Combined-Agent Vehicle Capabilities

Class	Minimum Water Capacity in Gallons (liters)	Minimum Auxiliary Agent in Pounds (kg)	Maximum Acceleration Time 0–50 mph (0–80 kmph) in Seconds
1	100 (400)	100 (45)	25
2	200 (800)	200 (90)	30
3	350 (1 400)	300 (135)	30

Notes:
(1) Class I: Not required to have turrets.
(2) Class II: Have turrets that flow 150 gpm (600 L/min) of foam.
(3) Class III: Have turrets that flow 250 gpm (1 000 L/min) of foam.

Figure 4.44 Foam pumper operated by an industrial fire brigade at a petroleum refinery. *Courtesy of Bob Norman.*

that either a fire or spill emergency occurs. Although they may be equipped to flow plain water on Class A fires, most industrial foam pumpers are primarily intended to produce large quantities of finished foam to attack Class B fires and spills. Industrial foam pumpers are built according to the standards provided in NFPA 1901, *Standard for Automotive Fire Apparatus* (1999).

Industrial foam pumpers may be equipped with around-the-pump, direct-injection, or balanced-pressure foam proportioning systems. Most large-scale industrial foam pumpers use some form of balanced-pressure proportioning system because of the reliability of the foam proportioning at large flows. The apparatus are equipped with fire pumps that range in capacity from 1,000 to 3,000 gpm (3 785 L/min to 11 356 L/min) or greater. Some have foam-proportioning systems with capabilities that exceed the pumping capacity of the apparatus itself. Most industrial foam pumpers have a large foam concentrate tank on board. These tanks range from 500 to 1,500 gallons (1 893 L to 5 678 L). The apparatus is typically equipped with a large fixed foam/water turret capable of flowing the entire capacity of the fire pump. The entire apparatus itself may be mounted on either a commercial or custom truck chassis.

Some industrial foam pumpers are equipped with a secondary extinguishing agent system. Halon and dry-chemical systems are the two most common types of systems used. Systems that contain 100 to 1,000 pounds (45 kg to 450 kg) of agent are common. This agent can be discharged through a fixed turret or a reeled handline.

Depending on the special needs of a particular facility, some facilities may have aerial apparatus equipped with large-scale foam systems. These apparatus generally have the same capabilities as industrial foam pumpers with the added ability to discharge finished foam from an elevated position. They may also be used for firefighter access to upper levels and other standard firefighting functions for which aerial apparatus are commonly used.

Municipal Fire Apparatus Foam Systems

Some municipal fire departments choose to equip fire department pumpers with fixed Class A and/or Class B foam systems **(Figure 4.45).** Class A foam systems are used on the standard types of fires to

Figure 4.45 Municipal pumper apparatus may also be required to provide foam capabilities at refineries and chemical plants. *Courtesy of Paul Valentine.*

which municipal firefighters commonly respond (vehicles, brush, structures, etc.). Class B foam systems allow firefighters to handle small-scale flammable/combustible liquid fires and spills. NFPA 1901 contains the requirements for these apparatus.

The Class A foam systems installed on municipal fire apparatus may be the high-energy or low-energy types described in Chapter 3, Foam Proportioning and Delivery Equipment. Regardless of the systems used, only certain discharges are provided with foam equipment and capable of flowing finished foam. Apparatus equipped with high-energy systems also require the installation of a sizable air compressor.

An around-the-pump proportioner is the most common type of Class B fixed foam proportioner used on municipal fire apparatus. These systems are relatively inexpensive and suitable for most small-scale incidents to which municipal firefighters commonly respond. Unlike Class A systems, Class B around-the-pump proportioners supply foam solution to all discharges on the apparatus. In some cases, this feature may be a disadvantage because of the apparatus' inability to flow plain water at the same time as finished foam. Fire departments that need that dual capability must have their apparatus equipped with a balanced-pressure or direct-injection system.

Municipal fire apparatus equipped with foam systems have foam tanks that supply the system. The most common size foam tanks for municipal fire apparatus range from 20 to 100 gallons (76 L to 379 L). These tanks are designed so that they can be refilled with 5-gallon (19 L) pails when necessary.

Foam Tankers (Tenders/Mobile Water Supply Vehicles)

Industrial and municipal fire departments that have the potential for large-scale Class B fires may have foam tankers (tenders/mobile water supply vehicles) to transport bulk quantities of foam concentrate to an emergency scene when needed **(Figure 4.46)**. These apparatus are similar in design and purpose to standard fire department tankers (tenders/mobile water supply vehicles) with the primary difference being that they haul foam concentrate rather than water. These

Figure 4.46 Industrial brigades may also own or have access to large-capacity foam tenders. *Courtesy of Conoco, Inc.*

apparatus are commonly found at petrochemical refineries and storage facilities, airports, and large fire departments.

The method by which the foam concentrate is unloaded depends on the standard operating procedures of the organization that uses it. Some tankers are equipped with small pumps to unload the concentrate. Others are equipped with discharges that allow the concentrate to be either pumped by an external pump or drafted. Still others are equipped with large dump valves that allow the concentrate to be quickly dumped into folding tanks.

Foam tankers may be single-chassis or tractor-trailer-type apparatus. Single-chassis apparatus may carry up to 4,000 gallons (15 142 L) of foam concentrate. Single-chassis apparatus that carry more than 1,500 gallons (5 678 L) require tandem rear axles. Tractor-trailer apparatus may carry as much as 8,000 gallons (30,283 L) of foam concentrate.

Foam Trailers

In some cases, foam equipment and bulk foam concentrate are carried on trailers that are pulled to the emergency scene by another vehicle **(Figure 4.47)**. The advantage of this arrangement is reduced costs when compared to a self-propelled foam vehicle. Foam-trailer design varies widely. Depending on the needs of the purchaser and the design of the trailer, foam trailers may contain foam master stream devices, foam handlines and nozzles, foam proportioning equipment, bulk foam concentrate supplies, or any combination of this equipment.

Figure 4.47 Foam trailers may be equipped with high-volume foam nozzles (such as the one shown), foam tanks, or both.

Wildland Apparatus

Wildland apparatus may be equipped with high- or low-energy Class A foam systems such as those described in Chapter 3, Foam Proportioning and Delivery Equipment **(Figure 4.48)**. Wildland apparatus are similar in function to standard fire department pumpers with a few exceptions. Wildland apparatus typically have smaller pumps and water tanks to reduce the size of the vehicle. This design allows for more maneuverability in off-road situations. These apparatus are also commonly mounted on a four-wheel-drive chassis. Additional information on wildland firefighting apparatus may be found in the IFSTA manual, **Wildland Fire Fighting for Structural Firefighters**, and from the National Wildfire Coordinating Group (NWCG).

Figure 4.48 Wildland apparatus are designed for off-road work and built on commercial all-wheel-drive chassis. The foam/water tank, pump, and hose reel may be permanently attached or part of a skid unit that can be easily removed from the vehicle.

Summary

While the design, installation, and maintenance of fixed foam distribution systems is the responsibility of the flammable and combustible liquids facility owner, knowledge of the system is the responsibility of all responding fire and emergency services personnel. Responders must be able to recognize the types of systems, the types of facilities that are protected, and how to effectively support the systems. The fixed systems are intended to control any leak or fire that may result from a leak. However, it is also possible that a fire may overwhelm the system, it may not function properly, or it may require additional foam concentrate to complete the control of the incident. It is the responsibility of fire and emergency responders to support the system in order to bring an incident to completion.

Courtesy of NIFC

Chapter 5

Class A Foam Fire Fighting

Job Performance Requirements

This chapter provides information that will assist the reader in meeting the following job performance requirements from NFPA 1001, *Standard for Fire Fighter Professional Qualifications*, 2002 edition; NFPA 1002, *Standard for Fire Apparatus Driver/Operator Professional Qualifications,* 1998 edition; NFPA 1021, *Standard for Fire Officer Professional Qualifications,* 1997 edition; NFPA 1081, *Standard for Industrial Fire Brigade Member Professional Qualifications,* 2001 edition; and NFPA 1051, *Standard for Wildland Fire Fighter Professional Qualifications,* 2002 edition. Colored portions of the standard are specifically addressed in this chapter.

NFPA 1001

5.3.8 Extinguish fires in exterior Class A materials, given fires in stacked or piled and small unattached structures or storage containers that can be fought from the exterior, attack lines, hand tools and master stream devices, and an assignment, so that exposures are protected, the spread of fire is stopped, collapse hazards are avoided, water application is effective, the fire is extinguished, and signs of the origin area(s) and arson are preserved.

(A) *Requisite Knowledge:* Types of attack lines and water streams appropriate for attacking stacked, piled materials and outdoor fires; dangers—such as collapse—associated with stacked and piled materials; **various extinguishing agents and their effect on different material configurations;** tools and methods to use in breaking up various types of materials; the difficulties related to complete extinguishment of stacked and piled materials; water application methods for exposure protection and fire extinguishment; dangers such as exposure to toxic or hazardous materials associated with storage building and container fires; obvious signs of origin and cause; and techniques for the preservation of fire cause evidence.

(B) *Requisite Skills:* The ability to recognize inherent hazards related to the material's configuration, operate handlines or master streams, break up material using hand tools and water streams, evaluate for complete extinguishment, operate hose lines and other water application devices, evaluate and modify water application for maximum penetration, search for and expose hidden fires, assess patterns for origin determination, and evaluate for complete extinguishment.

5.3.10 Attack an interior structure fire operating as a member of a team, given an attack line, ladders when needed, personal protective equipment, tools, and an assignment, so that team integrity is maintained, the attack line is deployed for advancement, ladders are correctly placed when used, access is gained into the fire area, effective water application practices are used, the fire is approached correctly, attack techniques facilitate suppression given the level of the fire, hidden fires are located and controlled, the correct body posture is maintained, hazards are avoided or managed, and the fire is brought under control.

(A) *Requisite Knowledge:* **Principles of fire streams; types, design, operation, nozzle pressure effects, and flow capabilities of nozzles;** precautions to be followed when advancing hose lines to a fire; observable results that a fire stream has been properly applied; dangerous building conditions created by fire; **principles of exposure protection;** potential long-term consequences of exposure to products of combustion; physical states of matter in which fuels are found; common types of accidents or injuries and their causes; and the application of each size and type of attack line, the role of the backup team in fire attack situations, attack and control techniques for grade level and above and below grade levels, and exposing hidden fires.

(B) *Requisite Skills:* The ability to prevent water hammers when shutting down nozzles; open, close, and adjust nozzle flow and patterns; apply water using direct, indirect, and combination attacks; advance charged and uncharged 1½-in. (38-mm) diameter or larger hose lines up ladders and up and down interior and exterior stairways; extend hose lines; replace burst hose sections; operate charged hose lines of 1½-in. (38-mm) diameter or larger while secured to a ground ladder; couple and uncouple various handline connections; carry hose; attack fires at grade level and above and below grade levels; and locate and suppress interior wall and subfloor fires.

5.3.19 Combat a ground cover fire operating as a member of a team, given protective clothing, SCBA if needed, hose lines, extinguishers or hand tools, and an assignment, so that threats to property are reported, threats to personal safety are recognized, retreat is quickly accomplished when warranted, and the assignment is completed.

(A) *Requisite Knowledge:* **Types of ground cover fires, parts of ground cover fires, methods to contain or suppress,** and safety principles and practices.

(B) *Requisite Skills:* The ability to determine exposure threats based on fire spread potential, protect exposures, construct a fire line or extinguish with hand tools, maintain integrity of established fire lines, and suppress ground cover fires using water.

6.3.2 Coordinate an interior attack line team's accomplishment of an assignment in a structure fire, given attack lines, personnel, personal protective equipment, and tools, so that crew integrity is established; attack techniques are selected for the given level of the fire (for example, attic, grade level, upper levels, or basement); attack techniques are communicated to the attack teams; constant team coordination is maintained; fire growth and development is continuously evaluated; search, rescue, and ventilation requirements are communicated or managed; hazards are reported to the attack teams; and incident command is apprised of changing conditions.

(A) *Requisite Knowledge:* **Selection of the nozzle and hose for fire attack given different fire situations;** selection of adapters and appliances to be used for specific fire ground situations; dangerous building conditions created by fire and fire suppression activities; indicators of building collapse; the effects of fire and fire suppression activities on wood, masonry (brick, block, stone), cast iron, steel, reinforced concrete, gypsum wall board, glass, and plaster on lath; search and rescue and ventilation procedures; indicators of structural instability; suppression approaches and practices for various types of structural fires; and the association between specific tools and special forcible entry needs.

(B) *Requisite Skills:* The ability to assemble a team, choose attack techniques for various levels of a fire (e.g., attic, grade level, upper levels, or basement), evaluate and forecast a fire's growth and development, select proper tools for forcible entry, incorporate search and rescue procedures and ventilation procedures in the completion of the attack team efforts, and determine developing hazardous building or fire conditions.

NFPA 1002

2-3.7 **Operate all fixed systems and equipment** on the vehicle not specifically addressed elsewhere in this standard, given systems and equipment, manufacturer's specifications and instructions, and departmental policies and procedures for the systems and equipment, so that each system or piece of equipment is operated in accordance with the applicable instructions and policies.

(a) *Requisite Knowledge:* Manufacturer specifications and **operating procedures,** policies, and procedures of the jurisdiction.

(b) *Requisite Skills:* The ability to deploy, energize, and monitor the system or equipment and to recognize and correct system problems.

6-1.1 Perform the specified routine tests, inspections, and servicing functions specified in the following list, in addition to those contained in 2-2.1, given a wildland fire apparatus and its manufacturer's specifications, so that the operational status is verified.

- Water tank and/or other extinguishing agent levels (if applicable)
- **Pumping systems**
- **Foam systems**

(a) *Requisite Knowledge:* Manufacturer specifications and requirements, policies, and procedures of the jurisdiction.

(b) *Requisite Skills:* The ability to use hand tools, recognize system problems, and correct any deficiency noted according to policies and procedures.

6-2.3 Produce a foam fire stream, given foam-producing equipment, so that the proper proportion of foam is provided.

(a) *Requisite Knowledge:* **Proportioning rates and concentrations, equipment assembly procedures, foam systems limitations,** and manufacturer specifications.

(b) *Requisite Skills:* The ability to operate foam proportioning equipment and connect foam stream equipment.

8-1.1 Perform routine tests, inspections, and servicing functions specified in the following list, in addition to those specified in the list in 2-2.1, given a fire department mobile water supply apparatus, so that the operational readiness of the mobile water supply apparatus is verified.

- Water tank and other extinguishing agent levels (if applicable)
- **Pumping system** (if applicable)
- Rapid dump system (if applicable)
- **Foam system** (if applicable)

(a) *Requisite Knowledge:* Manufacturer specifications and requirements, policies, and procedures of the jurisdiction.

(b) *Requisite Skills:* The ability to use hand tools, recognize system problems, and correct any deficiency noted according to policies and procedures.

NFPA 1021

2-6.2 Develop an initial action plan, given size-up information for an incident and assigned emergency response resources, so that resources are deployed to control the emergency.

(a) *Prerequisite Knowledge:* **Elements of a size-up, standard operating procedures for emergency operations,** and fire behavior.

(b) *Prerequisite Skills:* The ability to analyze emergency scene conditions, to allocate resources, and to communicate verbally.

NFPA 1081

6.2.3 Operating as a member of a team, attack an exterior fire, given a water source, an attack line, personal protective equipment, tools, and an assignment, so that team integrity is maintained, the attack line is properly deployed for advancement, access is gained into the fire area, appropriate application practices are used, the fire is approached safely, attack techniques facilitate suppression given the level of the fire, hidden fires are located and controlled, the correct body posture is maintained, hazards are avoided or managed, and the fire is brought under control.

(A) Requisite Knowledge. Principles of fire streams; types, design, operation, nozzle pressure effects, and flow capabilities of nozzles; precautions to be followed when advancing hose lines to a fire; observable results that a fire stream has been correctly applied; dangerous conditions created by fire; **principles of exposure protection;** potential long-term consequences of exposure to products of combustion; physical states of matter in which fuels are found; and the application of each size and type of attack line, the role of the backup team in fire attack situations, attack and control techniques, and exposing hidden fires.

(B) Requisite Skills. The ability to prevent water hammers when shutting down nozzles; open, close, and adjust nozzle flow and patterns; apply water using direct, indirect, and combination attacks; advance charged and uncharged 38 mm (1½ in.) diameter or larger hose lines; extend hose lines; replace burst hose sections; operate charged hose lines of 38 mm (1½ in.) diameter or larger; couple and uncouple various handline connections; carry hose; attack fires; and locate and suppress hidden fires.

7.2.1 Operating as a member of a team, attack an interior structural fire, given a water source, an attack line, personal protective equipment, tools, and an assignment, so that team integrity is maintained, the attack line is deployed for advancement, access is gained into the fire area, correct application practices are used, the fire is approached safely, attack techniques facilitate suppression given the level of the fire, hidden fires are located and controlled, the correct body posture is maintained, hazards are avoided or managed, and the fire is brought under control.

(A) Requisite Knowledge. **Principles of conducting initial fire size-up; principles of fire streams; types, design, operation, nozzle pressure effects, and flow capabilities of nozzles;** precautions to be followed when advancing hose lines to a fire; observable results that a fire stream has been correctly applied; dangerous building conditions created by fire; **principles of exposure protection;** potential long-term consequences of exposure to products of combustion; physical states of matter in which fuels are found; common types of accidents or injuries and their causes; and the application of each size and type of attack line, the role of the backup team in fire attack situations, attack and control techniques, and exposing hidden fires.

(B) Requisite Skills. The ability to prevent water hammers when shutting down nozzles; open, close, and adjust nozzle flow and patterns; apply water using direct, indirect, and combination attacks; advance charged and uncharged 38 mm (1½ in.) diameter or larger hose lines; extend hose lines; replace burst hose sections; operate charged hose lines of 38 mm (1½ in.) diameter or larger; couple and uncouple various handline connections; carry hose; attack fires; and locate and suppress hidden fires.

8.2.2 Develop an initial action plan, given size-up information for an incident and assigned emergency response resources, so that resources are deployed to control the emergency.

(A) Requisite Knowledge. **Elements of a size-up, SOPs for emergency operations,** and fire behavior.

(B) Requisite Skills. The ability to analyze emergency scene conditions, to allocate resources, and to communicate verbally.

NFPA 1051

5.1.1 The Wildland Fire Fighter I shall meet the job performance requirements defined in Sections 5.1 through 5.5.

(A) Requisite Knowledge. Fireline safety, use, and limitations of personal protective equipment, agency policy on fire shelter use, basic wildland fire behavior, **fire suppression techniques, basic wildland fire tactics,** the fire fighter's role within the local incident management system, and first aid.

(B) Requisite Skills. Basic verbal communications and the use of required personal protective equipment.

5.5.3 Recognize hazards and unsafe situations given a wildland or wildland/urban interface fire and the standard safety policies and procedures of the agency, so that the hazard(s) and unsafe condition(s) are promptly communicated to the supervisor and appropriate action is taken.

(A) Requisite Knowledge. Basic wildland fire safety, fire behavior, and **suppression methods.**

(B) Requisite Skills. None specified.

5.5.6 Describe the methods to reduce the threat of fire exposure to improved properties given a wildland or urban/interface fire, suppression tools, and equipment so that improvements are protected.

(A) Requisite Knowledge. Wildland fire behavior, wildland fuel removal, **structure protection methods**, and equipment and personnel capabilities.

(B) Requisite Skills. The application of requisite knowledge to protect structures.

5.5.7 Mop up fire area, given a wildland fire, suppression tools, and water or other suppression agents and equipment, so that burning fuels that threaten escape are located and extinguished.

(A) Requisite Knowledge. **Mop-up principles, techniques, and standards.**

(B) Requisite Skills. Use of basic tools and techniques to perform mop-up operations.

6.5.2 Select fireline construction methods, given a wildland fire and line construction standards, so that the technique used is compatible with the conditions and meets agency standards.

(A) Requisite Knowledge. **Resource capabilities and limitations, fireline construction methods,** and agency standards.

(B) Requisite Skills. None specified.

6.5.3 Effect the reduction of fire exposure to improved properties given a wildland or wildland/urban interface fire and available tools and equipment so that improvements are protected and the risk from fire is reduced.

(A) Requisite Knowledge. Knowledge of fire behavior in both wildland and improved properties, and **the effects of fuel modification to reduce the hazard.**

(B) Requisite Skills. The use of tools and equipment to protect the improved property.

7.5.2 Size up an incident to formulate an incident action plan, given a wildland fire and available resources, so that incident objectives are set and strategies and tactics are applied according to agency policies and procedures.

(A) Requisite Knowledge. Size-up procedures, fire behavior, **resource** availability and **capability, and suppression priorities.**

(B) Requisite Skills. Identification of values at risk, objective setting, and selection of correct wildland fire-suppression strategies.

7.5.5 Deploy resources to suppress a wildland fire, given an assignment, personnel, equipment, and agency policies and procedures, so that appropriate suppression actions are taken, and safety of personnel is ensured.

(A) Requisite Knowledge. Fireline location and construction techniques, burning out procedures, **capabilities of fire-fighting equipment** and personnel, radio communications capabilities and protocols, and **techniques of the proper and safe deployment of the assigned resources.**

(B) Requisite Skills. Capabilities of assigned personnel and equipment.

7.5.7 Evaluate incident conditions, given a wildland fire, so that progress, changes in fuels, topography, weather, fire behavior, and other significant events are identified and communicated to the supervisor and to assigned and adjoining personnel.

(A) Requisite Knowledge. Intermediate wildland fire behavior.

(B) Requisite Skills. Collect wildland fire weather, fuels, and topographic information.

7.5.12 Deploy resources to **mop up a wildland fire,** given a wildland fire, personnel, equipment, and agency policies and procedures, so that appropriate mop-up actions are taken.

7.5.13 Complete wildland fire suppression operations, given a wildland fire that has been controlled and mopped up and agency policies and procedures, so that the **fire area is extinguished and resources are returned to service.**

9.5.1 **Risk and Hazard Assessment.** Assess the actual and potential risks, hazards, and values at risk for the wildland/urban interface fire incident, given incident intelligence, predicted fire behavior, and agency policies, so that all risks, hazards, and values at risk are identified for planned mitigation efforts.

(A) Requisite Knowledge. Have an understanding of the factors that constitute a wildland fire hazard and risk and the impacts they will have on fire-suppression efforts.

(B) Requisite Skills. The ability to prioritize the various risks and hazards as a plan of operations is being developed.

9.5.6 **Structure Protection Plan.** Develop and monitor a structure protection plan, given incident intelligence, current and predicted fire behavior, community data, and available resources, so that various structures and other improvements that are or may be threatened during a wildland/urban interface incident are protected and the plan is modified as needed.

(A) Requisite Knowledge. Knowledge of the availability and capability of fire apparatus and equipment that can be involved in an incident, **the elements of a structure protection plan, incident objectives, and the weather forecast.**

(B) Requisite Skills. The ability to develop and implement a structure protection plan and to constantly evaluate the wildland fire situation and change and modify the structure protection plan accordingly.

Chapter 5
Class A Foam Fire Fighting

The majority of fires to which fire and emergency services personnel respond are those involving materials that are designated as Class A fuels. Depending on the service area, these fires may involve structures, wildlands, exterior concentrations of materials, or a combination. The National Fire Protection Association (NFPA) defines a *structure fire* as one that involves a building, enclosed structure, vehicle, vessel, aircraft, or like property. It also defines a *wildland fire* as an unplanned and unwanted fire requiring suppression action or an uncontrolled fire spreading through vegetative fuels. Exterior concentrations of materials may consist of stored lumber, piles of used tires or waste, or trash in landfills (**Figure 5.1**). All three Class A fire types may threaten exposures such as other structures, vegetation, stored materials, or other improvements. The use of Class A foam technology in combating these fires provides the fire and emergency services with improved capabilities over the use of water as the sole extinguishing agent.

Figure 5.1 Exterior concentrations of materials such as raw lumber or used pallets are examples of hazards that may require the use of Class A foams to control a fire.

This chapter provides information on tactical priorities for all emergency situations and the use of Class A foam solutions for structural fire fighting, wildland fire fighting, and incidents involving other exterior concentrations of Class A materials. Recent evaluations by major fire and emergency services organizations along with advancements in the design and production of Class A foam concentrates and proportioning equipment have provided validation for the use of Class A foam solutions in situations that were previously ignored. An overview of the basic considerations for Class A finished foam application precedes a discussion of the use of Class A foam solutions for interior and exterior structural fire fighting. This discussion is followed by a look at wildland fire fighting and ends with the use of Class A foam solutions on exterior concentrations of Class A materials.

Tactical Priorities

[NFPA 1021: 2-6.2; NFPA 1051: 5.1.1, 7.5.2; NFPA 1081: 6.2.3, 8.2.2]

Numerous tactical approaches are possible in firefighting operations. Regardless of the type of hazard or the type of extinguishing agent, some method for prioritizing the operation must be in place. According to Frank Bateman, Training Manager for Kidde Fire Fighting, the ultimate goal of any emergency responder at an emergency incident is *"the timely and efficient restoration of normal conditions within the limits of acceptable risk."* This goal applies to all types of incidents that fire and emergency services personnel respond to on a daily basis. It, therefore, can be applied to all operations that involve the use of foam concentrates as extinguishing agents.

To accomplish this goal, emergency responders must follow an established protocol that prioritizes the objectives that are essential to meeting that final goal. One of the many prioritization tools available was developed many years ago by Lloyd Layman and is referred to as the *RECEO* model (see sidebar). The acronym stands for *Rescue, Exposures, Confine, Extinguish,* and *Overhaul.* It is both a list of priorities and a sequence of operation that are used at emergency incidents. The Incident Commander implements the *RECEO* model within the structure of the Incident Management System.

Class A Foam Use Considerations

[NFPA 1001: 5.3.8; NFPA 1051: 5.1.1, 5.5.3, 6.5.2, 6.5.3, 7.5.2, 9.5.6; NFPA 1081: 6.2.3]

Before discussing the specifics of Class A foam fire fighting, fire and emergency services personnel should understand the basic considerations for using Class A foam concentrates. These considerations include the advantages and disadvantages of using Class A foam concentrates, concentrate selection issues, and

RECEO Model

The application of *RECEO* to all types of emergencies results in the efficient use of personnel, water, foam concentrates, and equipment and in the successful termination of an incident within the acceptable limits of risk. As with all crisis situations, emergency incidents are rarely linear, that is, follow a strict step-by-step pattern. Therefore, some steps take place simultaneously and in support of other steps. However, each step of the operation is accomplished in order, starting with Rescue, the highest priority, which is based on life safety.

- **Rescue** — Rescue those who are endangered. Life safety takes precedent at all emergencies, including the safety of emergency responders. The use of finished foam during the rescue phase can help reduce the heat within a structure, provide a foam blanket over a spill, provide a barrier between victims and the fire or hazardous spill, or provide a protected avenue of advancement to perform rescues.
- **Exposures** — Protect areas or structures that are threatened. A direct attack on the seat of the fire is the most effective means of exposure protection when an immediate attack is within the capabilities of the initial response units. If it is impossible to make such an attack, then protecting the exposures limits the overall loss to the initial burn area. The use of thick, dry finished foam can quickly and efficiently place a barrier between the fire or hazardous spill and any exposed area. This procedure is true in interior structure fires, wildland fires, flammable liquid spills, or flammable liquid fires. The use of foam concentrate requires less water than standard water curtain barriers.
- **Confine** — Limit the fire (or hazardous spill) to the smallest space within the area of origin. Confinement may take place in the corner of a room, in a patch of woods on a hillside, at an undiked spill adjacent to a vehicle accident, or at a single burning oil tank. Foam concentrates can be used to supplement other confinement techniques that involve diking or redirecting spills or burning materials. When used for confinement purposes, finished foam works best on static pools of unignited fuel.
- **Extinguish** — Control and extinguish the fire. Once the previous priorities have been met, fire extinguishment can take place. (Obviously, extinguishment does not occur in situations involving unignited flammable or combustible liquids. However, the term can be used to indicate the use of extinguishing agents to prevent ignition.) Finished foam cools, smothers, and blankets the fuel to both extinguish and prevent reignition. Minimum application rates are based on those given in NFPA 11, *Standard for Low-, Medium-, and High-Expansion Foam* (2002), and NFPA 11A, *Standard for Medium- and High-Expansion Foam Systems* (1999). Use the manufacturer's recommended rate if it is higher.
- **Overhaul** — Check for and extinguish hot spots or prevent reignition, investigate fire cause, and conduct salvage/loss control operations. This phase involves restoring conditions to normal or as close to normal as possible. Finished foam is used to extinguish hidden or remaining fires, soak into small confined spaces, prevent reignition or the release of vapors, and prevent any form of rekindle. Only when the hazard is completely removed should foam operations cease.

the requirements for logistics, staffing, training, safety, and personal protective clothing and equipment.

Advantages and Disadvantages

Using Class A foam concentrates (in the proper proportions) delivered through standard fog nozzles or a compressed-air foam system (CAFS) have certain advantages over water for extinguishing fires in structures, wildlands, and exterior stored materials. However, testing by Underwriters Laboratories Inc. (UL) for the National Fire Protection Research Foundation has revealed some disadvantages to the use of Class A foam and CAFS in these types of fires. Although the disadvantages must be considered, the advantages appear to outweigh them.

Advantages

- ***Reduces fire extinguishment or knockdown times*** — The inherently high surface tension of plain water impedes its ability to spread across a burning surface and absorb into fuels. Water molecules are attracted only to other water molecules, which causes most of the water used in a typical fire attack to bead and roll off surfaces. The addition of a Class A foam concentrate enhances the effectiveness of water as an extinguishing agent by working against the problems associated with water's high surface tension. The surfactant qualities of Class A foam concentrates reduce the surface tension of water, thereby improving knockdown and overhaul/mop-up time by allowing faster penetration into fuels. In both fireground operations and tests conducted by UL, the performance of Class A foam was found to be superior to or the same as plain water. No reports of inferior performance by the foam were indicated.
- ***Improves the water absorption rate of fuels*** — The addition of the hydrocarbon surfactant in Class A concentrates creates a solution that is chemically different from plain water. The resultant foam solution is one that has a high affinity for (attraction to) carbon particles, which is a major byproduct of the combustion process. This high affinity for carbon particles facilitates rapid water absorption and the wetting of fuels (**Figure 5.2**).

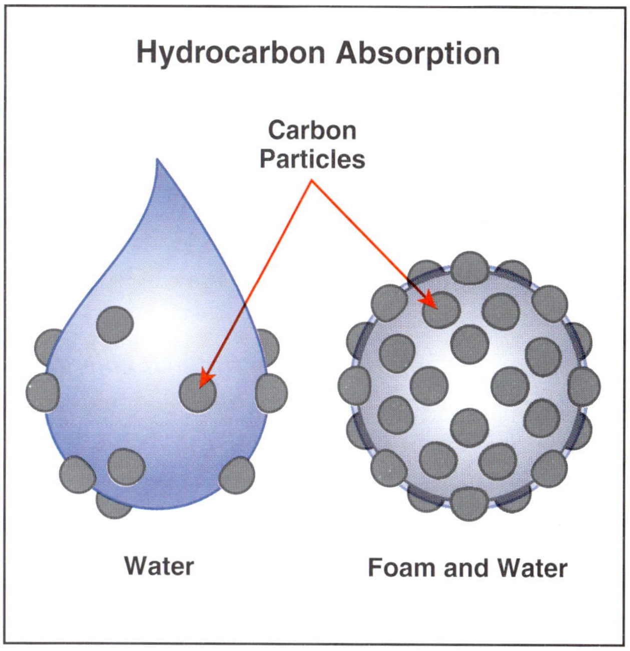

Figure 5.2 Carbon particles adhere to foam at a higher rate than they do to water.

- ***Reduces water damage*** — Class A finished foam acts as a vapor seal by adhering to burning Class A materials and insulating them, which prevents further combustion by creating a barrier that isolates the fuel from oxygen and heat. Lowering the water content of a foam solution through aspiration also enhances these insulating characteristics; that is, air injected into the foam solution expands the foam and water solution while using less water in the process. This characteristic results in decreased water damage to the unburned portions of the material.
- ***Reduces the amount of water required for extinguishment*** — A common misconception is that the addition of Class A foam concentrate to water actually increases the quantity of water. Some people believe that 200 gallons (757 L) of foam solution aspirated to a 10:1 expansion ratio actually gives 2,000 gallons (7 571 L) of water, but this is not accurate. Obviously, 200 gallons (757 L) of water is always 200 gallons (757 L) of water, regardless of the quantity of foam concentrate or the amount of air added to it. The addition of foam concentrate does not "create" more water, but it makes water more efficient at soaking into fuels and creating a vapor barrier to prevent reignition. Therefore, less water (in the form of foam solution) is required to extinguish a fire than

if plain water is used alone. Water conservation is most apparent when using a CAFS to apply Class A foam. Estimates based on field tests indicate that the efficiency of water, and therefore its conservation, may be as high as 5 to 10 times that of plain water in some applications. An additional advantage is the reduced water loading on fire-weakened structures, which may help prevent structural collapse.

- ***Reduces potential stress on fire and emergency services personnel at the incident*** — Field use of Class A foam, and particularly CAFSs, indicates that fire and emergency services personnel are exposed to less physical stress than encountered during operations using plain water hose streams. The reasons for the reduced stress are faster suppression times, reduced overhaul time required, and shorter incident times, which cause a more rapid return to quarters. Additionally, a CAFS decreases stress due to the lighter weight and easier maneuverability of the hoselines (**Figure 5.3**).
- ***Creates several specific advantages in wildland applications:***
  - — Opens the waxy coating of green vegetation, allowing water to penetrate into the surface
  - — Increases the relative humidity in the area of application
  - — Is visible from both the air and ground

Figure 5.3 Firefighters experience less stress when using compressed air foam system (CAFS) foam handlines because they are lighter in weight and more maneuverable than comparable water-charged hoselines. *Courtesy of Tom Ruane.*

- ***Forms a protective blanket*** — Like Class B foam, Class A foam forms a vapor barrier between the fuel and the oxygen in the air. This barrier also insulates unburned fuel from radiant heat and direct flame impingement. The blanket is effective for both exposure protection and preventing reignition.
- ***Can be produced at a relatively low cost*** — Class A foam concentrates are normally proportioned between 0.1 percent and 1 percent. Because less foam concentrate is required to produce an adequate foam blanket, the operational cost of the foam is less. This factor is an added benefit in training, allowing Class A foam to be used to simulate Class B foam during flammable and combustible liquids fire training.
- ***Can be used on flammable and combustible liquid fires*** — Although not documented in scientific testing, Class A foam can be used to suppress flammable and combustible liquid fires. Use of Class A foam in this situation should only occur when the situation requires rapid action and Class B foam is not available. Fire and emergency services personnel must realize that the Class A foam blanket deteriorates rapidly creating the possibility of reignition.
- ***Is visible during and after application*** — Visibility is particularly advantageous during wildland fire-fighting operations, allowing personnel to determine the areas that have been covered and those requiring additional coverage.
- ***Clings to most surfaces much longer than plain water*** — The ability of Class A foam to cling to most surfaces helps reduce water runoff, reduces water damage, and aids in exposure protection.
- ***May help in the preservation of evidence for fire cause determination*** — Because of the penetration of Class A foam into burned materials and the extinguishment of deep-seated fires, the need for manual overhaul is reduced. Overhaul reduction may create a situation where evidence that is normally disturbed or removed is left intact. This advantage is an unsubstantiated benefit based on field reports and not on scientific evidence.
- ***May provide long-term cost savings and reduced property damage*** — Although this advantage has not been thoroughly documented, it appears

from field reports provided to UL that some savings can be justified. Cost savings are encountered in the reduced incident times, lowered apparatus fuel costs from shorter operating times, and reduced property damage to both the equipment used by the organization and the fire scene itself.

Disadvantages

- ***Delays fire investigations*** — Some fire investigators have complained that the time that it takes for finished foam to degrade or dissolve prevents or delays them from surveying the fire scene interior of an involved structure. This complaint requires an adjustment in local operating procedures if it becomes an issue within the organization.
- ***Creates strong nozzle reactions and stream forces that can cause injuries or damage*** — The use of a CAFS creates very strong nozzle reactions. Additionally, loose objects and debris may blow around by the force of the foam stream, causing injuries to personnel and damage to surroundings. Use nozzles with pistol grips on CAFS hoselines, and wear full protective clothing, including face shields or respiratory protection (self-contained breathing apparatus), during all CAFS operations (**Figure 5.4**).

Figure 5.4 Strong nozzle reaction created by the added air pressure in compressed-air foam system (CAFS) hoselines requires nozzles with pistol grips and complete personal protective equipment. *Courtesy of Tom Ruane.*

- ***Obscures vision by coating facepieces and face shields*** — Finished foam can coat facepieces and face shields, obscuring vision. However, the foam can easily be wiped away with a gloved hand. The side benefit is the cleaning ability of finished foam, which removes soot from facepieces or face shields.
- ***Obscures tripping hazards*** — A finished foam blanket can obscure tripping hazards on floors or outside (including depressions, rocks, or limbs), which requires care in movement on the part of emergency personnel when walking in finished foam.
- ***Requires structure ventilation to remove products of combustion*** — Finished foam has little effect on reducing the temperatures at ceiling level. Therefore, proper ventilation is required in structures to remove the smoke, gases, and superheated air from this level.
- ***Creates slipping hazards*** — Finished foam can coat floors within structures and exterior surfaces made of concrete or asphalt and create slipping hazards. If this situation is a problem, hose the surfaces with water following overhaul/mop-up.
- ***May result in substantial equipment, concentrate, and training costs*** — The purchase and continuing maintenance of specialized foam nozzles, proportioners, eductors, and CAFS equipment can run from a few hundred dollars to as high as $50,000. Although Class A foam concentrate is diluted to generate large quantities of foam solution, continued use requires an ongoing replacement budget demand. For effective and efficient use of Class A foam, personnel must be adequately trained and reevaluated periodically.

- ***Is a corrosive detergent*** — Class A foam concentrates can damage the painted surfaces of apparatus, corrode metal, and remove lubricants from the valves and internal systems of pumps, proportioners, eductors, and nozzles. Care is required to prevent the corrosive affects, including cleaning up foam concentrate spills; flushing pumps, proportioners, eductors, and nozzles; and reapplying lubricants to valves and moving parts exposed to the foam. Additionally, Class A foam concentrate can damage personal protective equipment and have a drying affect on skin exposed to the concentrate.
- ***May have negative impact on the environment*** — While Class A foam solution may not have a negative affect on the environment, Class A foam concentrate may damage both plant and marine life. Studies on long-term impacts are still ongoing.
- ***May cause confusion with Class B foam containers/packaging*** — Class A foam concentrate containers have a similar appearance to Class B foam concentrate containers. This similarity can result in the intermixing of the two types of concentrates, which can lead to failure of the foam distribution system.
- ***Increases chance of equipment failures because of foam distribution equipment complexity*** — This complexity is particularly apparent with CAFS equipment. The potential for system failures can result in a loss of foam, water, or both at the nozzle and cause potential injury to personnel. Additional maintenance and resources are required to keep the equipment operational.
- ***Can restrict the flow of plain water when using CAFS discharge lines*** — Restrictions in the line that help produce the foam solution can reduce the gpm (L/min) output of the pump when plain water is flowing.
- ***May cause hose kinking when using CAFS hoselines*** — Although the UL report did not indicate that CAFS hoselines had a greater tendency to kink than hoses charged with plain water, some field reports indicate that they do. There is no scientific indication that hose kinking is a problem.
- ***Creates several specific disadvantages in wildland applications:***
  — Cannot maintain the integrity of the finished foam blanket in areas where personnel and equipment move continually

  — Cannot be used in areas where it is possible to pollute drinking water sources or environmentally sensitive sites

  — Will freeze when air temperatures are below freezing (32°F/0°C)

Concentrate Selection

Because Class A foam concentrates have a wide range of extinguishment and exposure protection capabilities, it is important that fire and emergency services personnel select the proper concentrate for the specific task. According to NFPA 1145, *Guide for the Use of Class A Foam in Manual Structural Fire Fighting* (2000), Class A foam concentrates are divided into low-, medium-, and high-expansion categories. Low-expansion Class A foam concentrates are then subdivided into three types: wet, fluid, and stiff/dry. These categories/types are characterized by expansion ratio, drain time, and consistency. These categories/types are referenced throughout the rest of this chapter. The characteristics for each category and type of foam concentrate appear in **Table 5.1**, Class A Foam Concentrate Categories and Characteristics. **Table 5.2**, Class A Foam Selection Matrix, gives a general matrix for the selection of Class A foam concentrates for the various types of nonwildland Class A fires (see p. 144).

Logistics

In planning the logistics for an efficient Class A foam operation, it is important to have sufficient amounts of foam concentrate available at any given time. Consider that a 0.5-percent foam solution requires ½ gallon (2 L) of concentrate for each 100 gallons (379 L) of water. This fact means that fire apparatus with a 500-gallon (1 893 L) water tank requires at least 2½ gallons (9 L) of foam concentrate. For extended operations, extra foam concentrate should be readily available to replenish empty apparatus foam concentrate tanks (**Figure 5.5**).

Table 5.1
Class A Foam Concentrate Categories and Characteristics

Foam Category	Low-Expansion			Medium-Expansion	High-Expansion
Foam Type	Wet	Fluid	Stiff/dry		
Expansion Ratio	1:5	5:10	10:20	20:200	200:1,000
Consistency	Watery, sloppy	Watery shaving lather, sloppy	Dry or stiff lather	Dry foam, medium to large bubbles	Very dry foam, large bubbles
25% Drain Time (in seconds)	Less than 30	30–90	90–120	Greater than 120	Greater than 300
Generator	Nonaspirating, aspirating, or CAFS	Aspirating or CAFS	CAFS	Large screened foam tubes	High-expansion generator
Usage	Penetration, cooling, blanketing, overhaul			Exposure protection, blanketing, overhaul, filling voids	Filling voids and spaces
Attack Mode	Direct or indirect, exposure protection			Indirect	

Based on NFPA 1145, *Guide for the Use of Class A Foam in Manual Structural Fire Fighting* (2000), Table 2.3.2.

CAFS = compressed-air foam system

Figure 5.5 Additional 5-gallon (19 L) foam concentrate containers are carried on some apparatus. *Courtesy of Harry Eisner.*

Staffing

Staffing requirements for the application of foam solutions on structural, wildland, and exterior concentrations of materials involved in fire are the same as those required for the application of plain water. Although the use of foam concentrate significantly improves the effectiveness of a fire attack, there is still the need for backup hoselines and personnel to operate them, rapid intervention crews (RICs), and personnel to support the overall operation. While only the driver/operator is required when foam concentrate is carried in a tank on the apparatus and proportioned by an onboard proportioner, foam concentrate that is carried in individual 5- and 55-gallon (19 L and 208 L) containers requires more personnel.

Training

Because the application techniques and the proportioning equipment vary just like the types of available Class A foam concentrates, training in the proper use of concentrates and equipment is essential. Simply purchasing foam concentrates and

Table 5.2
Class A Foam Selection Matrix

Class A Type Fire	Tactical Application	Exposure Type	Considerations and Foam Type: Primary Consideration (PC)	Foam Type (PC)	Secondary Consideration (SC)	Foam Type (SC)	Nozzle Type: Low-Energy Generator	High-Emergy Generator
Exterior Structure	Direct Attack		Knockdown Through Cooling	Wet or Fluid	None Given	None Given	Nonaspirating Conventional	Conventional Combination
	Exposure Protection	Porous	Raising Moisture Content	Wet	Resistance to Reignition	Dry	Low or Medium Aspirating	Largest Discharge Device Available
	Exposure Protection	Nonporous	Blanketing	Dry	Resistance to Reignition	Dry	Low or Medium Aspirating	Largest Discharge Device Available
	Overhaul		Surface Penetration and Cooling	Wet	None Given	None Given	Low or Medium Aspirating	Largest Discharge Device Available
Interior Structure	Direct Attack		Knockdown Through Cooling	Wet or Fluid	None Given	None Given	Nonaspirating Conventional	Conventional Combination
	Indirect Attack		Rapid Cooling of Combustion Products	Wet or Fluid	None Given	None Given	Nonaspirating Conventional	Conventional Combination
	Overhaul		Surface Penetration and Cooling	Wet	Blanketing	Dry	Nonaspirating Conventional	Largest Discharge Device Available
Exterior Material-Concentration	Exposure Protection	Porous	Raising Moisture Content	Wet	Resistance to Reignition	Dry	Nonaspirating Conventional	Conventional Combination
	Exposure Protection	Nonporous	Blanketing	Dry	Resistance to Reignition	Dry	Low or Medium Aspirating	Largest Discharge Device Available

Based on information found in NFPA 1145, *Guide for the Use of Class A Foam in Manual Structural Fire Fighting* (2000), Chapter 4.

proportioning equipment, issuing them to fire-fighting companies, and hoping that they will use them can cause improper application, ineffective results, and unfounded complaints about inadequate foam operations. Before the foam concentrates or proportioning equipment is issued, all personnel must receive thorough training on when to use it, how to use it, how to store it, how to replenish it, and how to clean the equipment following usage (**Figure 5.6**). Chapter 9, Foam Concentrate Use Training, covers the training necessary for using all types of foam concentrates and equipment.

Figure 5.6 Basic classroom training in the use of Class A foam is necessary before skill training. *Courtesy of Tom Ruane.*

Safety

Class A foam concentrates that meet NFPA 1150, *Standard on Fire-Fighting Foam Chemicals for Class A Fuels in Rural, Suburban, and Vegetated Areas* (1999), and have been approved by the U.S. Forest Service have undergone testing relative to biodegradability and corrosiveness. These factors are important when considering the hazards that Class A foam use pose to personnel and the environment. Ensure that each Class A foam manufacturer provides a material safety data sheet (MSDS) for its concentrate. Keep this document readily available to personnel involved in foam operations. Appendix E contains a sample MSDS sheet for a Class A foam concentrate.

Class A foam concentrates are biodegradable and relatively safe when used in small ratios of 0.1 to 1 percent. However, fire and emergency services personnel must clean up spills involving concentrate as soon as possible with an absorbent that can be shoveled onto the spill. This cleanup is particularly important if a concentrate is in danger of entering a waterway. It is not recommended practice to flush spilled foam concentrate into storm sewers; it could contaminate ground water.

Personnel exposed to foam concentrates can take the same precautions as those taken for contact with any high-strength cleanser that is irritating to the eyes and skin. Inhaling concentrate vapors may irritate the mucous membranes and the respiratory tract. Take the following safety precautions:

- Wear goggles, waterproof gloves, and rubber boots when mixing concentrates with water or replenishing apparatus foam concentrate tanks.
- Rinse clothing thoroughly that has been soaked in concentrate.
- Treat eyes splashed with concentrate or solution by flushing them with clean water for 15 minutes.
- Use hand lotion on hands to avoid chapping.
- Clean up spilled concentrate with an absorbent material to avoid slips and falls (**Figure 5.7**).
- Be careful when walking in a finished foam blanket to avoid hidden objects and holes.
- Use caution when walking on surfaces covered with foam solution. These surfaces can become very slippery.
- Avoid contaminating water sources with foam concentrates or solutions.
- Open nozzles slowly when using all types of foam solutions to avoid the strong resultant reaction that could knock down the operator (**Figure 5.8**).

Figure 5.7 Clean up foam concentrate spills immediately to prevent slipping and falls.

Figure 5.8 Open the nozzle slowly to avoid the affects of a strong nozzle reaction. *Courtesy of Tom Ruane.*

Personal Protective Clothing and Equipment

Because the primary hazard to fire and emergency services personnel is the fire, the use of personal protective clothing and equipment is mandated for personal safety. The type of personal protection is determined by the type of fire, whether structural, wildland, or exterior material concentrations. The minimum clothing consists of respiratory protection, eye protection, fire-fighting coat and pants, boots, helmet, gloves, and protective hood. These, in turn, protect the wearer from any Class A foam solution or concentrate that may be present. Clean personal protective clothing and equipment thoroughly following an incident to remove any hydrocarbon surfactant or finished foam residue. Cleaning should be in accordance with NFPA 1971, *Standard on Protective Ensemble for Structural Fire Fighting* (2000) and the equipment manufacturer's recommendation **(Figure 5.9).**

Figure 5.9 Wear full personal protective equipment including respiratory protection when using Class A foam. *Courtesy of Tom Ruane.*

Structure Fires

[NFPA 1001: 5.3.10, 6.3.2; NFPA 1081: 6.2.3, 7.2.1]

Since 1995, the use of Class A foam solutions to combat structure fires has continued to gain acceptance with the fire-fighting community. The results of field tests and actual usage have shown that many of the same advantages realized in wildland fire fighting are duplicated when applying Class A foam solutions to structure fires. In general, the tactical considerations for applying Class A foam solutions during structural fire operations follow the same guidelines as those for the use of untreated water. This section discusses the use of Class A foam solutions in each of the following four main areas of tactical application:

- Interior (offensive) attacks
- Exterior (defensive) attacks
- Exposure protection
- Overhaul operations

NOTE: An ongoing debate exists over the effectiveness for Class A foam and CAFS in interior structural fire fighting. While there does not appear to be any scientific evidence to support claims by field personnel that Class A foam is vastly superior to plain water, UL tests have indicated that foam is at least as effective and certainly not inferior to water.

Interior Fire Attacks

When used in structural fire fighting, Class A finished foam is most effective on interior fire attacks. In order for Class A finished foam to have maximum effect, however, it must be applied directly to the burning material and any uninvolved fuels immediately adjacent to the fire (exposures). This application is done with a solid or straight stream to get finished foam onto the burning material quickly and pretreat (provide a protective barrier against heat) surrounding exposures. (See Exposure Protection section.) The use of Class A finished foam also creates improved conditions for attack crews by increasing visibility **(Figure 5.10)**. As plain water travels through heated air en route to the burning material, it vaporizes and creates large amounts of steam. Water that does reach the burning fuel tends to roll off the fuel and further vaporize. Finished foam is somewhat insulated from heat; therefore, only a small percentage of it converts to steam as it travels through the fire to the fuel. Furthermore, the excellent penetration quality of finished foam stops combustion more quickly than water, reducing the amount of byproducts created by incomplete combustion that enter the atmosphere. The use of Class A finished foam also enables faster absorption of heat, thereby reducing temperatures in the firefighting environment. These factors provide benefits for the safety and operational effectiveness of emergency responders.

The use of foam solutions for interior attacks follows the same general pattern that water-only interior attack takes. The fire or emergency services personnel must determine the appropriate method of attack, know the correct flow rate and fire hose size, and select the correct hose and nozzle for the situation.

Figure 5.10 Class A foam, which is becoming increasingly popular for use in structure fires, can help to improve visibility in the structure.

Attack Methods

Three standard attack methods are used for interior fire fighting: direct, indirect, and combination. Either a direct attack or combination attack is used when using Class A foam streams to make an interior attack. A *direct attack* is a discharge of a foam stream directly onto the burning fuel, which allows a fire to be quickly extinguished. The wetting ability of finished foam allows the agent to soak into the fuel and thoroughly halt combustion **(Figure 5.11)**.

An *indirect attack* deflects extinguishing agent off the ceiling above the fire area and is generally not recommended when using Class A foam streams. However, finished foam clinging to or absorbed into

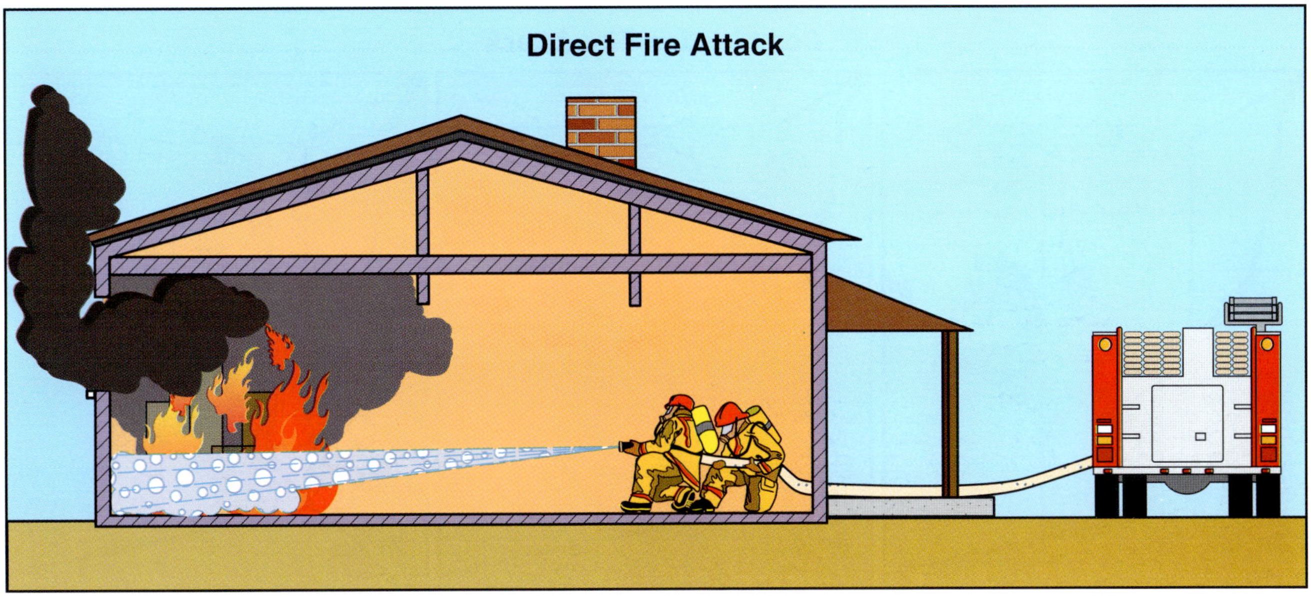

Figure 5.11 Direct foam attack in a compartment: Foam stream is dicharged directly on burning fuel.

the ceiling and other surfaces reduces the release of fuel vapors, which decreases the chance of flash-over (**Figure 5.12**).

A *combination attack* is one where the nozzle and fire stream are moved around the room in an *O, T,* or *Z* pattern. This movement allows agent to be applied to the fire and to the surrounding uninvolved fuel alternately. The combination attack is effective in extinguishing the burning fuel and also pretreating exposed fuels (**Figure 5.13**). One technique of the combination attack that could also be an indirect-attack technique involves "panel soaking": coating each of the interior surfaces with finished foam. The ceiling is coated first, followed by each of the wall surfaces. This foam coating reduces the potential for fire spread through radiation and flashover. Because the CAFS finished foam bubbles burst at 170°F (77°C), an immediate reduction in compartment temperature is realized, allowing for a more rapid advance into the area to reach the seat of the fire.

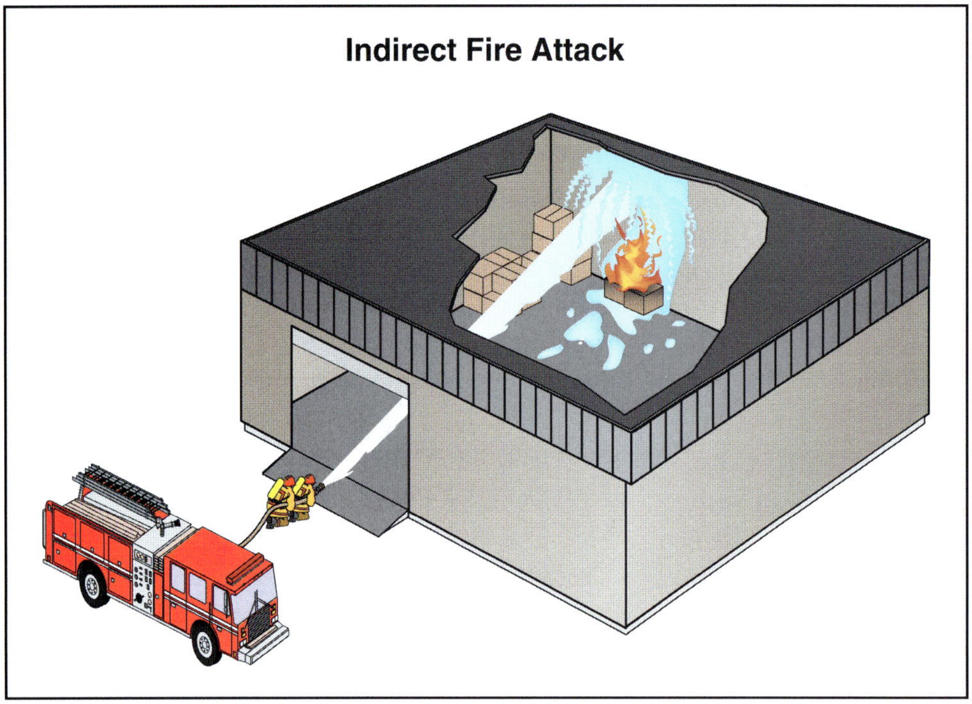

Figure 5.12 Indirect foam attack in a compartment: Foam stream is deflected off the ceiling onto the fire.

Figure 5.13 Combination fire attack: The foam stream is moved over the fire in patterns that form the letters *T, Z,* or *O.*

Flow Rates/Fire Hose Sizes

The topics of minimum acceptable flow rates and fire hose sizes for Class A foam streams used in structural attacks are subject to debates. Many unscientific tests using these streams on training fires show that fires are extinguished with Class A foam flow rates that are lower than those normally required for plain water. This condition is particularly true when using compressed-air foam system handlines. While a lower foam flow rate is very effective in extinguishing a fire, a safety concern arises in the event that either the foam proportioning system (on CAFS or low-energy systems) or air supply system (on CAFS) fails. The resultant foam stream or the flow of plain water, in itself, may not be enough to extinguish a fire or even provide protection for fire and emergency responders.

Until scientific testing proves otherwise, both IFSTA and several foam manufacturers recommend that flow rates used for Class A foam streams be the same as those used for plain water streams. Generally, the minimum accepted flow rate for an interior attack using water is 95 gpm (360 L/min). Low-energy Class A foam systems can easily provide this flow through 1¾-inch (45 mm) or larger hoselines. In order to achieve this flow with a CAFS system, a 2½-inch (65 mm) hoseline is required. Remember that the weight of the agent in the CAFS hoseline is considerably less than that of plain water. Thus, a 2½-inch (65 mm) foam hoseline handles much like a 1¾-inch (45 mm) plain water or low-energy foam line.

While the use of Class A foam solutions does not allow for reduced flow rates, the rapid fire knockdown associated with them reduces the time that a fire stream needs to be applied. The reduction in discharge time is the true factor in reduced water consumption when Class A foam solutions are used **(Figure 5.14)**. This factor means a reduced chance of structural collapse from the weight of added water and minimal water damage to the structure. While it could be possible to extinguish a fire with lesser flow rates than those used for plain water, it would take longer to effect the extinguishment. This situation increases the time that fire and emergency responders are exposed to hazardous conditions.

Figure 5.14 The increased extinguishment times provided by Class A foam result in reduced water damage in the structure. *Courtesy of Tom Ruane.*

Fire Hose Types/Nozzles

According to NFPA 1145, fire hose used by the fire and emergency services for the delivery of water in attack or supply lines is appropriate for use with low-energy foam systems. The hydraulic calculations used for flowing foam solutions are essentially the same as those used for flowing water **(Figure 5.15)**. Drier and stiffer foam developed by a CAFS may require longer hose lengths and hose with

Figure 5.15 To achieve an effective foam pattern (like the one shown), use hydraulic calculations for nozzle pressure that are the same as those for flowing water through hoselines. *Courtesy of Tom Ruane.*

rough interior surfaces to help agitate the foam solution. Attack hose that is resistant to kinking is required for continuous, unimpeded foam solution flow. Therefore, the use of reinforced rubber hose that is very kink resistant is recommended. Extruded-rubber hose and lined, woven-jacketed hose that are fairly resistant to kinking may also be used. A CAFS has not been shown to have a negative affect on any fire hose currently in use by the fire and emergency services.

Fog nozzles should be used on low-energy foam system hoselines. The turbulence created by these nozzles helps aerate the foam as it is discharged. Some fog nozzles are equipped with special clamp-on aspirating devices that improve the amount of aeration that takes place. Air-aspirating nozzles are not usually used for interior structural attacks due to their large size. Solid stream nozzles are most effective with CAFS foam streams.

Exterior Fire Attacks

Exterior fire attacks are employed in one of the two following scenarios:

1. A fire is "darkened down" or reduced in size from the exterior so that it is safe for fire and emergency responders to enter a structure in order to complete extinguishment.
2. Either a fire advances to a degree that it is not safe to enter the structure or not enough water can be discharged to effectively extinguish the fire.

The first scenario benefits greatly from the use of Class A foam operations. The use of Class A foam solutions for exterior attacks has proven quite successful in full-scale tests performed by personnel from Los Angeles (CA) County Fire Department, Morristown (NJ) Fire Bureau, and other departments/organizations worldwide. Employing the same fire stream techniques that were described previously, a fire can be knocked down quickly from a doorway or window. Of course, it is necessary to hit the seat of the fire from that position in order for this attack method to be successful (**Figure 5.16**). Only enough finished foam is discharged to reduce the size of the fire. Fire and emergency responders then wait a few moments for the heat to dissipate before entering the structure. This delay also allows the finished foam some time to penetrate the fuels and interior surfaces and further aids the extinguishment process. CAFS fire streams are especially well suited for this type of fire attack. The extended reach of the CAFS stream assures that finished foam is delivered deep into the structure and to the seat of the fire.

Testing of Class A foam operations in large-scale defensive fire attacks (scenario 2) has shown that foam concentrates do assist in improved fire extinguishment. The use of Class A finished foam allows the extinguishing agents to soak into the burning material, reach deep pockets of ignited materials, coat unburned materials, and blanket the fire while cooling the area. In these cases, large master streams apply the finished foam (**Figure 5.17**). However, if a fire has progressed beyond the emergency

Figure 5.16 Exterior attacks on interior structure fires with Class A foam and a compressed-air foam system (CAFS) allow the attacks to begin farther from the fire. *Courtesy of Mount Shasta (CA) Fire Protection District.*

Figure 5.17 Master stream appliances can be an effective means for applying Class A foams to wildland and urban-interface forest fires. *Courtesy of National Interagency Fire Center.*

organization's ability to apply adequate amounts of extinguishing agents, the use of Class A foam solutions may not be appropriate and complete flooding with water may be the only recourse.

Exposure Protection

Coating an exposure with finished foam provides a visible, lasting, protective barrier against the radiating heat of any fire. The ability of finished foam to enhance the soaking properties of water also allows greater penetration into an exposure's fuel to boost the moisture content and further reduce its ability to ignite. This technique is commonly referred to as *pretreating* the exposure. The advantage of using finished foam in this manner is that once the exposure is coated, emergency personnel may move on to the next exposure. When plain water is used, personnel must remain with the hoseline or monitor and continue to apply water to the exposure.

Dry finished foam is best for exposure protection because it clings to all horizontal and vertical surfaces (**Figure 5.18**). Finished foam is applied to all exterior surfaces of the exposure, including walls, eaves, roofs, porches, decks, columns, and any other threatened surfaces. The finished foam is lofted onto the surfaces from a distance to ensure that it clings and coats the surface, rather than bouncing off. The distance from which finished foam needs to be applied depends upon whether a nozzle-aspirated fire stream or a CAFS fire stream is used. CAFS fire streams are applied from a greater distance than nozzle-aspirated streams. CAFS streams have a 33 percent greater throw than regular water streams due to the increased pressure given

Figure 5.18 Class A foam adheres well to vertical surfaces as shown in this photo.

by the compressed air. Care must be taken not to damage the exterior of the structure with the force of a CAFS stream.

Apply at least ½ inch (13 mm) of finished foam to the exposure. When all other factors are the same, finished foam from a CAFS stream is expected to remain in place for about 1 hour. Finished foam applied from a nozzle-aspirated fire stream lasts about half that time. However, these figures are only general guidelines. Wind, fire, temperature, and humidity conditions affect the actual time that a foam blanket remains effective. On exposures where exterior finishes have extremely low moisture contents (such as shake shingles), it is effective to first wet the exposure with plain water before applying the finished foam coating.

Overhaul Operations

Class A foam exhibits some of its greatest benefits during overhaul operations. Using Class A finished foam drastically reduces the amount of water used, thereby limiting runoff and water damage to the structure. It is applied in the same fashion as water — by standard nozzles and techniques. An aspirating nozzle may also be used to spread a complete finished foam blanket over the material. The intent is the same in either case — blanket the fuel and give the water a chance to drain from the finished foam and penetrate the fuel (**Figure 5.19**). If the fire is deep-seated, the penetration of water alone may extend the length of time necessary for extinguishment (thereby causing additional damage) or not

Figure 5.19 Class A foam is an effective overhaul tool. However, the residual foam, as shown here, must also be cleaned up. *Courtesy of Harvey Eisner.*

extinguish the fire at all due to *thermal blocking:* a phenomenon that occurs when concealed hot spots contain enough heat to turn small amounts of penetrating water to steam. This situation prevents plain water from cooling sufficiently. However, the use of aspirated finished foam enables more water to defeat thermal blocking by increasing penetration and speeding the cooling process. Fire and emergency responders should make certain that finished foam reaches all heat pockets and hidden areas of a deep-seated fire. Because emergency responders can see where and how much finished foam has been applied, the tendency to apply more than necessary is reduced. When fire is located within the cavities of walls, holes must be cut to insert the foam nozzle. Class A finished foam has the advantage of filling the cavity and also of soaking into any insulation within the space.

Because Class A foam concentrates are hydrocarbon-based, some fire investigators have expressed concerns that finished foam residue provides false readings in tests for hydrocarbon accelerants at fire scenes. Laboratory tests indicate that finished foam residue does ***not*** contribute to any flammable liquids that may be found on a fire scene. In actual tests where structure fires were started with gasoline and extinguished using Class A finished foam, laboratory samples taken from the area where the gasoline was spread showed positive test results for the presence of gasoline. Samples where gasoline was not spread but where foam residue was present showed negative test results for the presence of gasoline. These results should eliminate any concern that fire investigators have in regard to this situation.

Wildland Fires

[NFPA 1001: 5.3.19; NFA 1002: 2-3.7, 6-1.1, 6-2.3, 8-1.1; NFPA 1051: 5.1.1, 5.5.3, 5.5.6, 5.5.7, 6.5.2, 6.5.3, 7.5.2, 7.5.5, 7.5.7, 7.5.12, 7.5.13, 9.5.1, 9.5.6]

The problem of wildland fires occurring is not a new one. Wildland fires have ravaged prairie and forest lands since the beginning of time. They have gained more attention in recent years because of their effects on human life and property. As the population of many communities has grown, structure development has spread into previously rural areas. Thus when a wildland fire occurs, in addition to worrying about the natural ground cover that is on fire, emergency personnel are faced with the challenge of protecting structures in the fire's path. The zone where an undeveloped wildland area meets a human development area is referred to as the *wildland/urban interface.*

The earliest and still largest use for Class A foam concentrates in the fire protection field is for the suppression of wildland fires. Class A finished foam acts as a suppressant by extinguishing the fire in the vegetation. Chemical retardants may also be used in conjunction with foam concentrates. The retardants (water mixed with diammonium phosphate) physically coat the fuel and remain in place for several days in the absence of rain. Before discussing wildland fire-suppression strategies and tactics, it is necessary to briefly review wildland fire behavior and the composition of these fires. Additional information can be found in the IFSTA **Wildland Fire Fighting for Structural Firefighters** manual and National Wildfire Coordinating Group documents.

Wildland Fire Behavior

Wildland fires involve burning materials such as weeds, grass, field crops, brush, forests, and similar vegetation. Wildland fires have characteristics that are not comparable to other forms of fire. Once a wildland fire starts, burning is generally rapid and continuous **(Figure 5.20).** Many factors affect wildland fire behavior, but the three most important ones are *fuel, weather,* and *topography.* Any one factor may dominate in influencing what an individual fire does, but usually the combined strength of all three factors dictates its behavior.

Fuel

Although all the materials are Class A combustibles, fuels found in wildland settings are generally classified by grouping those with similar burning characteristics together. Several systems are used for classifying wildland fuels such as the National Fire Danger Rating System (NFDRS) or by moisture content, position, and size. The method that uses the position of wildland fuels classifies them as subsurface, surface, and aerial (crown) fuels **(Figure 5.21)**. These descriptions are as follows:

Figure 5.20 Wildland fires can spread rapidly over all types of terrain. *Courtesy of National Interagency Fire Center.*

Figure 5.21 *(Below)* Types of wildland fuels classified by position: subsurface, surface, and aerial.

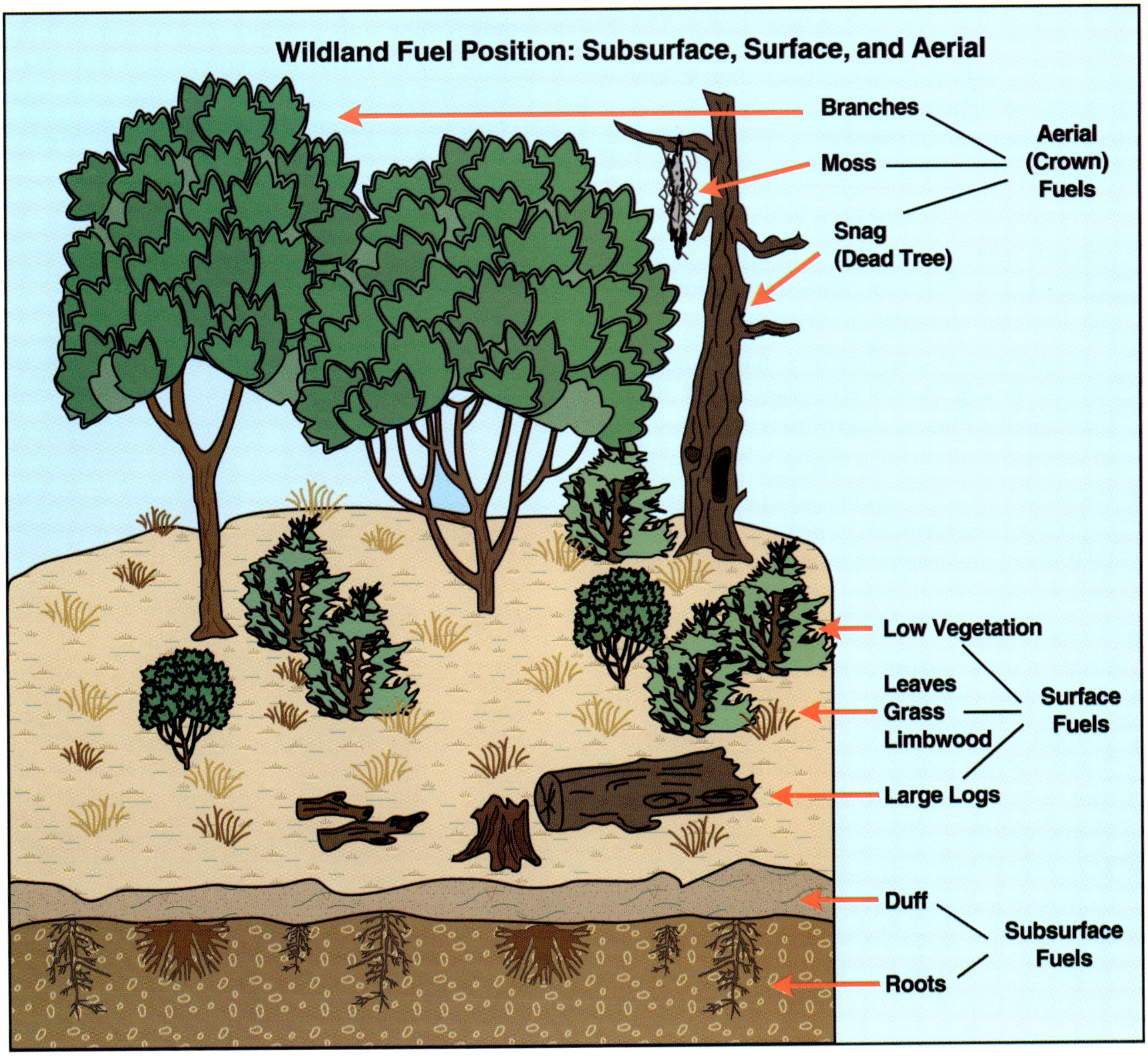

- ***Subsurface fuels*** — Roots, peat, duff, and other partially decomposed organic matter that lie under the ground's surface. These fuels do not burn rapidly but are difficult to extinguish completely.
- ***Surface fuels*** — Living surface vegetation including grass, brush, field crops, and other low vegetation; nonliving surface vegetation (small twigs, leaves, and needles); downed limbs, logging slash, and small trees. Small twigs, needles, or grass ignite easily and burn rapidly.
- ***Aerial (crown) fuels*** — Suspended and upright fuels physically separated from the ground surface to the extent that air can circulate freely around them, causing them to burn rapidly.

Fuel types vary among the regions of North America. Fire and emergency services organizations should be aware of the types within their service response areas and develop strategies and tactics accordingly. Regardless of the type of fuel, wildland fire behavior depends on certain fuel characteristics, including the following:

- ***Size*** — Fuel sizes are described as light, medium, or heavy. Small or light fuels burn faster.
- ***Compactness*** — Tightly packed fuel particles spread fire readily from piece to piece but generally burn slower than aerial types.
- ***Continuity*** — When fuels are spaced close together, the fire spreads faster because of radiant heat transfer. Fires in patchy fuels (those growing in clumps) may spread irregularly or not at all.
- ***Load (Volume)*** — The amount of fuel present in a given area determines a fire's intensity (rate of heat energy released) and the amount of water needed to perform extinguishment.
- ***Temperature*** — The temperature of the fuel is a result of the exposure to the rays of the sun. Fuels that are in direct sun can reach a temperature as high as 150°F (66°C). Higher temperatures make the fuel more likely to dehydrate and ignite by bringing them closer to their ignition temperatures.
- ***Moisture*** — The moisture content of fuels changes constantly in response to the amount of moisture that is available in the environment. Wind, temperature, and moisture in the atmosphere as well as in the soil can add or remove moisture from fuels.
- ***Position*** — How a fuel is located in relation to the ground differentiates the types of wildland fuels. A fuel's position (subsurface, surface, and aerial) greatly affect wildland fire behavior and how the fuel burns.

Weather

All aspects of the weather (state of the atmosphere over the earth's surface) have some effect upon the behavior of a wildland fire. Weather is the most changeable of the factors that affect wildland fire behavior. The effects of seasonal and daily weather cycles are considered along with individual weather elements. Some of these weather elements are as follows:

- ***Wind*** — Air in motion; it fans flames into greater intensity and supplies fresh air that speeds combustion. Rate and direction of fire spread are mostly functions of wind speed and direction. Medium- and large-sized fires may create their own winds, which may result in firestorms that can move above ground level **(Figure 5.22)**.

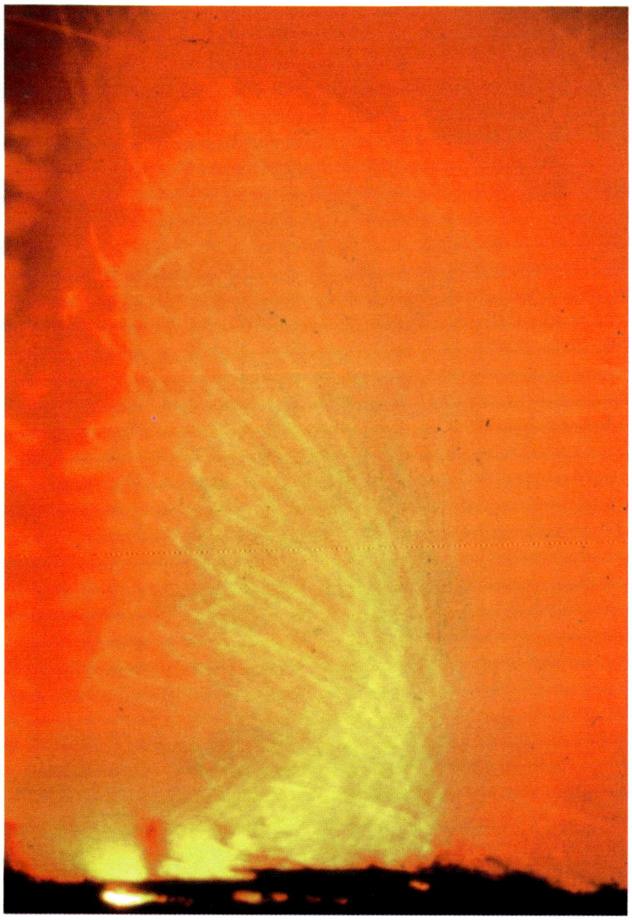

Figure 5.22 Wind adds fresh air to wildfires, increasing the speed of combustion and sometimes creating firestorms above the ground. *Courtesy of the National Interagency Fire Center.*

- ***Atmospheric temperature*** — Measure of the warmth or coldness of the air; has an effect on wind and is closely related to relative humidity (see next bullet); primarily affects the fuels as a result of long-term drying.
- ***Relative humidity*** — Percentage of moisture in air relative to the amount it can hold at a given temperature and atmospheric pressure. The greatest impact is on dead fuels that no longer have any moisture content of their own. However, relative humidity also impacts the rate of fire spread in live fuels. The lower the relative humidity, the drier the fuel, and the faster fire can spread.
- ***Precipitation*** — All forms of water particles that fall from the atmosphere; largely determines the moisture content of live fuels. Although dead fuels may dry quickly, large dead fuels will retain this moisture longer and burn slower.

Topography

Topography is the physical configuration of the land or terrain and has a decided effect upon wildland fire behavior (**Figures 5.23 a–d**). For example, the steepness of a slope affects both the rate and direction of a fire's spread. Because the fire dehydrates and preheats fuels on an upward slope, fires usually move faster uphill than downhill. Topographical features that influence wildland fire behavior include the following:

- ***Slope*** — Natural or artificial topographic incline; can range from slight to steep depending on the angle of elevation from the horizontal; the steeper the slope, the faster a fire moves.
- ***Aspect*** — Compass direction a slope faces. Full southern exposures (north of the equator) receive more of the sun's direct rays and therefore receive more heat. Wildland fires typically burn faster on southern exposures.

Figure 5.23a Topography example: lightly wooded grasslands.

Figure 5.23b Topography example: pasture land.

Figure 5.23c Topography example: rugged forests.

Figure 5.23d Topography example: thick tropical undergrowth typical of Hawaii. *Courtesy of Hawaii Department of Forestry and Wildlife.*

- ***Local terrain features*** — Terrain differences that directly affect air movement; may alter air flow and cause turbulence or eddies, resulting in erratic fire behavior. Canyons particularly can create turbulent updrafts that cause a chimney effect that increase wind velocity. Fires can then spread at extremely fast rates and become very dangerous. Some of these terrain features are as follows:
  — *V*-shaped drainages or gullies (chutes)
  — Depressions between two adjacent hilltops (saddles)
  — Obstructions such as ridges, trees, and even large rock outcroppings
  — Deep, narrow canyons

Wildland Fire Composition

The parts of a wildland fire are named according to their unique characteristics and positions in relation to the most active portion of the fire (**Figure 5.24**). Differences in fire behavior at various locations of a wildland fire can often be easily observed. The differences in fire behavior are consistently related to the primary direction of fire spread or most active portion of the fire. The most common names for the parts of a wildland fire are as follows:

- ***Origin*** — Area where the fire starts or originates and from which it spreads. The origin should be protected for further investigation of fire cause whenever possible.
- ***Head*** — Part with the greatest rate of fire spread. Wind and slope affect the direction of spread. Therefore, the head is either on the edge of the fire opposite the direction from which the wind is blowing or it is toward the upper part of a slope. The head of a fire may burn intensely and move with alarming speed. Controlling the head and preventing the formation of new heads are vital to suppressing a wildland fire.
- ***Fingers*** — Typically long, narrow strips of fire that extend from the fire's main body. They form when the fire burns in an area that has both light fuels and patches of heavy fuels. Light fuels burn faster than heavy fuels, which cause the "fingers" to form. If fingers are not controlled, they may form new heads.

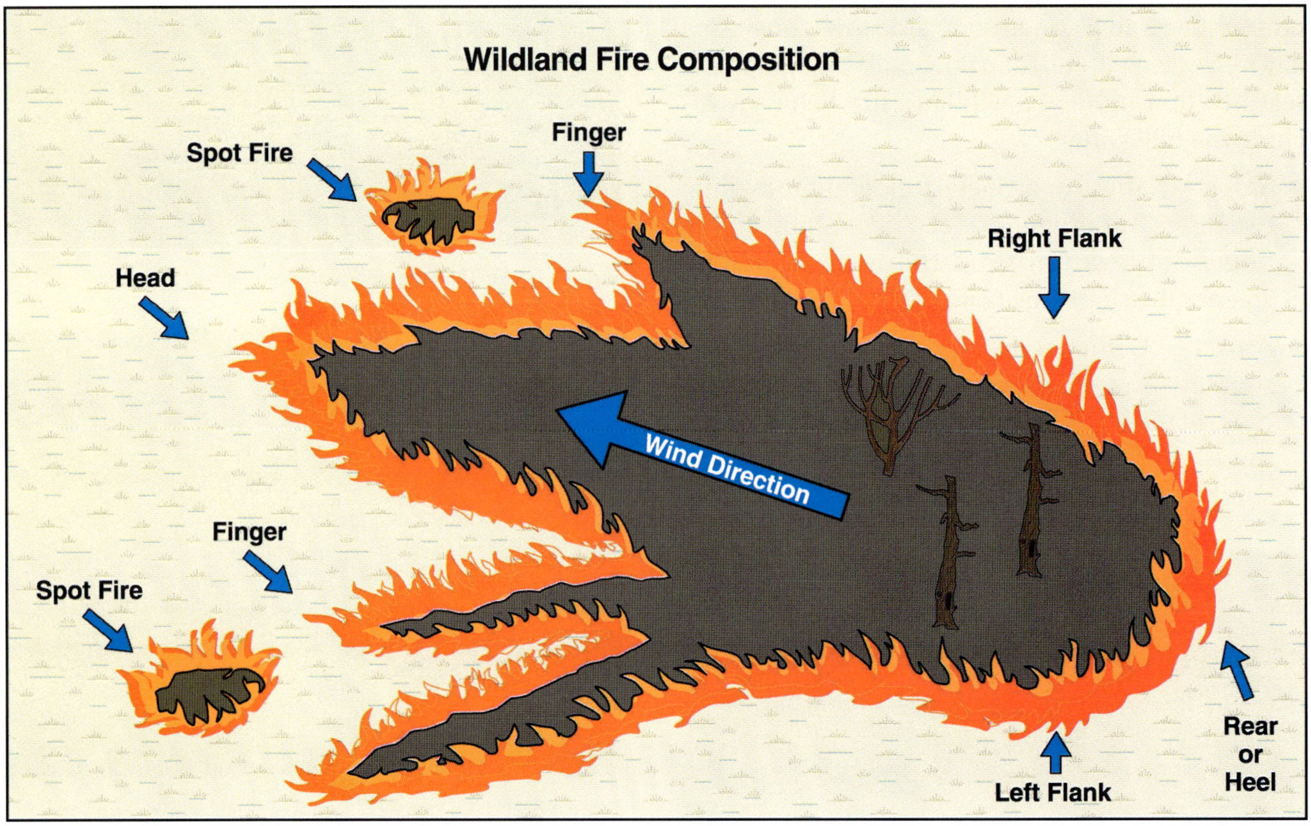

Figure 5.24 Wildland fire composition: the major parts of a wildland fire.

- ***Perimeter*** — Outer boundary or the distance around the outside edge of a burning or burned area (also called *fire edge*). Clearly, a fire's perimeter grows constantly until it is controlled and extinguished.
- ***Heel*** — Part opposite the head (also called *rear*). The heel usually burns with low intensity, has a low rate of spread, and is generally easier to control than the head. In most cases, the heel burns either downhill or against the wind.
- ***Flanks*** — Sides of a wildland fire roughly parallel to the main direction of fire spread. They separate the head from the heel and are identified as either left or right (looking from the heel of the fire facing the head). A shift in wind direction may change a flank into a head; therefore, flanks must be controlled as soon as possible.
- ***Islands*** — Unburned areas inside the fire perimeter. They are potential areas for more fire and must be patrolled and checked frequently. If these islands are close to a control line (barrier and treated fire edge), they may flare up later and throw embers across it.
- ***Spot fire*** — New fire caused by flying sparks or embers that land beyond the main fire and ignite fuel. Spot fires must be extinguished quickly to avert the possibility of new heads forming so that personnel and equipment are not trapped between two fires. Spot fires indicate that conditions are favorable for continued growth of the wildland fire.
- ***Slopover*** — Fire crossing a control line or natural barrier (also called *breakover*). A slopover differs from a spot fire in its location: Slopover occurs immediately across a control line; a spot fire occurs some distance from a control line.
- ***Green*** — Area of unburned fuels outside of but adjacent to the fire area (but not necessarily green in color); fuels that are not yet on fire but may be in the path of a fire.
- ***Black*** — Area in which the fire has consumed or "blackened" the fuels (opposite of green); can be a very hot and smoky environment.

Fire-Control Strategies

A critical initial decision when planning a wildland fire attack is whether the overall strategy is to be offensive (sufficient resources are available at the scene to control the fire), defensive (resources at scene are insufficient to extinguish a fire so actions are limited), or a combination. Defensive and offensive strategies are often combined when fighting fire in the wildland/urban interface when some structures are burning while others are merely threatened. Unlike structure fire-fighting protocols where the two strategic modes (offensive and defensive) are rarely combined, they are often combined when fighting wildland fires. In the offensive mode, the following two basic fire-control attack strategies are used to control wildland fires:

- ***Direct attack*** — Fighting the fire at its edge or closely parallel to it
- ***Indirect attack*** — Constructing a control line (using foam, chemical retardants, or brush-clearing techniques) some distance from the fire's edge and intentionally burning the uninvolved intervening fuels (know as *burning out*)

The principal difference between these two attack strategies involves the distance at which the control line is established from the fire's edge. Each has its own advantages as well as disadvantages, requiring careful consideration during application. While direct and indirect attacks are presented as two separate techniques, most wildland fire control activities use a combination of the two to take advantage of varying fuels and topography. Both strategies may involve the use of water or Class A foam concentrates in conjunction with other techniques, and both may be applied by ground forces using handlines or pump-can extinguishers or aerial forces using fixed- or rotary-wing aircraft.

Direct Attack

Direct attack is an operation where suppression action is taken directly on burning fuels. This attack method is accomplished by applying extinguishing agent (water, finished foam, or other extinguishing agents) directly onto a fire. Both water and finished foam smother the flames, cool the burning material, and penetrate deep into the fuel. The advantages of Class A foam concentrates that were mentioned earlier improve these extinguishing techniques and reduce the quantity of water required for the operation. A direct attack may be suitable in the following situations:

- Small fires or low-intensity sections of large fires **(Figure 5.25)**
- Hot spots that remain after a large fire has been extinguished
- Running (rapidly spreading) fires in light fuels (such as grass, leaves, small brush, duff, and field crops)
- Fires where the amount of area lost to the fire must be kept to a minimum
- Fires on high-value property such as commercial forests

The ability of finished foam to continue wetting and cooling fuels after fire and emergency responders have moved to other areas is one of the keys to its effectiveness. Wet finished foam is used for direct attacks because these types allow for maximum extinguishing, penetrating, and wetting of the fuel load. Applying the finished foam at the base of the fire front minimizes foam loss due to heat and thermal updraft. While attacking the edge of the fire itself, frequently moving the foam stream over yet unburned fuels assures that the fire's progress is halted. Apply finished foam only long enough to ensure that the fire is extinguished. As soon as steam is visible, move the foam stream to another area. Visible steam is an indication that extinguishment, cooling, and wetting have occurred. Finished foam that remains on the fuel after fire and emergency responders move to another area continues to cool and wet the fuel, thus making the job of mop-up considerably easier.

These increased abilities to extinguish fire more quickly do not preclude the need for fire and emergency responders to monitor areas that have already been extinguished to make sure finished foam is still visible and that no hot spots were missed. Another advantage is that finished foam is visible after extinguishment for a period of time — time that depends on the temperature, wind, fuel density, and

Figure 5.25 A direct attack with Class A foam is useful for small or low-intensity wildland fires. *Courtesy of Dan Thorpe.*

the degree to which the foam solution was expanded. If deeper penetration of the burning material becomes necessary, either increase the flow rate of the agent or discharge the agent closer to the fuel. Wet, fluid, and CAFS finished foams easily soak into most fuels. A CAFS is ideal for these situations because of the strong force of its fire stream **(Figure 5.26)**.

Once containment is reached, the foam stream is directed toward any remaining hot spots or other areas that are burning until the entire fire is totally extinguished. The mechanism for delivering the foam solution is capable of providing emergency personnel with a continuous foam stream during movement along the fire's edge. The hose length is kept as short as possible so that quick movement or escape can be made as well.

Some particular types of wildland fires are considered difficult to extinguish with plain water. Sample fires and suggestions for successful use of finished foam to extinguish them include the following:

- ***Fires within or at the tops of snags (standing dead trees)*** — It may be desirable to switch to stiff or dry finished foam so that the agent stays in place on the fuel.
- ***Log deck fires*** — These fuels require thinner finished foam that easily soaks into the fuel.
- ***Aerial (crown) fires*** — It may be desirable to switch to stiff or dry finished foam so that the agent stays in place on the fuel.
- ***Fires in heavy duff or dense fuel beds*** — These fuels require thinner finished foam that easily soaks into the fuel.

The safety of emergency personnel in a direct attack deserves careful evaluation. Working on the edge of a fire, as required in a direct attack, can expose personnel to intense heat and smoke. Thus it is important to be aware of the following advantages and disadvantages of a direct attack:

Advantages

- Emergency personnel are close to or working in the burned-out interior, which may be used as a safety zone during a direct attack. Escape routes must be maintained at all times in case winds or conditions change.

Figure 5.26 Class A foam soaks easily into most natural fuels when applied with a compressed-air foam system (CAFS). *Courtesy of National Interagency Fire Center.*

- Full advantage may be taken of dead, burned-out areas in constructing a control line.
- Minimal burning out is required to eliminate unburned fuels.
- The need for mop-up and standby is reduced because little unburned fuel remains inside the control line.

Disadvantages

- Emergency personnel are exposed to heat, smoke, and flame because they are working in the direct proximity of the fire.
- The length of the control line is usually extended because fires usually burn with irregular edges and fingers.
- Embers from hot spots may cause great risk of spot fires outside the control line.

Indirect Attack

In an indirect attack, a control line is constructed or located some distance from the edge of the fire and fuel between the two points is burned out. The distance from the control line to the fire's edge depends on the type and volume of fuel, the intensity of the fire and its rate of spread, and the topography. This attack may be useful if the fire is burning too intensely or spreading too rapidly to work safely and effectively on the fire's edge.

The control line of an indirect attack serves as a safety strip. It prevents further spread of fire once it has been brought under control. The success of an indirect attack depends, in part, upon the skillful choice of the location and the construction of the control line **(Figure 5.27)**. The line must be wide enough to prevent oncoming flame from jumping over to the fuel on the other side (generally two and a half times wider than the height of the flames). Accordingly, the location selected should be far enough ahead of the fire so that the line can be completed and burned out before the fire reaches it. The control line for an indirect attack must be adjacent to a safe, secure area (anchor point) where the attack begins. The anchor point (usually a natural or manmade barrier such as a road, body of water, or previously burned area) prevents the fire from outflanking crew members and dangerously surrounding them. Consider the following points in locating a control line:

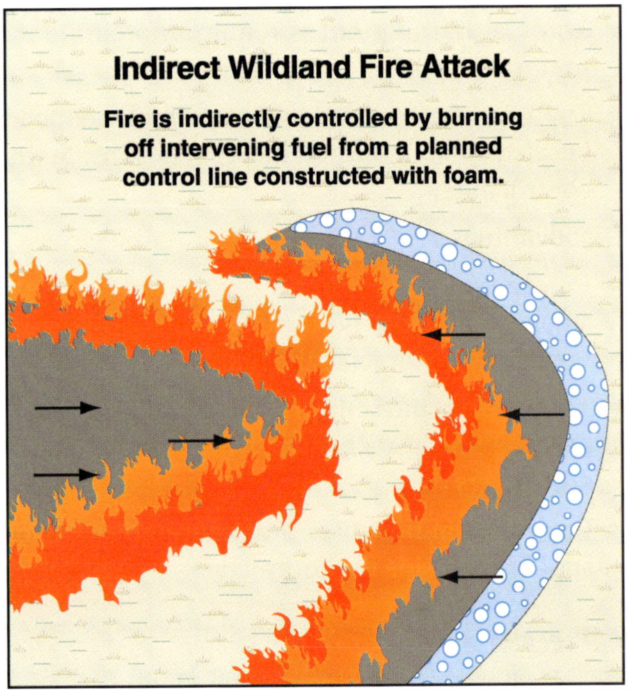

Figure 5.27 An indirect wildland fire attack using foam as a barrier.

- Anchor the line to a secure fire barrier.
- Take advantage of natural barriers.
- Avoid locating the line in dense fuels.
- Do not construct more line than crews can supervise.

Two main uses for Class A foam when conducting an indirect attack are to (1) construct a "wet line" barrier to an advancing wildland fire and (2) establish a line from which to start a backfire (intentional fire set to eliminate large areas of unburned fuel in the path of a wildland fire). When a foam line is used to stop an advancing wildland fire, the wetting action must be started far enough from the fire to allow fire and emergency responders time to construct a complete line before the fire reaches it **(Figure 5.28)**. If at all possible, the foam line is placed downslope from the advancing fire, and its width is two to three times the height of the flames expected to contact the line. All vegetation must be covered from its base to the top in order for the foam line to be effective.

Figure 5.28 Constructing a fire barrier with Class A foam. *Courtesy of National Interagency Fire Center.*

Depending on factors such as ambient temperature, wind, humidity, and fuel height, foam lines vary in effectiveness. However, they are expected to hold from three to five times longer than plain water lines. As the fire approaches the foam line, steam is generated. Emergency personnel should be prepared to apply additional finished foam to the line or make a direct attack in any areas where the fire appears to be overrunning the line. Off-road water supply apparatus are available that can lay a finished foam blanket for creating control lines.

When using finished foam to establish a barrier for backfire operations, apply it immediately ahead of the emergency personnel who are setting the backfire. Allow the foam blanket ample time to soak the fuel before lighting the backfire. The U.S. Bureau of Land Management reports success using Class A foam solutions through medium- and high-expansion equipment to establish a line from which to backfire.

The type of finished foam used during an indirect attack varies, depending on the fuel and weather conditions. Watery finished foams (wet and fluid types) are necessary when dealing with dense fuels (heavy duff and thick ground cover) and/or windy conditions. Thick finished foams (stiff or dry types) are used when fuels are either loose or elevated. The thick finished foam clings to elevated parts of trees and bushes. Regardless of what type of finished foam is used, continue to monitor foam lines and reapply the agent as necessary.

Continued size-up and modifications to the attack plan become extremely important in the successful application of an indirect attack. The advantages and disadvantages of an indirect attack are as follows:

Advantages

- A control line is usually shorter and easier to protect because it is not necessary to follow an uneven fire edge but can instead go directly from point to point.
- An indirect attack is generally less physically demanding on emergency personnel because they are building a control line away from the heat and smoky conditions found on the fire edge.
- The location of a control line is chosen so that it is easier to protect by picking sites where the fire is more easily controlled. Natural firebreaks such as rivers, streams, lakes, cliffs, or roads are used as a part of the control line.

Disadvantages

- Until fuel is burned out between the fire and the control line, a fire's intensity and rate of spread could increase drastically so that the control line cannot contain the fire, placing emergency responders in danger.
- A large area is sacrificed to the fire.
- Emergency personnel safety is generally decreased because they are not close to the burned-over interior of the fire, which may be used as a safety zone. Instead personnel are in the direct path of the fire rather than on the burned side.
- Unburned fuel between the fire's edge and the control line can allow a fire to develop erratic behavior, increasing the risk of spot fires. Removal of fuel between the control line and the fire's edge presents added risks, especially as the main fire approaches the burned-out area.
- A fire can change directions because of changes in wind conditions, rendering much of the previously completed work useless.

Fire-Control Tactics

Various tactical operations (including those used in direct or indirect attacks) are used in wildland fire operations. Depending on the size and speed of the fire, aerial attacks with extinguishing agents may be required. In addition, every fire requires mop-up to ensure that extinguishment is complete. Wildland/urban interface fires pose the extreme challenge of protecting homes and other structures that are in the potential paths of fire. Exposure protection can place fire and emergency responders in very dangerous positions. The use of Class A foam and CAFS to protect exposures reduces the chances of structures falling victim to fire.

Initial Attacks

Several initial fire-control tactics can be chosen. They may be used singly or in combination. The decision regarding which tactic to use takes into account the actual or potential life hazards, exposures, fuel burning, current and expected weather, and topography. Some of the more common tactics are as follows:

- ***Frontal attack*** — Attacks the head of a fire. The attack begins at or near the head of the fire and then proceeds to the flanks, which effectively limits the spread of the fire. However, either the intense burning of the head or other conditions such as fuel type, wind, or topography may make the frontal attack unsafe. This is especially true when fighting the fire from the unburned side. ***Frontal attacks can be very dangerous.*** Under high-intensity fire conditions, a flank attack may be a good alternative **(Figure 5.29).**
- ***Flank attack*** — Attacking the flanks simultaneously or successively. The attack begins at a secure anchor point on one or both flanks of the fire and works toward the head. Emergency personnel must keep watch on the rear to control or prevent fire from crossing the control line. Always take advantage of natural barriers such as rivers, streams, or roadways for flank control **(Figure 5.30).**
- ***Pincer attack*** — Simultaneous attack on two sides; similar to a flank attack. At times, a low-intensity fire is accessible from all sides. If sufficient personnel are available, an attack may be made simultaneously on the flanks, the fingers, and the head of the fire (sometimes referred to as *envelopment*). A similar method called *point and cut off* occurs when personnel attack several heads or fingers of the fire at the same time and then connect the short control-line segments. All of these simultaneous attacks require the use of several closely supervised groups of emergency personnel working in highly coordinated efforts **(Figure 5.31).**
- ***Mobile attack*** — Employing fire apparatus in a fast-moving direct attack (also known as *pump and roll*). Mobile attack is primarily used on grass fires and requires terrain that the apparatus can safely negotiate **(Figure 5.32).**
- ***Tandem Attack*** — Attack by engines, mechanized equipment, hand crews, or aircraft working together in a coordinated effort. Any combination of two resources can be used. The first units do a quick knockdown of the fire and the other units follow and do more thorough extinguishment and mop-up **(Figure 5.33).**

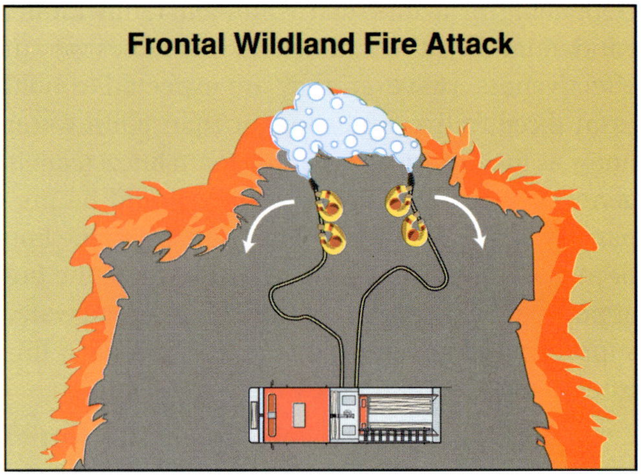

Figure 5.29 The frontal wildland fire attack starts at the furthest advance of the fire (the head) and then moves down the fireline in opposite directions.

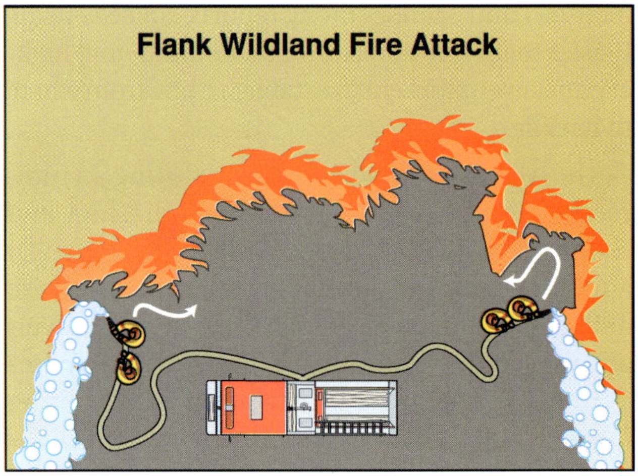

Figure 5.30 In a flank wildland fire attack, foam is applied to both ends of the firelines as crews work towards one another.

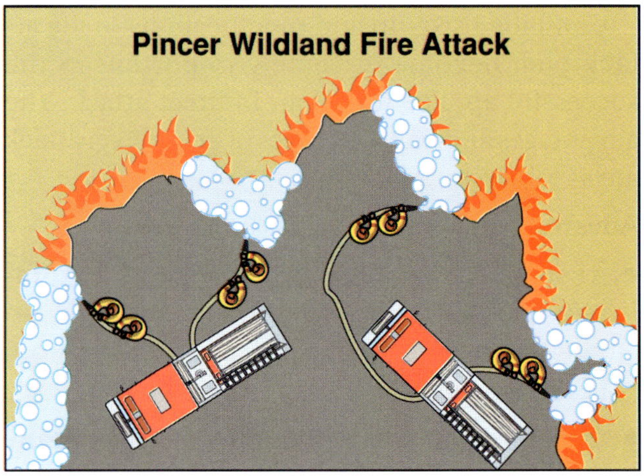

Figure 5.31 The pincer wildland fire attack and the point and cut-off attack methods are similar and involve attacking two sides of the fire at the same time.

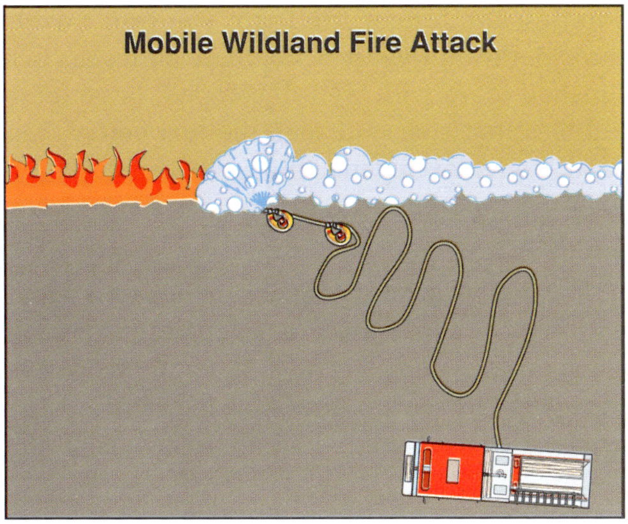

Figure 5.32 In a mobile wildland fire attack, firefighters operate from the burned side and move the foam stream between the fire and the unburned fuel.

Figure 5.33 A tandem wildland fire attack involves the use of multiple types of fire-fighting assets, in this case two engines and an aerial tanker.

Aerial Attacks

The use of aircraft as part of a coordinated wildland fire attack must be safe for the aircraft, air crews, and emergency crews on the ground. Favorable flying conditions and coordination with ground personnel are crucial elements. Various long-term agents (chemical suppressants and retardants), short-term agents (water and foam), or water/chemical solution agents (either long- or short-term) are used in aerial attacks. Class A foams are suitable for use in aerial attack (airdrop) applications on wildland fires from both helicopters and fixed-wing aircraft. Foam dropped from aircraft is used for both fire attack and mop-up operations **(Figure 5.34)**.

Figure 5.34 Both fixed-wing and rotary-wing aircraft can be very effective in the application of foam on wildland fires. *Courtesy of National Interagency Fire Center.*

Tests and actual experiences show that aerial foam attacks are effective on initial direct attacks on small or low-intensity fires. These fires involve ground cover fuels and have flame lengths of 8 feet (2.5 m) or less. In these situations, the finished foam is dropped on the leading edge of the fire. This attack knocks down the flame front and coats the immediately adjacent fuel, which aids in halting the advance of the fire. Aerial attacks must be immediately followed by a ground forces attack to complete the extinguishment and mop-up process. Airdropped foam is a short-term agent because it only holds the situation in check for a maximum of 5 to 15 minutes. If ground forces are not available, then repeated application of foam solutions might be necessary in addition to long-term retardant applications. On sloped terrain, the finished foam is dropped slightly uphill of the fire front and allowed to run down to the fire. In areas where there is heavy tree cover, tests indicate that as much as 40 percent of the foam hangs in the canopy of the trees.

The remainder strikes the ground fire. As the water drains from the foam blanket in the canopy, it cascades to the ground, providing additional extinguishing effects.

Airdropped foam is generally not very effective as a long-term fire retardant. In long-term situations, conventional fire retardants are more desirable. As well, airdropped foam is generally not effective in reapplication situations where the finished foam from the initial drop has been "burned through" (dehydrated). Emergency personnel on the ground are needed to extinguish those fires. In addition, wind direction and velocity can scatter the finished foam and make it ineffective. Plain water dropped on areas where finished foam has previously been dropped damages the foam blanket already in place.

Foam solution for airdrops is prepared using two proportioning methods: *direct injection* or *batch mixing.* Both fixed-wing and rotary-wing aircraft may be equipped with a direct injection proportioner that is operated by the pilot while in flight. The foam concentrate is injected into the tank or bucket and mixed with water before the drop. However, the simplest way to make foam solution is the batch-mixing method. After an aircraft or helicopter is loaded with water, foam concentrate is added to the tank or bucket. The amount of concentrate added depends on the size of the tank/bucket and the concentration desired. The desired concentration varies depending on the type of finished foam desired, the type of aircraft used, the height from which the foam solution will be dropped, and the flying speed of the aircraft. Foam concentrate must not be added to the tank before it is filled with water, otherwise excessive foaming can occur when water is added after the concentrate. The manufacturer of a particular foam concentrate may have specific concentrations that apply. General guidelines for aerial foam concentrations are as follows:

- ***Helicopters***
  - — Buckets: 0.1 to 1 percent
  - — Belly tanks: 0.2 to 1 percent
- ***Fixed-wing aircraft*** — 0.25 to 1 percent

The amount of aeration that takes place also depends on the altitude (height) of the drop and speed of the aircraft. More air is introduced into the foam solution the longer the foam solution travels through the air (**Figure 5.35**). More air results in a finished foam with well-developed, large foam bubbles that coat and soak the fuel better than small, underdeveloped bubbles that generate from a short aeration time (low altitude or slow velocity drops).

To gain the greatest effect from the aerial application of Class A foam concentrates, several elements are considered in addition to the cost of the operation. Each of the following elements has an effect on the pattern or footprint created by the airdropped foam:

- ***Aircraft velocity*** — Increased aircraft velocity produces drier finished foam, lengthens the placement pattern or footprint of the finished foam, and decreases the coverage level (foam blanket thickness).
- ***Drop height*** — The average altitude for airdrops is between 75 and 165 feet (23 m to 50 m) above the canopy. Drops lower than 75 feet (23 m) are considered unsafe and ineffective.
- ***Aircraft attitude*** — Attitude is the angle of the aircraft, front to rear, while in flight (**Figure 5.36, p.166**).
  - — *Level flight:* Produces a uniform, elliptical pattern
  - — *Diving:* Shortens the pattern and reduces the effective length; usually performed with rotary-wing aircraft
  - — *Lofting:* Elongates the pattern, reducing the concentration at the end of the run; usually performed with rotary-wing aircraft
  - — *Banking:* Throws the load into a part of the fire that the aircraft cannot fly over
  - — *Downhill run:* Tends to elongate the pattern
- ***Wind conditions*** — Wind direction can be used to advantage in distributing and spreading the finished foam. It can also be a hindrance and must be considered when planning the aerial attack (**Figure 5.37, p.166**).
  - — *Headwind:* Produces similar results to those found in diving (shortens the foam pattern)
  - — *Downwind:* Produces similar results to those gained through lofting (elongates the foam pattern)

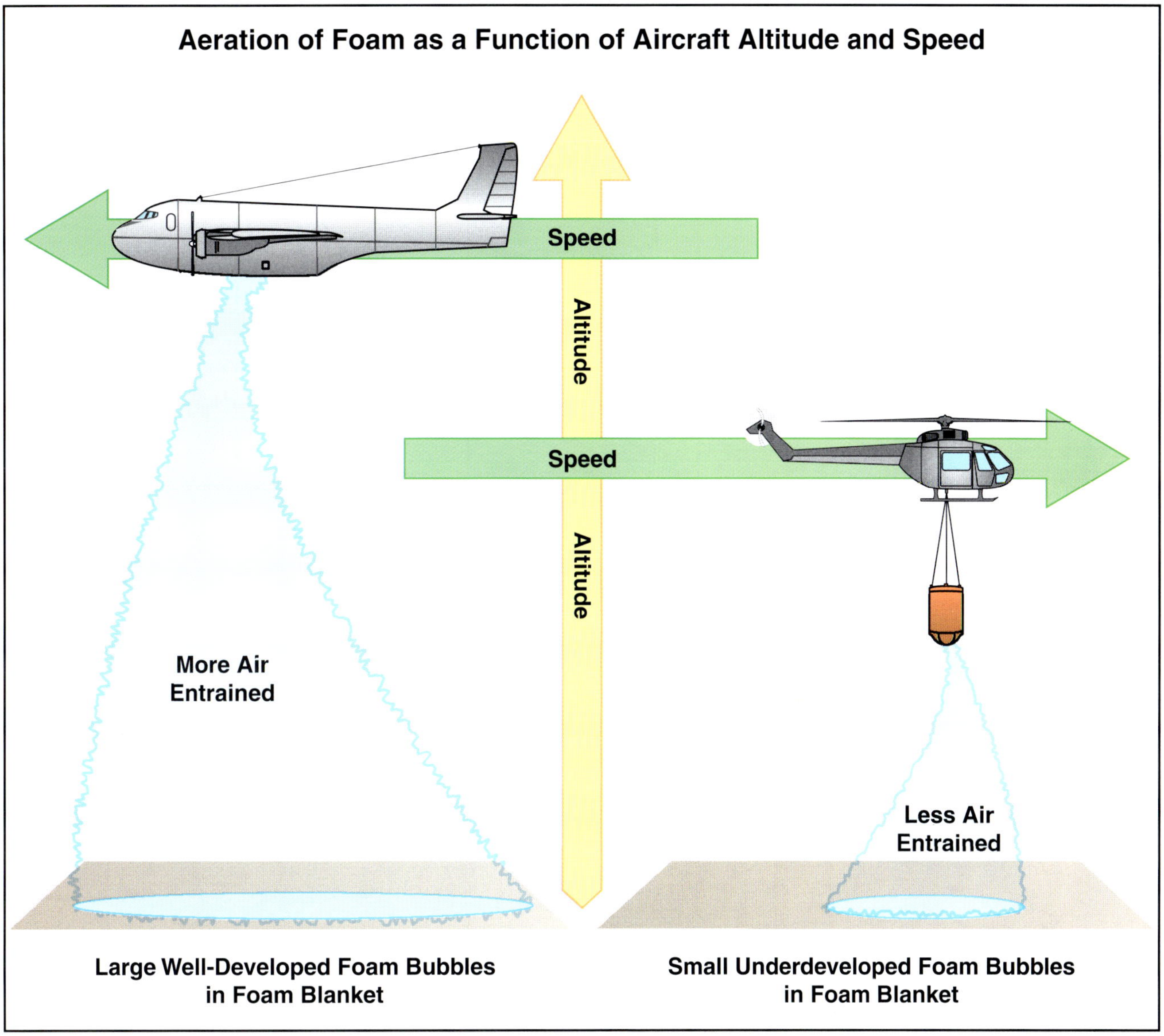

Figure 5.35 Aircraft altitude and speed have a direct affect on the development of a good foam blanket.

— *Crosswind:* Can have a significant effect on pattern placement; causes the foam solution to drift and disperse; can be used to spread finished foam in a line parallel to the flight path and to the side of the aircraft rather than directly beneath it

- ***Thermal updrafts*** — A thermal updraft (convection column) is created by high-intensity fires and can create unsafe operating conditions and reduce the effectiveness of the finished foam. To compensate for this condition, use a combination of the following: Use wetter foam concentrates (heat dries the solution during the drop), lower the drop height (lessens the time the foam solution is in the thermal/convection column), or use headwind drops (increases the core or quantity of the foam solution while it is in the thermal/convection column).

- ***Fuel arrangement and position*** — In forested areas, the tree canopy can prevent the foam solution from reaching the fuels on the forest floor. Wetter foam solution mixtures should be used to penetrate the canopy.

- ***System flow rate and volume*** — Flow rate is determined by the proportioning system and the foam concentrate that is in use. Rates will range from 0.1 percent to 0.9 percent, depending on the operational requirements. The volume of foam solution depends on the flow rate, tank or bucket capacity, and aircraft lift capacity.

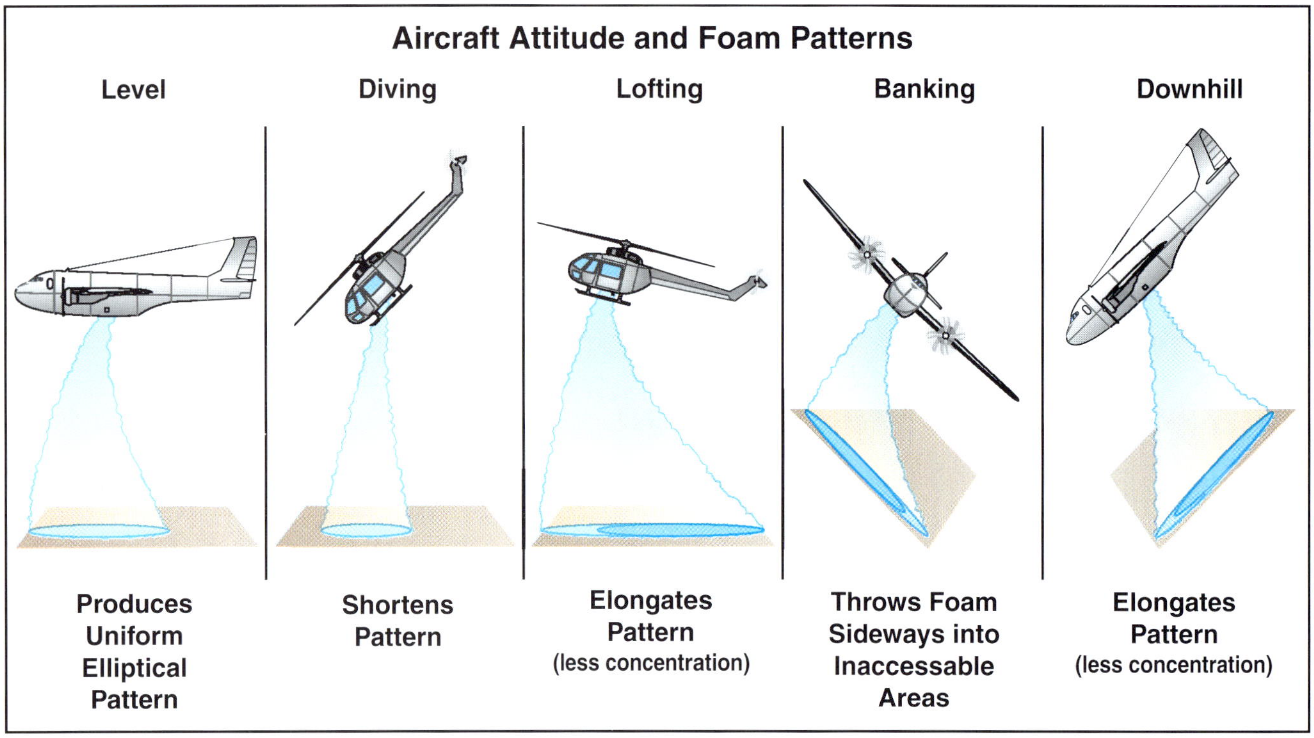

Figure 5.36 The position or attitude of the aircraft at the time foam is released influences the pattern of the foam blanket.

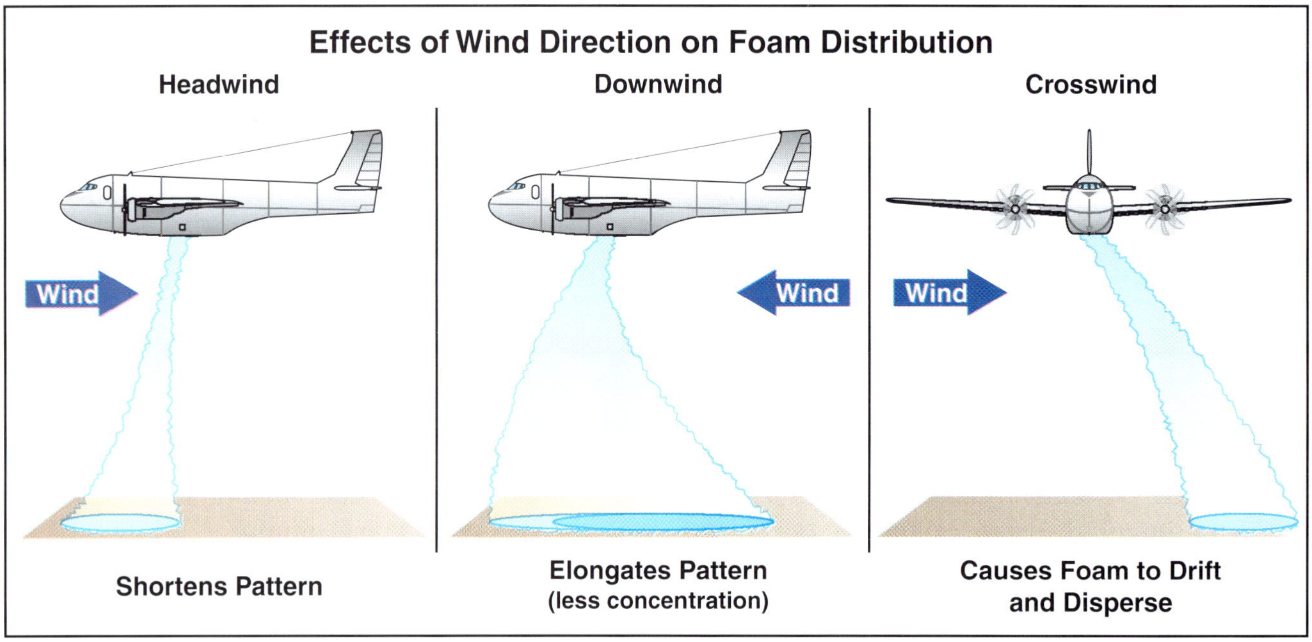

Figure 5.37 Wind direction at the time of a foam drop also influences the pattern of the foam blanket.

- *Operational cost* — The cost effectiveness of aerial operations is the cost of the foam concentrate, aircraft operational costs, aircrew costs, and fuel costs versus the potential loss due to the fire. It may be more cost effective to use the aerial attack for a rapid confinement and extinguishment than to use ground forces only and endure a long and hazardous operation.

Mop-Up Operations

Wildland mop-up operations involve completing extinguishment (extinguishing hot spots and hidden fires) after the main body of fire has been controlled. This step is important because deep-seated combustion is often missed when the fire is initially stopped. In time, these hot spots may develop into open burning and cause a rekindle of the original fire. Mop-up also involves making sure that the control line is scraped down to mineral soil (soil containing little or no combustible material) and is wide enough to keep any burning materials from blowing across it into unburned fuels.

Using finished foam during mop-up has several advantages — the primary one is increased emergency responder safety. Finished foam's ability to quickly extinguish remaining hot spots and blanket the area reduces the amount of smoke and vapors to which emergency responders are ordinarily exposed (**Figure 5.38**). Foam use reduces the amount of manual mop-up required, which saves wear and tear on equipment, reduces stress on personnel, and saves time. It also reduces the overall costs associated with the operation.

Figure 5.38 Additional foam may have to be added during mop-up operations to ensure that the fuel is completely saturated. Axes may be required to expose smoldering material to the foam. *Courtesy of National Interagency Fire Center.*

The standard method for mop-up involves a combination of finished foam application and manual mop-up procedures. In this operation, wet finished foam is used for maximum penetration capabilities. Finished foam application begins at the edge of the burned area and works inward until the entire area is covered. Spots where intense steaming is seen indicate areas that need to be manually overhauled by emergency responders using hand tools.

A second technique used for mop-up is to apply finished foam to the burned area with a medium-expansion foam nozzle. Allow the finished foam to remain in place, undisturbed, for about 30 minutes. Then check the area for spots where the foam blanket has dissipated or steam continues to rise. Either of these situations indicates areas where more finished foam may need to be applied or manual mop-up tactics must be employed.

Figure 5.39 Dry foam can be used to protect endangered structures in the wildland urban interface. *Courtesy of National Interagency Fire Center.*

Urban Interface Structure Protection

Class A finished foam is an excellent agent for protecting exposures. The principles concerned with protecting urban interface structure exposures in wildland fire situations are the same as those covered in the Structure Fires section of this chapter. Because exposure protection in wildland/urban interface fires may involve multiple structures, apparatus with CAFS-mounted equipment have an advantage. Due to the extended reach of CAFS streams, apparatus moving slowly down a street can successfully apply finished foam from structure to structure. This technique is suitable in situations where speed is of the essence. However, it is virtually impossible to coat all parts of a structure using this method. In addition, if a fire is advancing toward the side of a structure opposite the street, the foam agent cannot be adequately applied to the most severely exposed side (**Figure 5.39**). If time allows, complete the coating with handlines. As an added measure of protection, medium- or high-expansion foam nozzles can apply a barrier of finished foam around threatened structures that have already been coated with finished foam. This procedure will slow or halt the fire's progress as it approaches the structure.

Exterior Material-Concentration Fires

[NFPA 1001: 5.3.8; NFPA 1081: 6.2.3]

Fires in exterior concentrations of Class A materials can take many forms such as the following:

- Landfill refuse
- Used tires in storage
- Auto salvage yard materials **(Figure 5.40)**
- Junkyard materials
- Lumber in storage
- Hay bales in storage
- Items stored in open areas
- Construction site supplies **(Figure 5.41)**
- Vehicles
- Trash in Dumpsters® and other receptacles

The characteristics of Class A finished foam in soaking, cooling, and blanketing are advantages when extinguishing fires in these types of locations. The choice of finished foam type (wet, fluid, or dry) depends on the type of hazard, wind, humidity, and temperature, much like the hazards associated with wildland fires. In general, the procedure for fighting these types of fires is as follows:

- Select the type of foam concentrate depending on the type of hazard.
  - — *Wet foam for deep penetration:* Deep-seated fires in Dumpsters®, landfills, or tires
  - — *Wet/fluid foam for cooling and penetration:* Lumber, hay, open storage materials, and construction site supplies
  - — *Dry foam for complete coating and long-term protection:* Exposures
  - — *Wet/fluid foam for cooling and penetration and medium-expansion foam for filling voids:* Vehicle fires, auto salvage yard materials, or junkyard materials **(Figure 5.42)**

Figure 5.40 An auto salvage yard is an example of an exterior concentration of Class A materials.

Figure 5.41 New apartment complexes and other construction sites contain large quantities of raw lumber and other building materials.

Figure 5.42 Vehicle fires can be extinguished with Class A foam or CAFS-applied foam. *Courtesy of Tom Ruane.*

- Apply finished foam from the perimeter of the site working from the upwind side.
- Use standard overhaul methods for locating hidden or remaining fires and fill the voids with finished foam.
- Monitor the wind direction, velocity, and weather conditions that may reduce the effectiveness of the finished foam.

Finished foam can be applied with handlines or master stream appliances, depending on the size of the concentration of the involved materials. In the case of very large fires, master stream water attack may be necessary before the application of any finished foam or the use of a CAFS. Applying Class A finished foam or using a CAFS can then extinguish the deep-seated portion of the fire and blanket the smoldering materials. A master stream water attack may be necessary for the following reasons:

- The quantity of Class A foam concentrate is insufficient to handle the incident.
- The cost of the Class A foam concentrate needed would exceed the value of the burning materials.
- Water is just as effective and less costly as a CAFS in these types of incidents.

Some exterior materials concentrations such as those in auto salvage yards may contain limited quantities of flammable or combustible liquids. In small quantities, these liquids can be controlled with the use of Class A finished foam as it blankets the area. If larger quantities such as 55-gallon (208 L) drums are encountered, then Class B finished foam may be required to control the situation.

Summary

Although the use of Class A finished foam to extinguish wildland fires has been an accepted practice for many years, the advantages for using it on structure and exterior material-concentration fires has only been recognized recently. Thanks to the efforts of many fire departments, foam and nozzle manufacturers, and training organizations, the advantages have been quantified and are available to the fire and emergency services. The use of water as the primary extinguishing agent for Class A fires may never cease, but the use of Class A foam concentrates to improve and enhance the effectiveness of water will definitely increase.

Chapter 6

Class B Foam Use: Unignited Class B Liquid Spills

Job Performance Requirements

This chapter provides information that will assist the reader in meeting the following job performance requirements from NFPA 1021, *Standard for Fire Officer Professional Qualifications,* 1997 edition; NFPA 1081, *Standard for Industrial Fire Brigade Member Professional Qualifications,* 2001 edition; NFPA 472, *Standard for Professional Competence of Responders to Hazardous Materials Incidents,* 2002 edition; and NFPA 1051, *Standard for Wildland Fire Fighter Professional Qualifications,* 2002 edition. Colored portions of the standard are specifically addressed in this chapter.

NFPA 1021

2-6.2 Develop an initial action plan, given size-up information for an incident and assigned emergency response resources, so that resources are deployed to control the emergency.

(a) *Requisite Knowledge:* **Elements of a size-up, standard operating procedures for emergency operations,** and fire behavior.

(b) *Requisite Skills:* The ability to analyze emergency scene conditions, to allocate resources, and to communicate verbally.

NFPA 1081

6.3.4 Operating as a member of a team, extinguish an ignitable liquid fire, given an assignment, an attack line, personal protective equipment, a foam proportioning device, a nozzle, foam concentrates, and a water supply, so that the correct type of foam concentrate is selected for the given fuel and conditions, a correctly proportioned foam stream is applied to the surface of the fuel to create and maintain a foam blanket, fire is extinguished, re-ignition is prevented, and team protection is maintained.

(A) Requisite Knowledge. Methods by which foam prevents or controls a hazard; principles by which foam is generated; causes for poor foam generation and corrective measures; difference between hydrocarbon and polar solvent fuels and the concentrates that work on each; the characteristics, uses, and limitations of fire-fighting foams; the advantages and disadvantages of using fog nozzles versus foam nozzles for foam application; **foam stream application techniques;** hazards associated with foam usage; and methods to reduce or avoid hazards.

(B) Requisite Skills. The ability to prepare a foam concentrate supply for use, assemble foam stream components, master various foam application techniques, and approach and retreat from fires/spills as part of a coordinated team.

7.3.4 Operating as a member of a team, extinguish an ignitable liquid fire, given an assignment, an attack line, personal protective equipment, a foam proportioning device, a nozzle, foam concentrates, and a water supply, so that the correct type of foam concentrate is selected for the given fuel and conditions, a correctly proportioned foam stream is applied to the surface of the fuel to create and maintain a foam blanket, fire is extinguished, re-ignition is prevented, and team protection is maintained.

(A) Requisite Knowledge. Methods by which foam prevents or controls a hazard; principles by which foam is generated; causes for poor foam generation and corrective measures; difference between hydrocarbon and polar solvent fuels and the concentrates that work on each; the characteristics, uses, and limitations of fire-fighting foams; the advantages and disadvantages of using fog nozzles versus foam nozzles for foam application; **foam stream application techniques;** hazards associated with foam usage; and methods to reduce or avoid hazards.

(B) Requisite Skills. The ability to prepare a foam concentrate supply for use, assemble foam stream components, master various foam application techniques, and approach and retreat from fires/spills as part of a coordinated team.

NFPA 1051

9.5.5 **Hazardous Materials.** Analyze the potential involvement of various hazardous materials, given incident information and resources, so that hazardous conditions are identified and mitigated.

(A) Requisite Knowledge. A working knowledge of the types of hazardous materials that can be involved and the hazards they can pose to the public, fire-fighting personnel, and the environment; NFPA 472, *Standard for Professional Competence of Responders to Hazardous Materials Incidents,* First Responder level.

(B) Requisite Skills. None required.

NFPA 472

5.1.2.1 The first responder at the operational level shall be able to perform the following tasks:

(1) Analyze a hazardous materials incident to determine the magnitude of the problem in terms of outcomes by completing the following tasks:

(a) Survey the hazardous materials incident to identify the containers and materials involved, determine whether hazardous materials have been released, and evaluate the surrounding conditions.

(b) Collect hazard and response information from MSDS; CHEMTREC/CANUTEC/SETIQ; local, state, and federal authorities; and shipper/manufacturer contacts.

(c) Predict the likely behavior of a material as well as its container.

(d) Estimate the potential harm at a hazardous materials incident.

(2) Plan an initial response within the capabilities and competencies of available personnel, personal protective equipment, and control equipment by completing the following tasks:
   (a) **Describe the response objectives for hazardous materials incidents.**
   (b) **Describe the defensive options available for a given response objective.**
   (c) Determine whether the personal protective equipment provided is appropriate for implementing each defensive option.
   (d) Identify the emergency decontamination procedures.
(3) Implement the planned response to favorably change the outcomes consistent with the local emergency response plan and the organization's standard operating procedures by completing the following tasks:
   (a) Establish and enforce scene control procedures including control zones, emergency decontamination, and communications.
   (b) Initiate an incident management system (IMS) for hazardous materials incidents.
   (c) Don, work in, and doff personal protective equipment provided by the authority having jurisdiction.
   (d) Perform defensive control functions identified in the plan of action.
(4) Evaluate the progress of the actions taken to ensure that the response objectives are being met safely, effectively, and efficiently by completing the following tasks:
   (a) Evaluate the status of the defensive actions taken in accomplishing the response objectives.
   (b) Communicate the status of the planned response.

5.3.2 Identifying Defensive Options. Given simulated facility and transportation hazardous materials problems, the first responder at the operational level shall identify the defensive options for each response objective and shall meet the following requirements:

(1) **Identify the defensive options to accomplish a given response objective**.
(2) **Identify the purpose for, and the procedures, equipment, and safety precautions used with, each of the following control techniques:**
   (a) Absorption
   (b) Dike, dam, diversion, retention
   (c) Dilution
   (d) Remote valve shutoff
   (e) Vapor dispersion
   (f) **Vapor suppression**

5.4.4 Performing Defensive Control Actions. Given a plan of action for a hazardous materials incident within their capabilities, the first responder at the operational level shall demonstrate defensive control actions set out in the plan and shall meet the following related requirements:

(1) **Using the type of fire-fighting foam or vapor suppressing agent and foam equipment furnished by the authority having jurisdiction, demonstrate the effective application of the fire-fighting foam(s) or vapor suppressing agent(s) on a spill or fire involving hazardous materials.**
(2) **Identify the characteristics and applicability of the following foams:**
   (a) **Protein**
   (b) **Fluoroprotein**
   (c) **Special purpose**
      i. **Polar solvent alcohol-resistant concentrates**
      ii. **Hazardous materials concentrates**
   (d) **Aqueous film-forming foam (AFFF)**
   (e) **High expansion**
(3) **Given the required tools and equipment, demonstrate how to perform the following defensive control activities:**
   (a) Absorption
   (b) Damming
   (c) Diking
   (d) Dilution
   (e) Diversion
   (f) Retention
   (g) Vapor dispersion
   (h) **Vapor suppression**
(4) Identify the location and describe the use of the mechanical, hydraulic, and air emergency remote shutoff devices as found on cargo tanks.
(5) Describe the objectives and dangers of search and rescue missions at hazardous materials incidents.
(6) Describe methods for controlling the spread of contamination to limit impacts of radioactive materials.

8.3.4.1 Performing Response Options Specified in the Plan of Action. Given an assignment by the incident commander in the employee's individual area of specialization, the private sector specialist employee B shall perform the assigned actions consistent with the organization's emergency response plan and standard operating procedures and shall meet the following related requirements:

(1) Perform assigned tasks consistent with the organization's emergency response plan and standard operating procedures and the available personnel, tools, and equipment (including personal protective equipment), including the following:

(a) **Confinement activities**

(b) Containment activities

9.1.2.2 In addition to being competent at the awareness, operational, and technician levels, the hazardous materials branch officer shall be able to perform the following tasks:

(2) **Plan a response within the capabilities** and competencies **of available** personnel, personal protective equipment, and **control equipment** by completing the following tasks:

(a) Identify the response objectives for hazardous materials incidents.

(b) **Identify the potential action options (defensive, offensive, and nonintervention) available by response objective.**

9.3.2 Developing a Plan of Action. Given simulated facility and transportation hazardous materials incidents, the hazardous materials branch officer shall develop a plan of action consistent with the local emergency response plan and the organization's standard operating procedures that is **within the capability of the** available personnel, personal protective equipment, and **control equipment** and shall meet the following related requirements:

(3) Given the local emergency response plan or the organization's standard operating procedure, identify procedures to accomplish the following tasks:

(e) Coordinate with fire suppression services as it relates to hazardous materials incidents.

(f) **Coordinate hazardous materials branch control, containment, or confinement operations.**

9.5.1 Evaluating Progress of the Plan of Action. Given simulated facility and transportation hazardous materials incidents, the hazardous materials branch officer shall evaluate the progress of the plan of action to determine whether the efforts are accomplishing the response objectives and shall meet the following related requirements:

(3) Determine the effectiveness of the following:

(a) Hazardous materials response personnel being used

(b) Personal protective equipment

(c) Established control zones

(d) **Control, containment, or confinement operations**

(e) Decontamination process

Chapter 6
Class B Foam Use: Unignited Class B Liquid Spills

Incidents that involve ignited flammable and combustible liquids (Class B fuels) tend to be very spectacular and gain a great deal of media and emergency responder attention (see Chapter 7, Class B Foam Fire Fighting: Class B Liquids at Fixed Sites). However, fire and emergency response personnel are more likely to encounter incidents involving unignited spilled fuels in their day-to-day activities. The term *fuel spill* describes the spill of any flammable, combustible, acidic, basic, or otherwise hazardous liquid fuel. Although the appearance of an incident involving unignited fuels is not nearly as spectacular as that of a fire incident, an unignited spill has the potential to be extremely dangerous and life threatening. The sudden ignition of vapors surrounding a spill can catch emergency responders off guard and lead to either serious injuries or deaths.

No jurisdiction is immune from the threat of a large spill of flammable or combustible liquids. A fuel spill may occur at fixed facilities such as petrochemical refining and handling facilities, large industrial plants, maritime facilities (including shipboard), automobile service stations, and airports. A fuel spill may also be encountered as the result of a motor vehicle accident involving a flammable or combustible liquid transport vehicle or simply a ruptured fuel tank on an automobile (**Figure 6.1**).

The information in this chapter pertains to those substances that are normally found in a liquid state at atmospheric pressure. Compressed and liquefied gases are more difficult to deal with when they are spilled or released, and the information in this chapter may not apply to them. This chapter outlines the activities that must take place to control a flammable or combustible liquid spill, the types of foam concentrates used, and how finished foam is applied to minimize the production of vapors and to reduce the potential for ignition of the fuel. Vapor suppression operations and incident termination are discussed, and a sample tank vehicle incident is given. General strategic and tactical considerations used when dealing with unignited Class B liquid spills is also given.

Figure 6.1 The majority of fuel spills to which fire and emergency services personnel respond are small-quantity spills such as the one shown. *Courtesy of Martin Grube.*

General Strategic and Tactical Considerations

[NFPA 1051: 9.5.5]

In the context of the *RECEO* model, the primary objective for fire and emergency services personnel at an unignited flammable or combustible liquids spill is to prevent the ignition of the vapors that are released by the liquid and stop its flow. The first arriving units have to suppress the vapors through either the use of a water fog or through the application of a foam blanket. A foam blanket should be applied as soon as possible with the *RECEO* model introduced in Chapter 5, Class A Foam Fire Fighting, as a consideration. Additional considerations are as follows:

- Correct type of foam concentrate is available
- Sufficient foam concentrate is available to completely blanket the spill
- Sufficient water supply is available to develop the foam solution

If these requirements cannot be met immediately, then other actions must be taken to prevent ignition or otherwise mitigate the situation while waiting for the arrival of additional personnel, equipment, foam concentrate, and technical support. Stopping the flow of the liquid is only attempted when leaks are small and emergency personnel have the proper materials, personal protective equipment, and training (**Figure 6.2**).

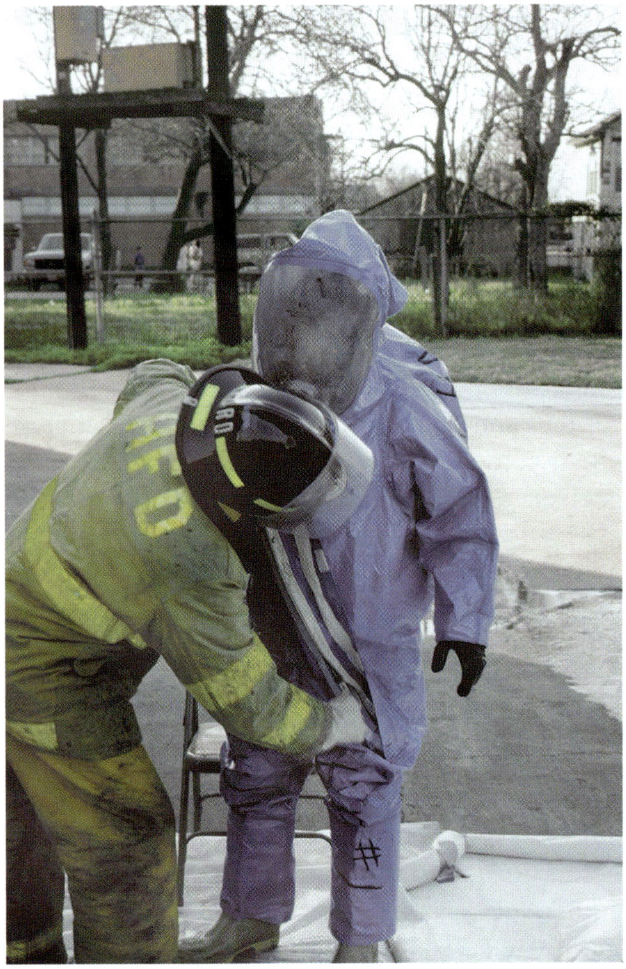

Figure 6.2 Only personnel properly trained and equipped should attempt to stop a large leak from a large container.

As is the case with Class A structures and Class B flammable/combustible liquid fires, general strategic and tactical considerations must be made in a logical sequence to ensure that spills are safely and efficiently controlled. These considerations include hazard assessments, incident preplanning, size-up (including product identification and characteristics, ignition sources, and wind direction and topography), and tactical priorities (including confine and contain).

Hazard Assessments

A hazard assessment is developed by a fire or emergency services organization to provide an understanding of the types and frequency of hazardous situations to which it may respond. This assessment includes the number and types of industrial complexes, high-value target hazards, high life-loss hazards, or transportation systems within the jurisdiction. Each of these hazards are in turn evaluated for the individual type of hazard such as flammable and combustible liquids storage, required water flow to extinguish a fire in the hazard, number of personnel required to handle an incident, type of equipment required, and type of foam concentrate necessary to combat a situation. Once this information is gathered, the pre-incident planning phase begins. At an actual incident, on-site hazard assessments are made during initial size-up and this information is combined with the earlier hazard assessments to develop incident action plans.

Pre-incident Planning

Most fire and emergency services personnel are familiar with pre-incident planning. It takes place daily with conducting site inspections and surveys, reviewing material safety data sheets, testing and flowing hydrants, determining potential travel routes (including locating barriers to site access and staging areas), and training on the use of various extinguishing agents and equipment (**Figure 6.3**). Specifically, it is planning for the worst possible incident involving a hazard located within the jurisdiction. Based on the hazard assessment, the responding emergency units establish the necessary steps needed to resolve an incident. Specific duties are assigned to the first, second, and subsequent arriving units. Types of equipment and

Figure 6.3 Fire and emergency services personnel should become familiar with a facility's types of fire protection equipment (such as this portable foam trailer) during site surveys. *Courtesy of Paul Valentine.*

extinguishing agent are designated for specific incidents. Personnel then drill on their assigned duties and equipment. Pre-incident planning prepares emergency responders for the likelihood that an incident will occur at a target hazard or involve a general type of hazard such as a rail tank car spill (**Figure 6.4**).

Size-Up

Following the establishment of command, the first priority upon arrival at any emergency incident is *size-up*: the mental process of considering all available factors that will immediately affect an incident during the course of an operation. Information gained from the size-up determines the strategy and tactics that are applied to the incident based on a

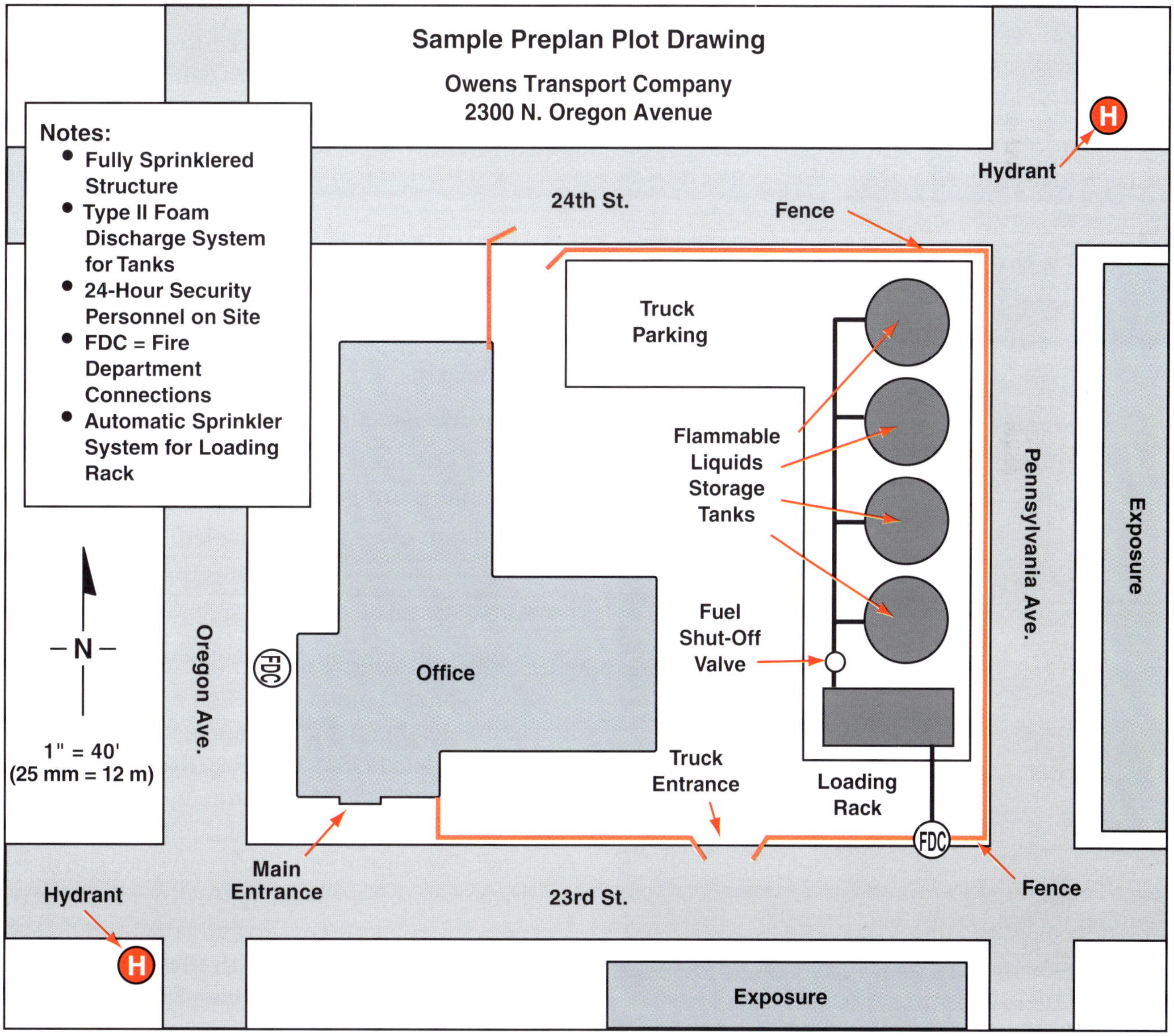

Figure 6.4 A sample preplan plot drawing. Such drawings are normally carried on emergency response apparatus.

model such as *RECEO* introduced in Chapter 5, Class A Foam Fire Fighting. (See Tactical Priorities section.) Size-up is a continual evaluation. It begins with pre-incident planning, extends through the receipt of an alarm, and continues throughout the incident. Upon arrival at an incident scene, the first incident commander (IC) conducts an extensive on-site hazard assessment of the existing conditions. During the course of the operation, the incident commander periodically conducts additional hazard assessments to ensure that progress is being made in resolving the incident **(Figure 6.5)**.

At the time of an alarm, specific information is obtained for size-up, including the nature of the call, its location, type of equipment responding, time of day, and weather information (such as wind direction and speed, temperature, and humidity). Emergency responders en route to an incident can review the following information to assist in the size-up of an incident:

Figure 6.5 The size-up of a spill begins as the incident commander mentally compares the response information with a visual review of the incident scene.

- Response route
- Pre-incident plans and sketches
- Arrival times of other responding units
- Exposure types and their distances from the original incident
- Hydrant and water supply conditions
- Access to the scene
- Preliminary plans for apparatus placement at the scene
- Additional information from the dispatcher/telecommunicator
- Need for additional units

CAUTION

None of the review procedures to assist size-up should interfere with the safe operation of apparatus traveling to an emergency scene.

The incident commander evaluates the following conditions on the scene of an unignited Class B liquid spill for on-site hazard-assessment purposes:

- Life hazards
- Unusual signs (leaking material, vapor clouds, etc.)
- Material(s) involved
- Condition of a spill: Is it pooled (static) or running (dynamic)?
- Path of travel of a running spill
- Actions already taken by people on the scene

The incident commander then adds the on-site hazard-assessment information to the information made available before arrival. With this combined information, the incident commander develops a formal plan of action for the unignited Class B liquid spill. Overall, the initial plan must focus on three strategic priorities: life safety, environmental protection, and property conservation **(Figure 6.6)**. The IC evaluates the situation with these priorities in mind and determines the potential for successfully accomplishing them.

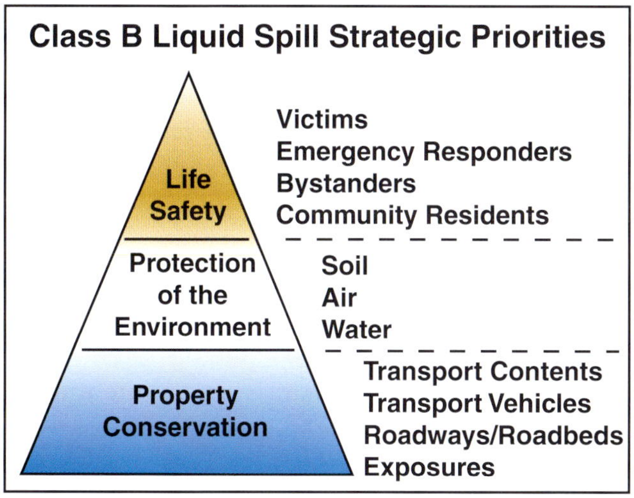

Figure 6.6 The three strategic priorities of a Class B liquid spill are life safety, environmental protection, and property conservation.

When dealing with unignited fuel spills, several parts of the standard size-up routine become crucial. Product identification is the first critical factor, and once the incident commander identifies the product, the next factor is to obtain all possible information on the chemical and physical characteristics of the product. Any potential ignition sources in the area of the spill must also be considered along with wind direction and the topography surrounding the incident site.

Product Identification and Characteristics

Emergency responders must use all available means to identify the product involved and determine the characteristics of the fuel. For more detailed information on product identification, see the IFSTA **Hazardous Materials for First Responders** manual. However, some ways to identify a product are as follows:

- Signs, placards, or labels (**Figure 6.7**)
- Type of container, tank, or trailer
- Shipping documents
- Pre-incident plans
- Information provided by eyewitnesses, vehicle drivers, or company employees
- Obvious physical indicators such as odor, color, etc.

The location of shipping documents is mandated by the U.S. Department of Transportation and Transport Canada for vehicle, rail, air, and maritime cargo carriers. On vehicles, the bill of lading or

Figure 6.7 Tank trucks that carry hazardous materials in North America are required to display the U.S. Department of Transportation/Transport Canada (DOT/TC) warning placards.

manifest must be within reach of the driver/operator, usually on the passenger seat or in a pocket in the driver's door. Manifests for rail cargo are located in the cab of the lead engine near the engineer. On cargo aircraft, the papers are on or near the flight deck. For maritime transport, manifests are located on the bridge while at sea or near the gangway when in port. Barge cargo manifests are located on the bridge of the towboat or tug.

Some sources that may give product information and characteristics once it is identified are as follows:

- Computer databases
- Chemical indexes
- *Emergency Response Guidebook,* current edition (**Figure 6.8, p.180**)
- CHEMTREC®/CANUTEC (see textbox on next page)
- Material safety data sheets or other manufacturer-supplied information

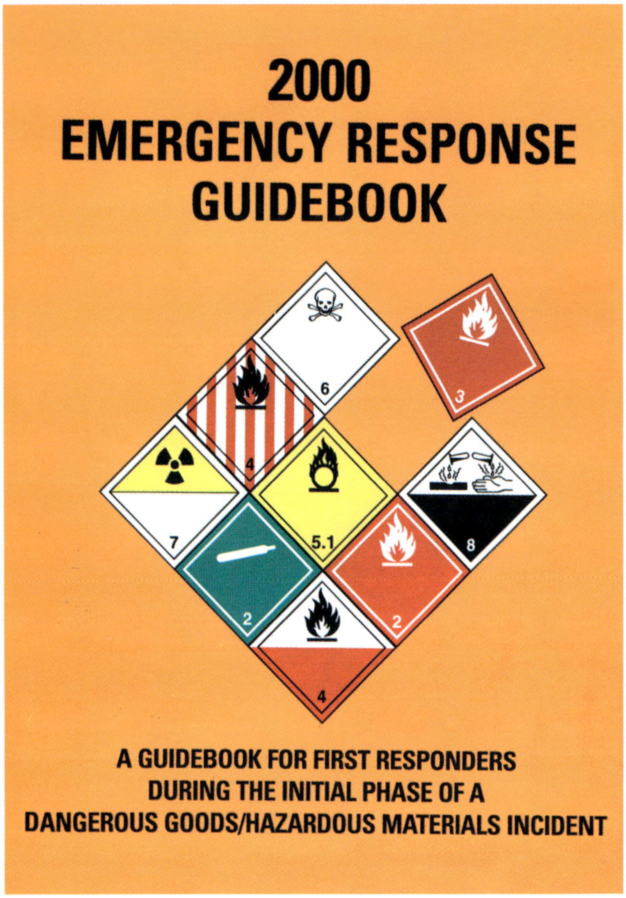

Figure 6.8 The *2000 Emergency Response Guidebook* published by the U.S. Department of Transportation is available in both English and Spanish language versions.

CHEMTREC®/CANUTEC

The Chemical Transportation Emergency Center (CHEMTREC®) was established by the chemical industry in the United States in 1971 to provide fire and emergency services personnel with information and assistance with incidents involving chemicals and hazardous materials. CHEMTREC® also helps shipping companies comply with U.S. Department of Transportation regulations regarding the manner in which these materials are shipped. CHEMTREC® provides a 24-hour emergency call center that can answer questions that on-scene fire and emergency services personnel have regarding the best way to handle spilled materials at an incident.

The Canadian counterpart to CHEMTREC® is CANUTEC, the Canadian Transport Emergency Centre, which is operated by Transport Canada, a department of the national government. CANUTEC provides the same services to fire and emergency services personnel in Canada that CHEMTREC® does in the United States.

Fire and emergency services personnel who deal with fuel spills have a particular interest in the following characteristics of the product: flash point, vapor density, flammable/explosive ranges, and rate of vaporization. Knowledge of these characteristics assists in determining the tactics used to control an incident, level of personal protective equipment required, and potential hazards that are created if a spill ignites. These characteristics were covered in Chapter 1, Fire Behavior and Extinguishment. However, the following characteristics are mentioned here as a refresher:

- ***Flash point*** — Minimum temperature at which a liquid fuel releases enough vapors to form an ignitable mixture with air near the liquid fuel's surface. At this temperature, the vapors flash but will not continue to support combustion.
- ***Vapor density*** — Comparison of the density of a gas or vapor to the density of air. Air has a vapor density of 1. Vapors with a density greater than 1 are heavier than air and therefore sink, accumulating in low spots in the terrain. Vapors with a density less than 1 rise and rapidly dissipate in the open. Within structures, the vapors accumulate in the upper portions of the structure, requiring ventilation.
- ***Flammable/explosive range*** — Percentage of a gas or vapor concentration in the air that burns if ignited. Below the flammable range (lower flammable limit or LFL) or lower explosive limit (LEL), a gas or vapor concentration is too lean to burn (not enough fuel and too much oxygen). Above the flammable range (upper flammable limit or LFL) or upper explosive limit (UEL), the gas or vapor concentration is too rich to burn (too much fuel and not enough oxygen). Within both the upper and lower limits, the gas or vapor concentration burns rapidly if ignited **(Figure 6.9)**.
- ***Rate of vaporization*** — Speed at which a liquid evaporates or vaporizes; is affected by the physical properties of the fuel, temperature, and wind speed. This rate determines the amount of flammable vapors released from a fuel spill. The rates at which any fuel involved in a spill vaporizes and the extent to which these vapors spread are difficult to determine. Computer models are available to assist fire and emergency services personnel with gathering this

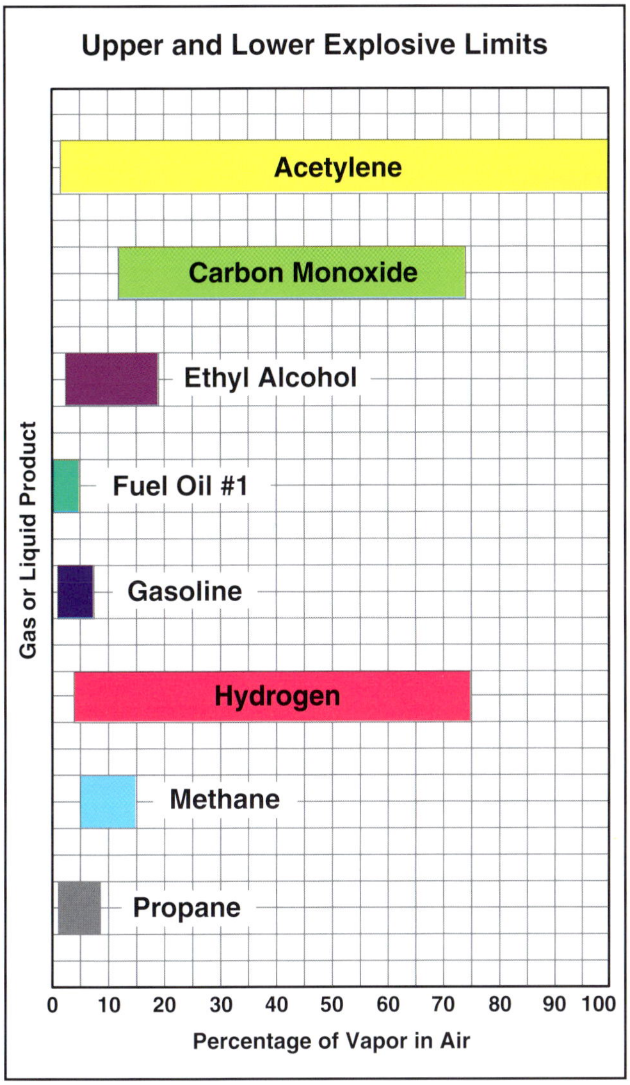

Figure 6.9 Examples of the flammable ranges of common gas or liquid products.

information and predicting rates of vaporization. Fire and emergency services personnel must remember that the longer a spill is left unattended, the more vapors are released.

> **CAUTION**
>
> **Consider any vapor concentration above 10 percent of the lower explosive limit (LEL) as a serious ignition potential that requires special consideration. Ventilating concentrations that are above the upper explosive limit (UEL) result in explosive mixtures. In areas that have concentrations too rich to burn, there are zones near the edge of the vapors that are within the flammable range.**

Ignition Sources

Early in a fuel spill incident, the incident commander must consider any potential ignition sources in the area so they can be neutralized. Depending on the location of the incident, an endless number of potential ignition sources may be found in the area. Some of the sources include the following: electrical equipment, motor vehicles, static electricity, pilot lights, power tools, and open flames (**Figure 6.10**).

Figure 6.10 Potential ignition sources include commercial welding units such as this one. *Courtesy of Conoco, Inc.*

Wind Direction and Topography

The incident commander must pay particular attention to the wind direction and the topography surrounding the incident site. It is important to note those ignition sources that are downwind and downhill of the spill and in the path of any vapors or runoff that are released (**Figure 6.11**). This awareness prevents responding units from approaching the incident from a downwind direction and risking the chance of an apparatus becoming an ignition source. It also prevents the apparatus from being parked in the path of a fuel spill. Preferably, the dispatch/telecommunications center can provide information on atmospheric conditions and topography to the units en route to the incident.

The ground surface also impacts the tactics that are used to control a spill. On hard, nonporous surfaces such as parking lots and roadbeds, the liquid fuel generally remains on the surface. Very little of the fuel soaks into the concrete, although some fuels may soften and react with asphalt or tar. On porous surfaces such as the soil alongside the roadway, gravel that composes the roadbed for railroads,

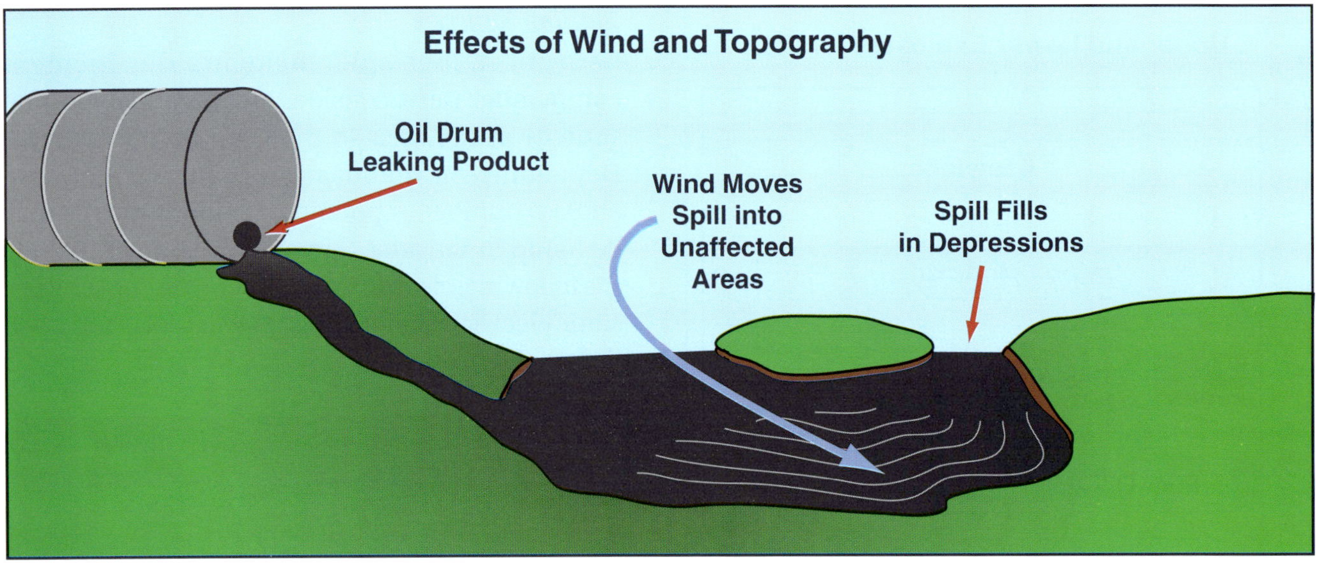

Figure 6.11 The spread, depth, and size of spills are affected by both wind and topography.

or just gravel or dirt roads, the fuel spill will soak into the soil creating the potential for a fire. This contaminated soil (sometimes to great depths) has to be blanketed with foam to prevent vapor release. Then the soil must be removed by hazardous waste companies and disposed of according to local laws. When responding to aircraft incidents away from an airport, soft or muddy soils may stop heavy apparatus and equipment, steep slopes may be difficult to traverse or climb, low or down-slope areas may become saturated with fuel, and rough or rocky terrain may be impassable.

Tactical Priorities

[NFPA: 1021: 2-6.2]

As discussed in Chapter 5, Class A Foam Fire Fighting, the *RECEO* model (rescue, exposures, confine, extinguish, and overhaul) can be applied to any emergency incident, including those involving unignited flammable and combustible liquid spills.

Rescue

A rescue operation, the initial step in the *RECEO* model, varies depending on the size of the spill, ignition sources, location and condition of victims, and resources available to the initial response units. If the first response unit has no foam or a very limited supply, it may be necessary to use water fog handlines to sweep the spill away from victims and protect personnel who are making the rescue. Class A foam, if available, provides an additional benefit in this instance because of its low solution ratio. Fire and emergency services personnel must be trained in the use of water fog and Class A foam during rescue operations just as they are trained to use Class B foam for spill control.

Figure 6.12 In order to prevent emergency services personnel from becoming victims during rescue operations, full personal protective clothing with respiratory protection must be worn. In this example, properly attired personnel train with a hoseline during a simulated rescue incident. *Courtesy of Tom Ruane.*

As part of the rescue step, all emergency personnel must be properly equipped with personal protective equipment including the appropriate level of respiratory protection (**Figure 6.12**). If large quantities of flammable or combustible liquids are spilled, then the surrounding area should be evacuated. Additional emergency personnel are required for this effort.

Exposures

The severity of exposure hazards depends on the location of the incident and can range from none to extreme. On a rural stretch of road, the consideration may only be the responding apparatus in the event of a change in wind direction. However, it is more likely that exposures will be at the other end of the severity scale. Roadway surfaces, overpasses, adjacent lanes of divided highways, multiple pipelines, buildings, and other structures, to name a few, have to be protected. Taking the following actions may provide this protection:

- *Remove the exposure* — Pull or tow vehicles, railway cars, ships, aircraft, or other mobile items out of the spill area.
- *Cover the exposure with foam* — Provide an effective barrier on exposures with Class A foam in case the spill ignites.
- *Confine or contain the spill* — Prevent the spill from flowing closer to the exposures or turn off the leak and contain the fuel to provide exposure protection.

Confine

When dealing with flammable or combustible liquids spills, the *confine* step (or *confinement*) is the process of controlling the flow of a spill and capturing it at some specified location. Confinement is primarily a defensive action. The objectives of confinement operations are to capture the moving materials as quickly as possible in the most convenient location and to recover the materials with the smallest amount of exposure to people, the environment, and other property. Spilled fuels and the water and finished foam used to control them must be kept out of water sources, storm sewers, and other bodies of water to prevent contamination of these water sources or the environment. Spilled materials may be confined by the following methods:

- Building dams or dikes near the source
- Catching the material in another container
- Directing (diverting) the flow to a remote location for collection **(Figure 6.13)**

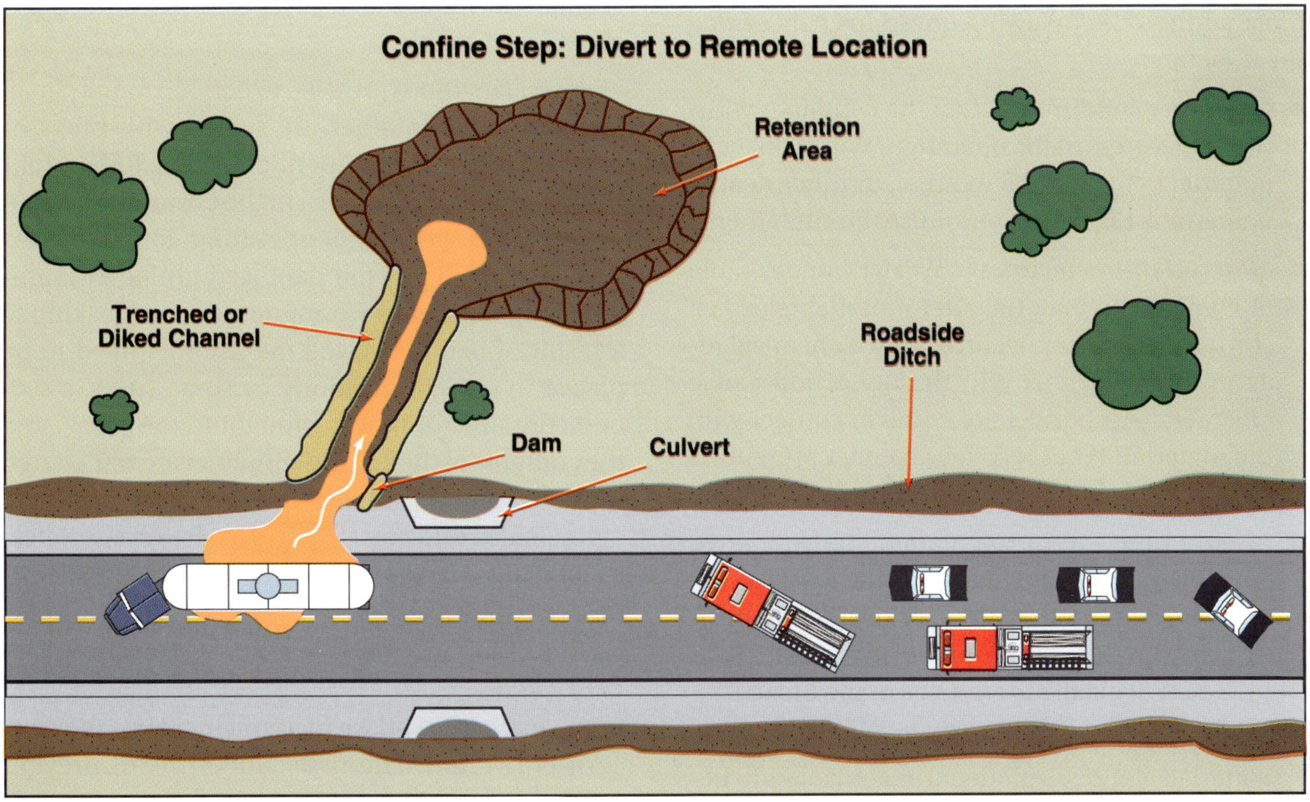

Figure 6.13 In some cases it may be possible to divert the flow of a product to a safe location for retention and collection. Soil contamination may be a result of this action and should be considered.

Generally, the following tools that can be used for confinement purposes are carried on fire apparatus: shovels for building earthen dams, salvage covers for making catch basins, and hoselines that are deployed and charged with water for creating diversion channels. Before using these tools to confine spilled materials, incident commanders should seek advice from technical sources to determine whether the spilled materials will affect the tools. Large or rapidly spreading spills may require the use of heavy construction-type equipment, floating confinement booms, or special sewer and storm drain plugs. Pre-incident plans can identify where these sources of help can be obtained when needed.

The *containment* step (included with the confine step when dealing with hazardous liquids) is the act of stopping the further release of a spilled material from its container. Fire and emergency services hazardous materials technicians usually perform containment procedures, although an outside contractor or trained personnel supplied by the owner of the spilled material may perform this task **(Figure 6.14)**. Several factors affect the ability to contain a spilled material, and the incident commander must address the following questions concerning these factors:

- ***Container condition*** — Can the container withstand the stress of the operation? Will changes in material behavior, the weather, or other nearby operations further compromise the container?
- ***Material characteristics*** — Will it be undesirable to contain the material after a partial release if contamination has occurred or if a chemical reaction has started? Will the operation cause emergency responders to come in contact with the material? What risk does contact present? Is the material changing physical states?
- ***Release rate*** — Can the material be contained before the vessel is empty? Will the leaking stop in a short amount of time because of the position of the breach (rupture)? Are sufficient resources available for the size of the breach?
- ***Incident hazard assessment*** — Will the leaking materials ignite, explode, or react violently before or during the operation? Will containment devices and the container remain intact until recovery operations end?

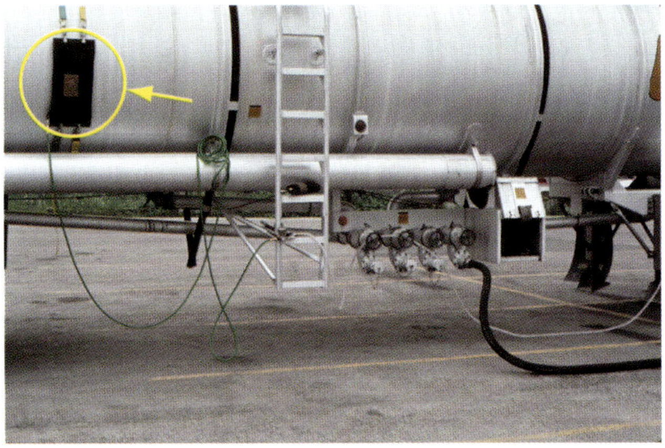

Figure 6.14 When possible, contain liquid flows in their place of origin. In this photograph, an air bag is used to seal a puncture in the side of a tank trailer. *Courtesy of Mike Hildebrand.*

Of course, the location of the incident, the type and number of exposures, the number and capabilities of responding forces, and a complex variety of other concerns affect the decision to contain unignited Class B liquid leaks. Confinement of the spilled material is often performed initially to protect exposures, and then the leak-containment process is started. Confinement and containment are started simultaneously if personnel and equipment are available.

Several methods are used to contain a spill. Small leaks from a container are sealed with plugs or patches. Most containment efforts include establishing a confinement area in the event that a patch or plug fails. Personnel operating control valves upstream from leaks in damaged piping or manifolds may stop leaks. Containers that are leaking as a result of being inverted may simply need to be placed in upright positions. In some cases, a second container or tank is brought to the scene, and the contents of the leaking container are transferred into the second container. All of these options are complicated and must be handled by appropriately trained technicians who wear proper personal protective equipment.

Extinguish

In the *extinguish* phase of a spill incident, the concern is twofold. The first concern is to prevent ignition. Removing potential ignition sources and/or providing a vapor barrier over the spill with foam will accomplish this task. The second concern is to extinguish the fire if the spill ignites. Chapter 7, Class B Foam Fire Fighting: Class B Liquids at Fixed Sites, and Chapter 8, Class B Foam Fire Fighting: Transportation Incidents, address this situation.

Overhaul

In a spill situation, *overhaul* (removing the hazard and preventing ignition) is usually the responsibility of a hazardous waste removal company. Fire and emergency services personnel can provide support by continuing to apply foam to the blanket or providing standby water fog and foam handlines in the event of an ignition of the spill.

Liquid Spill Control

[NFPA 472: 5.1.2.1 (2a, b), 5.3.2 (1, 2), 5.4.4, 8.3.4.1 (1a), 9.1.2.2 (2b), 9.3.2 (3f), 9.5.1 (3d); NFPA 1081: 6.3.4, 7.3.4]

Foam concentrates that are designed for the extinguishment of Class B fires are valuable when dealing with unignited flammable and combustible liquid spills. The foam concentrate characteristic of smothering provides a complete barrier to vapor generation and a barrier between the fuel and oxygen and ignition sources. However, fire and emergency services personnel must realize that foam concentrates that are designed for extinguishing fire have different considerations when they are used for vapor suppression. It is also important to remember that training in the use of foam concentrates for fire extinguishment is not sufficient to qualify personnel to use the same foam concentrate for vapor suppression **(Figure 6.15)**.

Figure 6.15 Fire and emergency services personnel must be trained in the use of foam for unignited spill control as well as flammable liquids fire control. Personnel in the photo are wearing proximity personal protective clothing that allows them to advance closer to the fire than would standard structural personal protective clothing. *Courtesy of Conoco, Inc.*

Vapor suppression operations include learning about vapor suppression concentrates, application rates and foam blanket application, and reapplication procedures. Proper incident termination procedures are required to successfully resolve an unignited Class B liquid spill incident. A sample tank vehicle incident involving an unignited spill is given at the end of the section.

Vapor Suppression Operations

Vapor suppression is the action taken to reduce the emission of vapors at a fuel spill. This suppression, in turn, lessens the effects of the vapors on people in the area and lessens the chance of the vapors igniting. Class B foam concentrates are the most common agents used for suppressing vapors at liquid fuel spills. Fire and emergency services personnel must understand the factors related to the use of foam concentrates for vapor suppression including the following:

- Types of vapor suppression foam concentrates
- Types of spilled materials (see Product Identification and Characteristics section)
- Reaction characteristics of spilled materials
- Equipment considerations
- Application rates
- Procedures for applying the foam blanket
- Reapplication procedures

NOTE: Vapor pressure rises as the fuel temperature increases. Hot diesel fuel or gasoline can pose significant postfire or spill-security problems because ignitable fuel vapor invades an otherwise secure-looking foam blanket.

Vapor Suppression Concentrates and Spilled Materials

Aqueous film forming foam (AFFF) and film forming fluoroprotein (FFFP) Class B foam concentrates are the most common types used for vapor suppression today. The standard forms of these foam concentrates suppress vapors at hydrocarbon fuel spills. The alcohol-resistant (AR) forms of these concentrates are suitable for most types of hydrocarbon and polar solvent fuels. Some fuels such as gasoline have elements of both hydrocarbons (the gasoline itself) and polar solvents (various additives in the gasoline). This combination may have an adverse effect on concentrates designed solely for hydrocarbons.

High-expansion Class B foam concentrates are also used to successfully suppress vapors from flammable and combustible liquid spills and can be effective in indoor or confined-space applications. Unlike low-expansion foam, high-expansion foam can suppress vapors from unignited liquefied natural gas. This action reduces the vaporization that occurs as a result of the extreme temperature difference between the fuel and the water in finished foam **(Figure 6.16)**.

If a spill consists of acids or bases, Class A and Class B fire-fighting foam concentrates may not be effective because most of them have a limited range of pH (measure of acidity of an acid or level of alkaline in a base) tolerance. Therefore, most of the protein, flouroprotein, AFFF, and high-expansion foam concentrates are not compatible with acids and bases. Some alcohol-resistant foams (such as AR-AFFF) that have a good pH tolerance may be useful for vapor suppression of both acids and bases in liquid form. The manufacturer's recommendation must be followed in all cases.

Emulsifying agents (surfactants added to water to allow it to mix with and dilute flammable or combustible liquids) may be used on some fuel spills. They are recommended for use on lightly distributed diesel fuel spills but not on pooled diesel spills or gasoline spills of any depth. The emulsifying agent is mixed with water as a 1-percent solution (5 gallons [18.92 L] of emulsifier to 50 gallons [189.27 L] of water) **(Figure 6.17)**.

The introduction of blended or reformulated gasoline for use in vehicles in both Canada and the United States in the 1990s has created a further challenge for fire and emergency services personnel. Government-mandated fuel additives such as ethanol, methanol, and oxygenated compounds have increased the combustibility of gasoline while de-

Figure 6.16 High-expansion foam can be used to suppress natural gas vapors from leaking pipes. In this photograph, the foam stream is directed into the vapor cloud above the leak.

Figure 6.17 Emulsifiers are designed for small spills and are mixed with water at a 1-percent ratio.

creasing the emission of harmful particulates into the air. At the same time, the blended or reformulated gasoline has become more difficult to cover with a foam blanket when it is spilled or extinguish when on fire because it interferes with the film-formation process of AFFF. NFPA 11, *Standard for Low-, Medium-, and High-Expansion Foam* (2002), states that alcohol-resistant foam concentrates must be used on fires involving gasoline that contains 10-percent oxygen additive. See sidebar.

Finally, no single foam concentrate is effective against all types of volatile hazardous liquids. Class A foam concentrates are generally not used on flammable or combustible liquid spills or other hazardous chemicals.

Chemical Reactions

Reactivity is a very important factor to consider when deciding to use foam concentrate for vapor suppression. Besides flammable and combustible liquids, numerous manmade liquid chemical compounds may react violently when mixed with water and create a more hazardous situation for emergency responders and the community than originally existed in the beginning of an incident. These combined compounds may become reactive with either water or the foam solution (**Figure 6.18, p. 188**).

A good resource for emergency responders to use when responding to liquid hazardous materials spills is the *Emergency Response Guidebook* (current edition). However, this resource only covers known manmade chemical compounds. Emergency responders must be very careful to recognize that they may be dealing with a mixture of chemicals from several leaking containers thus creating a completely different compound that may have completely different reactivity characteristics with water or foam solutions than the original chemicals in their separate states.

Blended or Reformulated Gasoline

The development of gasoline additives in the 1990s was motivated by a desire to achieve the following goals:

- Protect the ozone layer of the earth
- Maintain a healthy, breathable atmosphere
- Reduce petroleum consumption
- Produce more efficient fuels for transportation

To meet these goals, the petroleum industry in North America has pursued two different approaches through the development of two types of additives: ethanol and methyl tertiary-butyl ether (MTBE). Both have their proponents and detractors and are in use in both Canada and the United States.

Ethanol is a gasoline additive that is created from organic matter such as corn. Proponents claim that it burns cleaner and slightly cooler than gasoline without additives. It is available in two types: high-level and low-level. The high-level type (called *E85*) is a mixture of 85-percent ethanol with 15-percent gasoline. It requires some adjustment to a vehicle's engine and fuel system before it is used. The low-level mixture is blended between 5- and 10-percent ethanol to gasoline and requires no adjustment to a vehicle's engine. Ethanol must be added to gasoline near the point of distribution because it is easily contaminated by water, which makes the gasoline useless. Opponents argue that it requires more energy to manufacture ethanol than its use saves and that government subsidies will be required to make it profitable for farmers to produce the corn that goes into it.

Methyl tertiary-butyl ether is a chemical additive that has been in use as an octane enhancer and replacement for lead since the 1970s. MTBE may be found in approximately 30 to 50 percent of all gasoline sold in the United States. It is clean burning and exceeds all pollution-reduction goals according to proponents. Opponents claim that it is not as efficient as ethanol.

Both ethanol and MTBE add oxygen to gasoline to increase the efficiency of combustion of the fuel. Both destroy a nonalcohol-resistant foam blanket. Both require the use of alcohol-resistant foam concentrates to prevent ignition when they are spilled or extinguish when on fire. Other alcohol oxygenates that may be encountered are as follows:

- Ethers
  - — Tertiary-amyl methyl ether (TAME)
  - — Ethyl tertiary-butyl ether (ETBE)
  - — Diisopropyl ether (DIPE)
- Alcohols
  - — Methanol
  - — Tertiary-butyl alcohol (TBA)

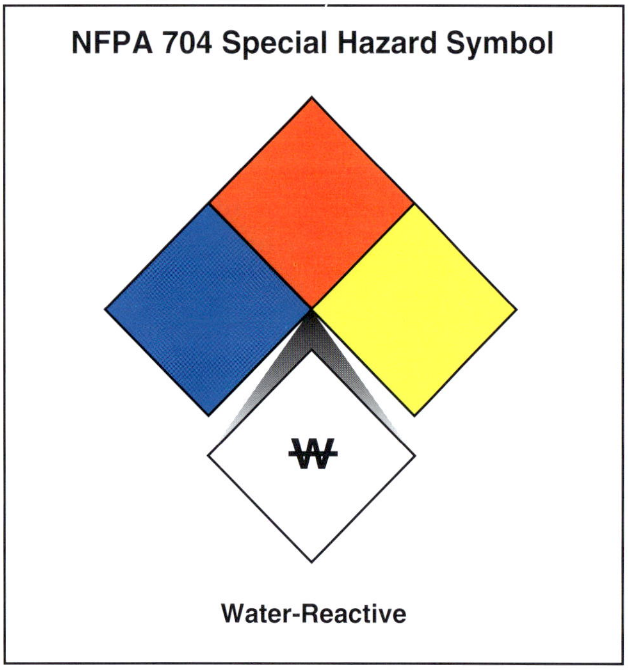

Figure 6.18 This is an example of the NFPA 704 symbols. The special hazard space indicates one of two situations: ₩ (contents will react with water) as shown here or OX (oxidizer) as shown earlier. The other three spaces indicate the health hazard rating (blue), the flammability hazard rating (red), and the instability hazard rating (yellow).

The use of foam concentrates in these types of emergencies needs to be monitored very closely to ensure that the desired results are being accomplished. Emergency responders must remember that foam concentrates that work for vapor suppression for individual liquid chemicals may not work on these uncontrolled chemical mixtures. Applications of finished foam in these situations may cause adverse conditions.

Equipment Considerations

The type of equipment used to apply finished foam to fuel spills is the same as those types discussed earlier in this manual. In general, the particular type of proportioning system used on unignited or ignited spills is not critical; however, it must operate properly. The main piece of equipment that may vary at different fuel spills is the nozzle. Depending on the type of equipment that is available and feasible to use, fire and emergency services personnel may use fog nozzles or low-, medium-, or high-expansion air-aspirating foam nozzles (**Figure 6.19**).

Aspirated finished foam is the most desirable type to use for vapor suppression at flammable and combustible liquid spill incidents. Aspirated finished foam provides a longer lasting cover than does nonaspirated finished foam. Low-expansion foam concentrates may be applied through low- or medium-expansion aspirating foam nozzles to form effective vapor suppression foam blankets.

Figure 6.19 Once the appropriate type of foam delivery system is determined, the nozzle is attached to the hoseline.

Figure 6.20 Standard fog nozzles may be affectively used to apply foam in either a straight stream or fog pattern.

Standard fog nozzles can be used to apply effective foam blankets when they are used in conjunction with AFFF or FFFP foam concentrates (**Figure 6.20**). Foam solutions applied from standard fog nozzles are not aspirated as much as foam solutions applied from foam nozzles. Because of this limitation, the need for more frequent applications is required. However, foam streams from fog nozzles have a longer reach than those from foam nozzles. Fog nozzles have adjustable stream patterns that can protect fire and emergency services responders in the event a spill ignites.

Application Rates

To control any type of incident (fire or fuel spill), it is crucial that an adequate flow of finished foam is discharged onto the fuel. Suggested minimum application rates and durations are established by NFPA 11; NFPA 11A, *Standard for Medium- and High-Expansion Foam Systems* (2002); and the foam concentrate manufacturer's recommendations. The recommendations in NFPA 11 are for spills that are on fire. Although it may be possible to perform vapor suppression at flow rates lower than those required for fire suppression, lower rates are not recommended. Flowing the rates recommended in NFPA 11 provides protection for fire and emergency responders in case the spill ignites.

CAUTION

When specifying and purchasing foam concentrates, take care to select foams that meet Underwriters Laboratories (UL), Factory Mutual (FM), or military specification (milspec) requirements and have the appropriate listing for the type of hazard. Be cautious of unreasonable claims concerning the ability of foams to protect against all types of hazards.

NFPA 11 recommends minimum flow rates that are maintained for at least 15 minutes on ignited spills of hydrocarbon fuels, 1 inch (25 mm) or less in depth in nondiked areas. These minimum flow rates for ignited spills are as follows:

- ***Protein and fluoroprotein foam concentrates:*** 0.16 gpm/ft^2 (6.5 [L/min]/m^2)
- ***AFFF and FFFP foam concentrates:*** 0.10 gpm/ft^2 (4.1 [L/min]/m^2)

NFPA does not specify a minimum flow rate for hydrocarbon or polar solvent fuel spills. Instead, it refers emergency responders to the manufacturer of the concentrate for directions based on the particular fuel being covered. If adequate foam concentrate and water supply are available at the spill site, then the application rate for ignited spills may be used to provide a foam blanket. If the spill ignites, use the 15-minute rate recommended by NFPA 11. See Chapter 7, Class B Foam Fire Fighting: Class B Liquids at Fixed Sites, for the use of foam on ignited flammable and combustible liquids.

To determine the amount of foam concentrate required to make an NFPA 11-specified application, emergency responders need to know the following facts:

- Size of spill
- Type of foam concentrate, type of delivery nozzle, concentrate proportioning ratio, and recommended application rate
- Recommended discharge duration

To increase the quarter-life time of foam solution, two methods may be followed: First, lower the nozzle discharge pressure when using medium-expansion foam nozzles or medium-expansion nozzle attachments. Foam solution leaving a medium-expansion nozzle at 50 to 75 psi (345 kPa to 517 kPa) improves finished foam's quarter-life. The second method of increasing finished foam's quarter-life time is to double the proportioning rate. For example, use a 3-percent foam concentrate at 6 percent or a 1-percent foam concentrate at 3 percent. Increasing the amount of concentrate (and therefore the amount of foaming chemical) in the solution creates a more rigid, slower draining foam blanket. Normally, fast draining AFFF, FFFP, and fluoroprotein foams hold water and air proportionally with the amount of foam concentrate (foaming chemical) introduced into the water stream.

The longest quarter-life results when foam solution is richly proportioned, nozzle pressure is lowered, and a medium-expansion nozzle or nozzle attachment is employed. These nozzles and attachments are often referred to as *haz-mat nozzles.*

The following examples (in both English System and SI System units) calculate the amount of concentrate needed to make one initial NFPA-recommended application of finished foam (see textbox, p. 190). Rarely would this amount be enough to actually control an incident. Additional finished foam is needed for reapplication when the initial blanket breaks down or burns through.

When using AFFF or FFFP concentrates, a general guideline is that hydrocarbon incidents with a square footage that is 10 times the rated flow of the foam eductor can be managed. For example, a 100-gpm (379 L/min) eductor provides enough finished

Concentrate Amount Calculations

English System Units

Calculate how much foam concentrate would be needed to cover a fuel spill of about 1,000 square feet given the following parameters:

- 3-percent AFFF foam (0.10 gpm/ft^2 application rate)
- 15-minute discharge time

Step 1: Determine the total required flow rate. Multiply the size of the spill by the application rate:

1,000 ft^2 X 0.10 gpm/ft^2 = 100 gpm required flow rate

Step 2: Once the flow rate has been established, determine the amount of concentrate required. Multiply the flow rate, the proportioning rate, and the discharge time together:

100 gpm X 0.03 X 15 minutes = 45 gallons of concentrate

SI System Units

Calculate how much foam concentrate would be needed to cover a fuel spill of about 100 square meters given the following parameters:

- 3-percent AFFF foam (4.1 L/min per square meter application rate)
- 15-minute discharge time

Step 1: Calculate required flow rate:

100 m X 4.1 (L/min)/m^2 = 410 L/min

Step 2: Calculate amount of concentrate required:

410 L/min X 0.03 X 15 minutes = 185 L of concentrate

foam to cover a 1,000-square-foot (93 m^2) spill. Another general guideline is that polar solvent incidents with a square footage that is 5 times the rated flow rating of the foam eductor can be managed. When the spill consists of polar solvents, an area five times the rated flow of the foam eductor is adequate. For example, a 100-gpm (379 L/min) eductor provides enough finished foam to cover a 500-square-foot (46 m^2) polar solvent spill.

Foam Blanket Application

Begin foam blanket application on an unignited spill as soon as possible. This action lessens the overall amount of vapors released by the spill and reduces the chance of ignition and exposure to fire and emergency services personnel. However, gather the following information before beginning the application:

- ***Identify the spilled fuel*** — Ensures that the foam concentrate available is compatible with the fuel
- ***Determine approximate size of the spill*** — Allows emergency responders to determine if they have a large enough supply of foam concentrate and foam delivery equipment to handle the spill **(Figure 6.21)**
- ***Calculate required flow rate and amount of foam concentrate needed*** — Allows emergency responders to make calculations (see the Application Rates section) to ensure concentrate and flow rate requirements are met to create the desired foam blanket
- ***Determine quantity of foam delivery equipment on the scene*** — Helps determine the ability of the available proportioning and distribution equipment (both fixed and portable) to provide the application rate based on the required flow rate for the hazard
- ***Determine wind direction*** — Helps determine the safest location for apparatus, emergency personnel, and foam delivery equipment **(Figure 6.22)**

Fire and emergency services personnel approach and handle vapor suppression at unignited spills with the same caution that they use at incidents involving ignited materials. The incident commander establishes zones of control around the incident and designates the area closest to the spill as the *hot zone*. When entering this zone, all emergency personnel must wear full personal protective clothing including respiratory protection such as self-contained breathing apparatus. When possible, emergency personnel should approach the spill and apply finished foam from the upwind side of the spill (that is, with the wind at their backs). Applying finished foam from the upwind side of a spill exposes emergency responders to a minimum amount of vapors. It is also a safer position should the spill ignite. It is much easier to apply the foam blanket if it is traveling with the wind rather than into the wind.

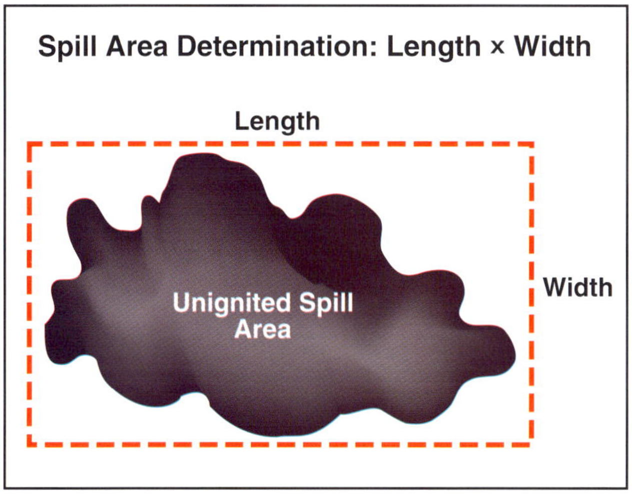

Figure 6.21 To determine the approximate size of a liquid spill, determine the length and width at the greatest points and multiply those two numbers together.

Depending on the needs of the incident and the equipment available, the foam application hoseline is equipped with either a fog or an air-aspirating foam nozzle. A backup hoseline crew protects each foam attack hoseline crew. The backup crew is responsible for providing the protective cover necessary for the foam attack hoseline crew to escape harm in the event the spill ignites while the foam attack hoseline is in service **(Figure 6.23)**. The backup hoseline may be either a foam hoseline or a plain water hoseline. If it is a foam hoseline, it should be equipped with a fog nozzle that gives a better protective pattern than a straight stream or foam delivery nozzle.

For finished foam to achieve maximum effectiveness, it is applied gently to the surface without plunging into the fuel. The three principal techniques for applying finished foam to the surface of a spill are as follows:

Figure 6.23 A backup crew in complete protective equipment provides protective cover for the attack hoseline crew.

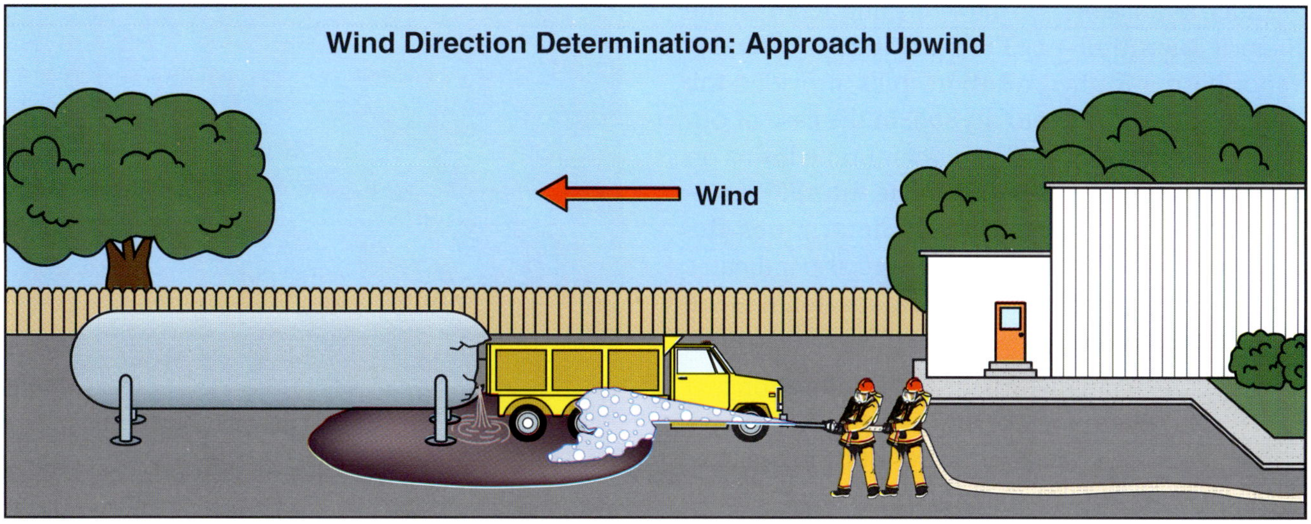

Figure 6.22 When possible, approach spills from an upwind position.

- ***Roll-on method (bounce)*** — Direct the foam stream on the ground near the front edge of the spill so finished foam rolls across the surface of the fuel. Continue applying finished foam until it has spread across the entire surface of the spill. It may be necessary to move the stream to different positions along the edge of the spill to cover it entirely **(Figure 6.24)**.
- ***Bank-down method (deflect)*** — Direct the foam stream off an object that is elevated above the spill, and allow finished foam to run down onto the surface of the fuel. Bank finished foam off a wall, support column, vehicle, or any other similar object in the area of the spill. As with the roll-on method, it may be necessary to direct the foam stream off various points around the spill area to achieve total coverage **(Figure 6.25)**.
- ***Rain-down method (raindrop)*** — Direct the foam stream into the air above the spill and allow finished foam to float gently down onto the surface of the fuel. Sweep the foam stream back and forth over the entire surface of the spill until it is completely covered **(Figure 6.26)**.

Fire and emergency services responders should attempt to achieve a 4-inch (102 mm) foam blanket over the entire surface of a spill to provide a good vapor barrier and give a cushion to permit a drainage time that negates the need to continually apply more finished foam **(Figure 6.27)**. It may be difficult to achieve a 4-inch (102 mm) foam blanket when using a fog nozzle because of the lesser degree of aeration that occurs. When this situation is the case, simply try to cover the surface of the spill as much as possible. The film-forming action of the foam (either AFFF or FFFP) provides an effective vapor barrier for hydrocarbon spills, provided the fuel has not soaked into the soil. In the case of polar solvent spills, a polymeric membrane forms on the surface of the spill when using AR-AFFF. Constant air sampling of the spill determines the effectiveness of the foam blanket and also indicates the need for reapplication of additional foam as the blanket dissolves.

Take care to not disturb the foam blanket once it is in place. Disturbance of the foam blanket can result in gaps in the vapor barrier, resulting in a release of vapors and possible ignition of the spill **(Figure 6.28)**. In the past, fire and emergency services responders have been injured and killed as a result of

Figure 6.24 The roll-on method is performed by directing the foam stream at the front edge of the spill.

Figure 6.25 The bank-down method relies on a vertical surface from which to deflect the foam stream.

Figure 6.26 Master stream appliance applying foam in the rain-down method.

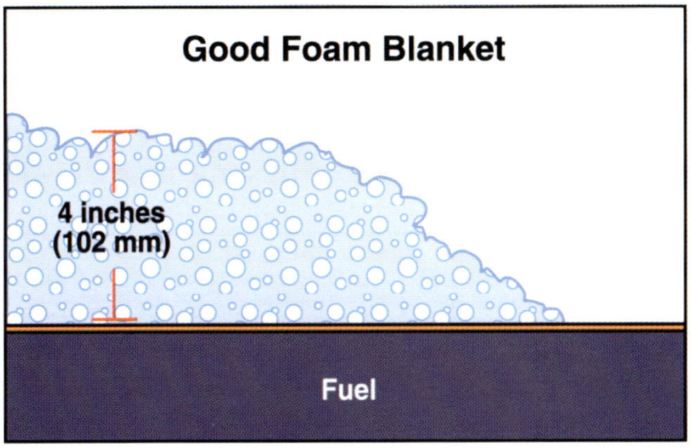

Figure 6.27 A 4-inch (102 mm) foam blanket should be established and maintained over the surface of the spill.

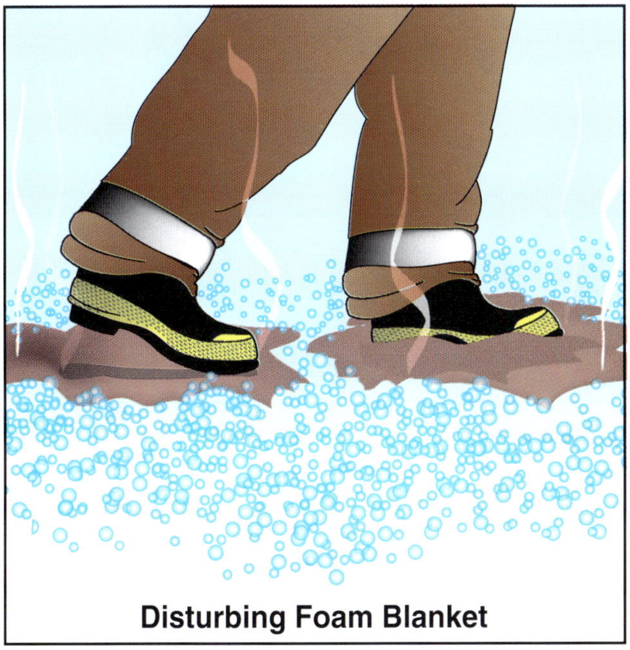

Figure 6.28 Walking in foam blankets creates holes and releases flammable vapors. Avoid this situation if at all possible. If walking in the blanket is necessary, reapply foam quickly.

this type of incident. If it becomes absolutely necessary (such as in a rescue situation) for emergency responders to walk though the foam blanket, apply a continuous stream of finished foam while they are in the blanket. Use a combustible gas detector to monitor any area where flammable vapors are suspected or potentially present. Fire and emergency services responders must be fully protected with full personal protective equipment when performing this monitoring task.

NOTE: The following factors should be taken into consideration for fuel spills: First, road surfaces may be well in excess of 140°F (60°C). Aqueous films may not be effective on hot, pooled hydrocarbon spills when the fuel is in excess of 140°F (60°C). Second, film-forming foams may not be a reliable vapor suppressant when the fuel has soaked into the soil.

CAUTION

The use of self-contained breathing apparatus (SCBA) is critical because many flammable liquids present a toxicity hazard well below their lower flammable range. Personal protective clothing that is designed for structural fire fighting does not provide adequate protection from liquids that may be absorbed through the skin. Hazardous materials protective clothing may be required in these cases.

Reapplication Procedures

Rarely will a single application of finished foam provide adequate vapor protection for the entire duration of an incident. Foam blankets begin to break down or dissolve from the moment they are put into place. The amount of time it takes finished foam to break down is commonly referred to as the *drainage time, drainage rate,* or the *drainage drop-out rate.* The speed at which a foam blanket dissolves depends on the following variables:

- Exposure to heat
- Wind conditions
- Expansion ratio of the foam concentrate
- Aerating equipment or nozzle
- Chemical compatibility between the fuel and the foam concentrate

Determining when it is time to reapply finished foam to the original blanket is, at best, an ill-defined science. Several methods can be used to determine when to reapply finished foam. A fairly reliable method is air sampling of the spill. As the blanket dissolves, flammable vapors are released. The air-monitoring instruments provide an immediate alarm indicating that additional foam is required. On large spills, air sampling may require that a firefighter enter the spill area to make the sample readings. It is important to ensure that the individual is fully protected by personal protective equipment, all sources of ignition are secure, and a charged foam line is ready.

A second method relies on common sense. Obviously, the odor of fuel vapors indicates that there is a leak in the foam blanket. Additionally, when the foam blanket begins to look thin or when holes appear, apply more finished foam to that area. In reality, the filming action of AFFF or FFFP finished foams is probably still covering visible fuel spills. However, applying more finished foam ensures maximum security.

The third method relies on the foam concentrate manufacturer's recommendations for reapplying finished foam. Each foam concentrate has been tested for the amount of time it takes for the finished foam to drain. Most manufacturers recommend reapplying finished foam when 25 percent of the original solution in the blanket has drained (quarter-drain time, 25-percent drainage time, or quarter-life). Most foam manufacturers are able to provide the quarter-life statistics that rely on NFPA foam quality test criteria. Failing all other methods of determining spill scene security, fire and emergency responders should reapply finished foam when the quarter-drain time expires (**Figure 6.29**). In some cases, it will

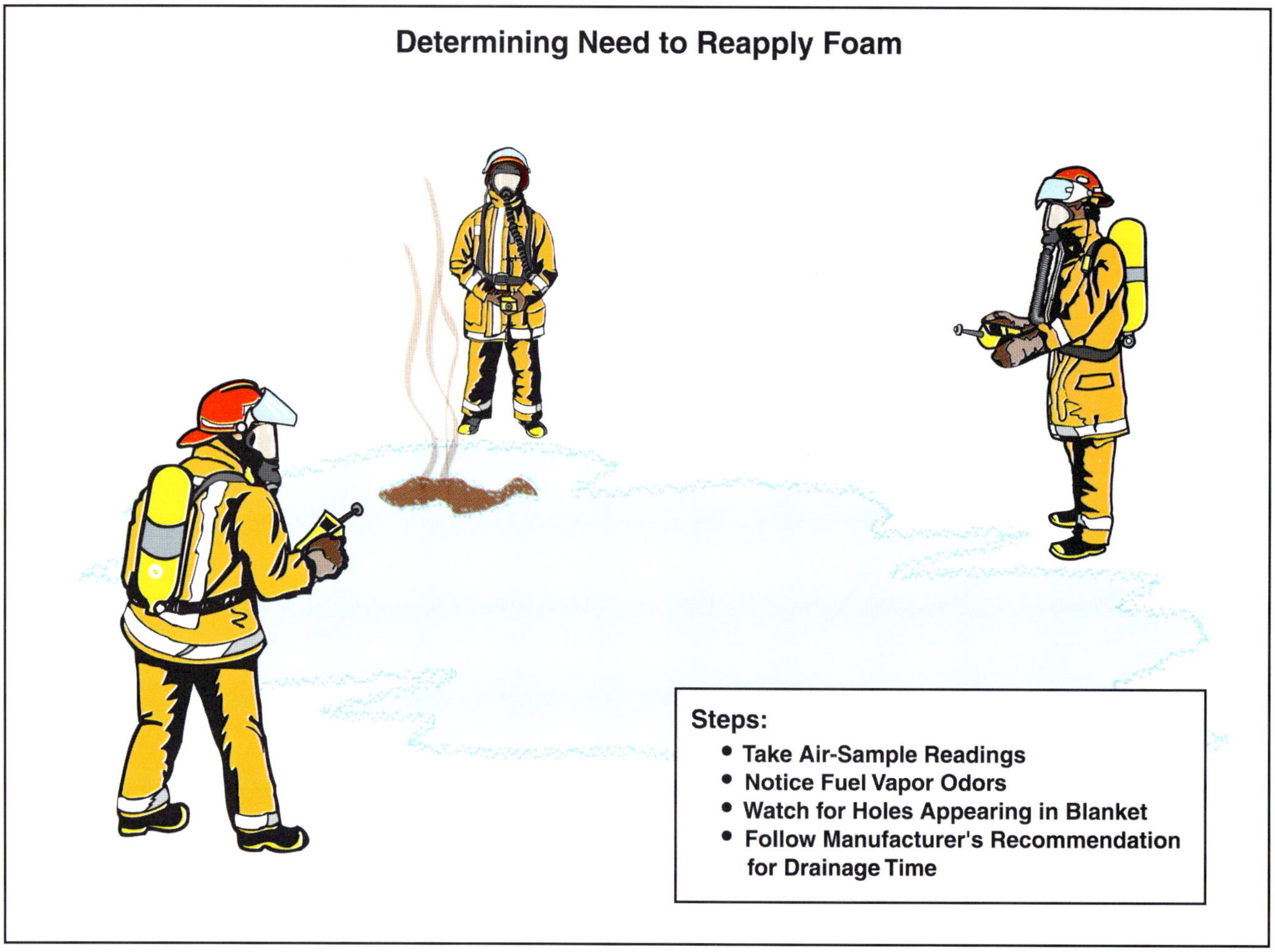

Figure 6.29 This is an illustration of methods for determining the need for reapplying a foam blanket to a spill. Monitor the spill with air-sampling instruments for the release of fuel vapors, watch the surface of the foam blanket for visual signs of holes, and follow the manufacturer's recommendation for reapplication based on the drainage time of the foam.

be obvious that it is time to reapply finished foam before the quarter-drain time has expired. Do not delay reapplication.

Incident Termination

When vapors are suppressed and an incident is stable, fire and emergency services personnel are usually relieved of any further responsibilities for the liquid spill. Hazardous materials technicians (either specially trained members of the emergency services organization or employees of a contractor hired for the purpose) begin the appropriate procedures for product recovery and incident termination. These activities include transfer of the spilled material into storage or transport tanks, disposal of contaminated soil or water and equipment, and other associated activities. These activities are performed in accordance with local standard operating procedures and state/province or federal ordinances for termination of hazardous materials incidents. Usually, fire and emergency services personnel are not responsible for performing this portion of the incident. However, fire and emergency services personnel may be required to remain on the scene and assist, stand by, or apply more finished foam if needed (**Figure 6.30**).

Figure 6.30 During the cleanup operation, fire and emergency services personnel may be required to remain on site in the event of an accidental ignition of the spilled liquid. *Courtesy of Martin Grube.*

The incident is not over until fire and emergency services personnel and their equipment are ready to return to service. This situation does not happen until personnel, protective clothing, hoses, pumps, proportioners, eductors, nozzles, and other equipment are properly decontaminated. Personnel must follow the equipment and foam manufacturer's recommendations for decontaminating the equipment and the environmental requirements for disposing of the cleaning solutions or water.

Finally, foam concentrate tanks on apparatus must be refilled to avoid sloshing and related air entrapment that can reduce proportioning accuracy. Replace empty pails with full containers of concentrate. It is essential that foam concentrate be replaced with the same type of foam. The mixing of Class A, AFFF, and AR-AFFF foam concentrates can cause AR-AFFF to solidify into a rubberlike mass in the tank and to clog proportioning valves, eductors, and nozzles. For more information on unignited Class B liquid spill incident termination, see Fire Protection Publication's *Hazardous Materials: Managing the Incident* manual.

Sample Tank Vehicle Incident

Sometimes tank vehicles are damaged in a collision or rollover that results in a slow leak, the complete failure of a single compartment, or the failure of the entire transport tank (**Figure 6.31, p. 196**). Fire and emergency services personnel must treat these incidents as potential fire situations. Therefore, they need to remember the following guidelines:

- Respond and approach from the upwind and uphill side of the incident.
- Do not drive or walk into vapor clouds.
- Operate from a safe distance — determined by type of hazard (toxicity and flammability), wind direction and velocity, and topography. Approach from upwind and uphill.
- Wear full personal protective equipment.
- Determine the type of tank vehicle and the type and quantity of liquid cargo.

As soon as possible, determine if there is a rescue situation. If so, take actions to rescue trapped persons or protect them someway. Use available foam on the first arriving apparatus to blanket

Figure 6.31 The semitrailer tank truck in this photograph is leaking fuel into an area that is below the level of the roadway. The likelihood of an ignition is lessened by the location of the tank. *Courtesy of Jim Reneau.*

spilled liquid cargo and prevent ignition if possible. Do not enter a flammable liquid area until an adequate foam blanket is covering the entire spill surface area. Even then, only the minimum number of essential personnel should enter hazardous areas. Continually reapply foam to maintain the blanket as needed. If personnel are working in the foam blanket, consider a continual foam application around their immediate area. Use combustible monitoring equipment to determine the effectiveness of the foam blanket **(Figure 6.32)**.

Figure 6.32 Combustible gas monitors (like the units shown here) are essential for use around both unignited and ignited fuel spills.

Eliminate or prevent ignition sources, especially downwind. The vehicle's batteries can be turned off at the disconnect switch. Continually identify the spread and location of flammable vapor concentrations with monitoring equipment. Flammable vapors can travel a considerable distance from a spill as well as collect in low areas.

Contain runoff by diking, damming, or redirecting **(Figure 6.33)**. Establish a berm of sand or other dry absorbent around the vehicle to contain the foam blanket. Prevent spill material from entering storm drain system or waterways. If spilled liquid has already entered the storm drains, it may be possible to contain it in that system. Consult storm drain maps and open street-access covers ahead of the flow of the material. Drop dry absorbents, sand, dirt, or sandbags down the access holes to contain the flow at that location. This procedure may also be performed at the storm sewer outfall (outlet) location. Apply foam into the storm drain to blanket spilled liquid in that location.

Figure 6.33 Spilled fuel can be confined with earthen berms, sand, foam, or absorbents.

During all operations involving a leaking tank vehicle, prepare for the possibility that it may ignite. Have enough resources, equipment, and foam supplies ready to extinguish a tank vehicle fire. Portable and apparatus-mounted high-volume foam monitors or turrets from aircraft rescue and fire fighting (ARFF) vehicles are very effective on spill fires. They can also provide protection for personnel working in hazardous areas.

Removing cargo remaining in the tank vehicle, uprighting the vehicle, and retrieving spilled cargo is the responsibility of the owner of the vehicle or contract services. However, fire and emergency services personnel may be required to remain at the scene to apply foam as long as there is the possibility of ignition of the remaining liquids.

Summary

The use of foam concentrates to mitigate a flammable or combustible fuel spill should be apparent to the experienced fire or emergency services personnel. The same foam concentrate that is used to extinguish fires in Class B fuels can also be used to prevent their ignition through the blanketing and smothering process. Fire and emergency responders must be trained to make rapid and accurate evaluations of an incident and to deploy the correct type of foam concentrate, foam delivery system, and personnel to prevent ignition of a fuel spill. Training and pre-incident planning are essential to developing these types of skills.

Chapter 7

Class B Foam Fire Fighting: Class B Liquids at Fixed Sites

Job Performance Requirements

This chapter provides information that will assist the reader in meeting the following job performance requirements from NFPA 1081, *Standard for Industrial Fire Brigade Member Professional Qualifications,* 2001 edition; NFPA 1003, *Standard for Airport Fire Fighter Professional Qualifications,* 2000 edition; NFPA 472, *Standard for Professional Competence of Responders to Hazardous Materials Incidents,* 2002 edition; and NFPA 1051, *Standard for Wildland Fire Fighter Professional Qualifications,* 2002 edition. Colored portions of the standard are specifically addressed in this chapter.

NFPA 1081

5.3 Site-Specific Requirements. The following job performance requirements shall be considered as site-specific functions of the incipient industrial fire brigade member. The management of the industrial fire brigade shall determine the requirements that are applicable to the incipient industrial fire brigade members operating on their site. The process used to determine the site-specific requirements shall be documented, and the job performance requirements identified shall be added to those identified by this standard.

5.3.1 Attack an incipient stage fire, given a hose line flowing up to 473 L/min (125 gpm), appropriate equipment, and a fire situation, so that the fire is approached safely, exposures are protected, the spread of fire is stopped, agent application is effective, the fire is extinguished, and the area of origin and fire cause evidence are preserved.

(A) Requisite Knowledge. **Types of handlines used for attacking incipient fires, precautions to be followed when advancing hose lines to a fire, observable results that a fire stream has been properly applied,** dangerous building conditions created by fire, principles of exposure protection, and dangers such as exposure to products of combustion resulting from fire condition.

(B) Requisite Skills. The ability to recognize inherent hazards related to the material's configuration; operate handlines; prevent water hammers when shutting down nozzles; open, close, and adjust nozzle flow; advance charged and uncharged hose; extend hose lines; operate hose lines; evaluate and modify water application for maximum penetration; assess patterns for origin determination; and evaluate for complete extinguishment.

5.3.3 Utilize master stream appliances, given an assignment, an extinguishing agent, and a master stream device, so that the agent is applied to the fire as assigned.

(A) Requisite Knowledge. **Safe operation of master stream appliances, uses for master stream appliances, tactics using fixed master stream appliances,** and property conservation.

(B) Requisite Skills. The ability to put into service a fixed master stream appliance, and to evaluate and forecast a fire's growth and development.

6.3.3 Utilize master stream appliances, given an assignment, an extinguishing agent, and a master stream device and supply hose, so that the appliance is set up correctly and the agent is applied as assigned.

(A) Requisite Knowledge. **Correct operation of master stream appliances, uses for master stream appliances, tactics using master stream appliances, selection of the master stream appliance for different fire situations,** the effect of master stream appliances on search and rescue, ventilation procedures, and property conservation.

(B) Requisite Skills. The ability to correctly put in service a master stream appliance and evaluate and forecast a fire's growth and development.

6.3.4 Operating as a member of a team, extinguish an ignitable liquid fire, given an assignment, an attack line, personal protective equipment, a foam proportioning device, a nozzle, foam concentrates, and a water supply, so that the correct type of foam concentrate is selected for the given fuel and conditions, a correctly proportioned foam stream is applied to the surface of the fuel to create and maintain a foam blanket, fire is extinguished, re-ignition is prevented, and team protection is maintained.

(A) Requisite Knowledge. **Methods by which foam prevents or controls a hazard; principles by which foam is generated; causes for poor foam generation and corrective measures; difference between hydrocarbon and polar solvent fuels and the concentrates that work on each; the characteristics, uses, and limitations of firefighting foams; the advantages and disadvantages of using fog nozzles versus foam nozzles for foam application; foam stream application techniques; hazards associated with foam usage; and methods to reduce or avoid hazards.**

(B) Requisite Skills. The ability to prepare a foam concentrate supply for use, assemble foam stream components, master various foam application techniques, and approach and retreat from fires/spills as part of a coordinated team.

7.3.4 Operating as a member of a team, extinguish an ignitable liquid fire, given an assignment, an attack line, personal protective equipment, a foam proportioning device, a nozzle, foam concentrates, and a water supply, so that the correct type of foam concentrate is selected for the given fuel and conditions, a correctly proportioned foam stream is applied to the surface of the fuel to create and maintain a foam blanket, fire is extinguished, re-ignition is prevented, and team protection is maintained.

(A) Requisite Knowledge. **Methods by which foam prevents or controls a hazard; principles by which foam is generated; causes for poor foam generation and corrective measures; difference between hydrocarbon and polar solvent fuels**

and the concentrates that work on each; the characteristics, uses, and limitations of fire-fighting foams; the advantages and disadvantages of using fog nozzles versus foam nozzles for foam application; foam stream application techniques; hazards associated with foam usage; and methods to reduce or avoid hazards.

(B) Requisite Skills. The ability to prepare a foam concentrate supply for use, assemble foam stream components, master various foam application techniques, and approach and retreat from fires/spills as part of a coordinated team.

NFPA 1003

3-3.4 Extinguish a three-dimensional aircraft fuel fire, given PrPPE, an assignment, and ARFF vehicle hand line(s) using primary and secondary agents, so that a dual agent attack is used, the agent is applied using the proper technique, the fire is extinguished, and the fuel source is secured.

(a) Requisite Knowledge: The fire behavior of aircraft fuels in three-dimensional and atomized states, physical properties and characteristics of aircraft fuel, **agent application rates and densities, and methods of controlling fuel sources.**

(b) Requisite Skills: Operate fire streams and apply agents, secure fuel sources.

NFPA 1051

9.5.5 Hazardous Materials. Analyze the potential involvement of various hazardous materials, given incident information and resources, so that hazardous conditions are identified and mitigated.

(A) Requisite Knowledge. A working knowledge of the types of hazardous materials that can be involved and the hazards they can pose to the public, fire-fighting personnel, and the environment; NFPA 472, *Standard for Professional Competence of Responders to Hazardous Materials Incidents,* First Responder level.

(B) Requisite Skills. None required.

NFPA 472

5.1.2.1 The first responder at the operational level shall be able to perform the following tasks:

(1) Analyze a hazardous materials incident to determine the magnitude of the problem in terms of outcomes by completing the following tasks:

(a) Survey the hazardous materials incident to identify the containers and materials involved, determine whether hazardous materials have been released, and evaluate the surrounding conditions.

(b) Collect hazard and response information from MSDS; CHEMTREC/CANUTEC/SETIQ; local, state, and federal authorities; and shipper/manufacturer contacts.

(c) Predict the likely behavior of a material as well as its container.

(d) Estimate the potential harm at a hazardous materials incident.

(2) Plan an initial response within the capabilities and competencies of available personnel, personal protective equipment, and control equipment by completing the following tasks:

(a) **Describe the response objectives for hazardous materials incidents.**

(b) **Describe the defensive options available for a given response objective.**

(c) Determine whether the personal protective equipment provided is appropriate for implementing each defensive option.

(d) Identify the emergency decontamination procedures.

(3) Implement the planned response to favorably change the outcomes consistent with the local emergency response plan and the organization's standard operating procedures by completing the following tasks:

(a) Establish and enforce scene control procedures including control zones, emergency decontamination, and communications.

(b) Initiate an incident management system (IMS) for hazardous materials incidents.

(c) Don, work in, and doff personal protective equipment provided by the authority having jurisdiction.

(d) **Perform defensive control functions identified in the plan of action.**

(4) Evaluate the progress of the actions taken to ensure that the response objectives are being met safely, effectively, and efficiently by completing the following tasks:

(a) Evaluate the status of the defensive actions taken in accomplishing the response objectives.

(b) Communicate the status of the planned response.

5.3.2 Identifying Defensive Options. Given simulated facility and transportation hazardous materials problems, the first responder at the operational level shall identify the defensive options for each response objective and shall meet the following requirements:

(1) Identify the defensive options to accomplish a given response objective.

(2) Identify the purpose for, and the procedures, equipment, and safety precautions used with, each of the following control techniques:

(a) Absorption
(b) Dike, dam, diversion, retention
(c) Dilution
(d) Remote valve shutoff
(e) Vapor dispersion
(f) **Vapor suppression**

5.4.4 Performing Defensive Control Actions. Given a plan of action for a hazardous materials incident within their capabilities, the first responder at the operational level shall demonstrate defensive control actions set out in the plan and shall meet the following related requirements:

(1) Using the type of fire-fighting foam or vapor suppressing agent and foam equipment furnished by the authority having jurisdiction, demonstrate the effective application of the fire-fighting foam(s) or vapor suppressing agent(s) on a spill or fire involving hazardous materials.

(2) **Identify the characteristics and applicability of the following foams:**
(a) **Protein**
(b) **Fluoroprotein**
(c) **Special purpose**
i. **Polar solvent alcohol-resistant concentrates**
ii. **Hazardous materials concentrates**
(d) **Aqueous film-forming foam (AFFF)**
(e) **High expansion**

(3) Given the required tools and equipment, demonstrate how to perform the following defensive control activities:
(a) Absorption
(b) Damming
(c) Diking
(d) Dilution
(e) Diversion
(f) Retention
(g) Vapor dispersion
(h) **Vapor suppression**

(4) Identify the location and describe the use of the mechanical, hydraulic, and air emergency remote shutoff devices as found on cargo tanks.

(5) Describe the objectives and dangers of search and rescue missions at hazardous materials incidents.

(6) Describe methods for controlling the spread of contamination to limit impacts of radioactive materials.

6.2.4.2 The hazardous materials technician shall **identify the impact of the following fire and safety features on the behavior of the products** during an incident at a bulk storage facility and explain their significance in the risk assessment process:

(1) **Fire protection systems**
(2) Monitoring and detection systems
(3) Product spillage and control (impoundment and diking)
(4) Tank spacing
(5) Tank venting and flaring systems

8.3.4.1 Performing Response Options Specified in the Plan of Action. Given an assignment by the incident commander in the employee's individual area of specialization, the private sector specialist employee B shall perform the assigned actions consistent with the organization's emergency response plan and standard operating procedures and shall meet the following related requirements:

(1) Perform assigned tasks consistent with the organization's emergency response plan and standard operating procedures and the available personnel, tools, and equipment (including personal protective equipment), including the following:
(a) **Confinement activities**
(b) Containment activities

9.1.2.2 In addition to being competent at the awareness, operational, and technician levels, the hazardous materials branch officer shall be able to perform the following tasks:

(2) **Plan a response within the capabilities and competencies of available personnel,** personal protective equipment, and **control equipment** by completing the following tasks:
(a) Identify the response objectives for hazardous materials incidents.
(b) **Identify the potential action options (defensive, offensive, and nonintervention) available by response objective.**

9.3.2 Developing a Plan of Action. Given simulated facility and transportation hazardous materials incidents, the hazardous materials branch officer shall develop a plan of action consistent with the local emergency response plan and the organization's standard operating procedures that is within the capability of the available personnel, personal protective equipment, and control equipment and shall meet the following related requirements:

(3) Given the local emergency response plan or the organization's standard operating procedure, identify procedures to accomplish the following tasks:
(e) **Coordinate with fire suppression services as it relates to hazardous materials incidents.**
(f) **Coordinate hazardous materials branch control, containment, or confinement operations.**
(h) Coordinate on-scene decontamination.

9.5.1 Evaluating Progress of the Plan of Action. Given simulated facility and transportation hazardous materials incidents, the hazardous materials branch officer shall evaluate the progress of the plan of action to determine whether the efforts are accomplishing the response objectives and shall meet the following related requirements:

(3) **Determine the effectiveness of the following**:

(a) Hazardous materials response personnel being used

(b) Personal protective equipment

(c) Established control zones

(d) **Control, containment, or confinement operations**

(e) Decontamination process

Chapter 7
Class B Foam Fire Fighting: Class B Liquids at Fixed Sites

Only a small percentage of incidents involving flammable and combustible liquids actually involve ignited and burning fuel (Class B fires). Most incidents involving the spill or release of hazardous materials are mitigated without ignition; therefore, these incidents attract little media or public attention (see Chapter 6, Class B Foam Use: Unignited Class B Liquid Spills). However, those rare incidents that involve either ignited flammable or combustible fuels or other hazardous materials are quite spectacular (**Figure 7.1**). The inherent volatile properties of these fuels produce fire conditions with large quantities of flame and smoke; therefore, these fires have a great impact on the community and attract much attention from both the media and public. These fires also pose problems for the fire and emergency services.

Because these fires are not common in most jurisdictions, most fire and emergency services are not prepared to combat sizable Class B fires. They may lack the training, equipment, or past experience that is necessary to handle such an incident. In addition, U.S. Environmental Protection Agency (EPA) regulations and state/province clean air and water resources laws often prohibit many fire and emergency services organizations and jurisdictions

Figure 7.1 Flammable or combustible fuel fires can be very dramatic and radiate high levels of heat.

from conducting extensive (or even adequate) Class B foam fire fighting training exercises on a regular basis. The cost of foam concentrates and flammable and combustible liquids used in training also prevents regular exercises in some cases.

Because Class B fires usually occur outdoors and involve liquids that generate extensive amounts of heat, flame, and toxic smoke, the principles associated with Class B fire attack are quite different from those used for the more common Class A structure, wildland, or exterior material-concentration storage fires (see Chapter 5, Class A Foam Fire Fighting). Therefore, all fire and emergency services personnel must be familiar with the principles of Class B fire attacks, use of foam concentrates, and use of foam application equipment. This knowledge assists them in making the appropriate decisions that result in an effective attack on Class B liquid fires.

Like the unignited flammable and combustible spills discussed in Chapter 6, Class B Foam Use: Unignited Class B Liquid Spills, incidents involving ignited flammable and combustible fuels or other hazardous materials can occur at any time of the day or night at any location. Chapter 8, Class B Foam Fire Fighting: Transportation Incidents, discusses transportation incidents that occur on roadways (involving motor vehicles and transport trailers), on railroad right-of-ways (involving tank cars), on waterways (involving boats, barges, and ships), with pipelines, or with aircraft both on and off airport property. This chapter focuses on the general uses of foam handlines and monitors at all flammable and combustible fuel incidents and specifically addresses the extinguishment of fires at fixed-site locations such as refineries, fuel storage tanks, loading racks, service/filling stations, and electrical transformer vaults. The topics discussed include basic decisions that fire and emergency services personnel must make early in any Class B fire incident and the considerations that affect those decisions, tactical priorities for all Class B fires, safety issues, basic foam application techniques, and specific information for some of the more common Class B fires in fixed-site facilities (fuel facilities, storage tanks, three-dimensional/pressurized fuel situations, and electrical transformer vaults).

Decision-Making Considerations

[NFPA 472: 5.4.4 (2) (3h), 6.2.4.2 (1), 9.1.2.2. (2b), 9.3.2 (3e & f), 9.5.1 (3d); NFPA 1003: 3.3.4; NFPA 1051: 9.5.5; NFPA 1081: 5.3.1, 6.3.4, 7.3.4]

At the beginning of a Class B fire incident, the incident commander (IC) on the scene has to make a crucial strategic decision. Does the fire and emergency services organization have the ability to extinguish the fire (through an offensive attack) or should it contain the incident (through a defensive attack) and allow the fire to consume the fuel (burn itself out) **(Figure 7.2)**? The outcome of this decision determines the tactics used for the duration of the incident. Eight main areas are considered in this decision-making process. When fire and emergency services personnel have determined the size of the fire, identified the type of fuel involved and its depth, calculated the appropriate application rate and foam concentrate amount required, inventoried the available concentrate quantity, calculated the correct delivery rate, and determined the amount of water available and the means of distributing it, the IC can develop and implement the rest of the strategy and tactics. The eight areas are as follows:

- Size of fire
- Type of fuel
- Depth of the fuel
- Required application rate for appropriate foam concentrate
- Amount of foam concentrate available on the scene and the ability to maintain enough supply to extinguish a fire
- Ability to deliver and sustain the required application flow rate
- Availability of an adequate water supply
- Availability of water distribution systems

Size of Fire

A standard Class A structure fire typically begins in a small portion of a building and continues to spread if it is not extinguished. However, Class B fires almost always involve the entire surface of an exposed fuel from the beginning, and they remain

Figure 7.2 It may be necessary to allow some flammable or combustible liquids fires to consume all of the available fuel.

that size throughout the duration of an incident. For example, fire and emergency services personnel arriving at the scene of a storage tank fire involving the entire surface of a fuel will probably face a fire of the same magnitude hours later unless the fuel is completely consumed or the fire is extinguished. It is likely that the fire will not grow larger unless either the storage tank fails (allowing the contents to spread) or other exposures such as tanks, structures, or piping systems become involved. Thus, the size of a fire may remain fairly constant throughout the course of an incident.

Determining the size of a fire involves either simple mathematical calculations or an accurate estimation of the exposed fuel area. In the case of a fire in a circular tank, the fire size is calculated by determining the area of the top of the tank. Use either of the following formulas to determine the area of a circle:

$$A = \frac{\pi d^2}{4} \qquad (1)$$

where: A = *area in square feet (square meters)*

π = *the constant pi* = *3.1416*

d = *diameter of the tank in feet (meters)*

OR

$$A = (0.8)(d^2) \qquad (2)$$

where: A = *area in square feet (square meters)*

0.8 = *a constant*

d = *diameter of the tank in feet (meters)*

A slight difference in the results occurs depending on which formula is used. This difference is not significant to the overall outcome of the incident. If the fire involves either a defined square or rectangular area such as a dike area surrounding a fuel loading rack, the area can be determined by multiplying the length and width of the dike area. For example, a dike area that measures 15 feet by 40 feet (5 m by 13 m) contains 600 square feet (65 m^2). A fire in an area that is not clearly defined in size requires an accurate estimation by either fire and emergency services personnel or the IC on the scene. One way is to estimate the dimensions of the greatest width and length of the fire area and develop a fire flow calculation based on a rectangle of that size (see Figure 6.21, Chapter 6, Class B Foam Use: Unignited Class B Liquid Spills). This size estimation results in a figure that slightly exceeds what is actually required; however, it is better to err on the side of excess in this case.

Type of Fuel

The next consideration is to determine the type of liquid fuel involved in the fire. The identity of the fuel may be determined from the sources mentioned in Chapter 6, Class B Foam Use: Unignited Class B Liquid Spills. Another common method for identifying the type of fuel involved in an incident is to observe the physical characteristics of the fire. If the flames are reddish orange and the fire is releasing large quantities of black smoke, it is most likely a hydrocarbon fuel. A polar solvent fuel might be involved if the flames have a clean blue color or are not visible to the naked eye or they are producing very little smoke. An odor that is sweet, fruity, or pungent coming from the fuel is also an indicator of a polar solvent fuel, and the odor can also indicate a specific type of polar solvent because each emits its own fragrance. The product material safety data sheet (MSDS) usually provides clues to the type of odor to expect with various polar solvents.

If the fuel type cannot be identified, the IC will have to make the decision to implement either an offensive or defensive mode to control the incident. Deploy the first foam handlines using the available concentrate that is on hand. Apply foam to the nearest edge of the spill or fire and observe the results for effectiveness. If the results are satisfactory, continue application until the foam blanket is complete and the spill or fire is controlled

(**Figure 7.3**). If the foam does not appear to be having a satisfactory effect, then take a defensive mode, protect exposures, and allow the fire to burn out. If other types of foam concentrate are available from other sources, they may be applied to determine effectiveness.

When the identity of the fuel is established, the IC can determine the important properties of the fuel that impact the strategy used to control the incident. Knowledge of fuel properties permits the IC to select the appropriate type of foam to control the situation. These fuel property concerns are as follows:

- ***Whether the liquid is a hydrocarbon or polar solvent*** — By knowing this information, the IC can determine if the foam concentrate at the scene is compatible with the particular fuel. Even if compatible foam concentrate is available, the type of fuel also affects whether the proper proportioning rate can be established.
- ***Whether or not the liquid is toxic*** — Some incidents occur where it is safer to let toxic fuels burn until they are consumed. This strategy reduces the hazard level of the cleanup operations and also eliminates the chance of spreading contamination as a result of runoff from the fire attack. Generally, technical information provided from emergency information sources such as CHEMTREC® or CANUTEC tells the IC when a nonintervention approach is best.
- ***Whether or not a fuel is water-reactive*** — A burning liquid that is water-reactive could have violent, explosive results from any attack using foam concentrates or plain water. If the reactivity of the fuel were uncertain through standard means of material identification, the approach would be to make a gradual application of water or foam from a safe distance, starting from the edge of the fire/spill. Should a reaction become obvious, the fire stream could be immediately moved from the fire/spill. Although fires involving combustible liquids can be attacked using plain water, IFSTA recommends the use of foam on all Class B liquid fires.

Depth of Fuel

Application rates vary according to the depth of the liquid fuel that is burning. NFPA 11, *Standard for Low-, Medium-, and High-Expansion Foam* (2002), recommends an application rate for a spill fire based on the type of foam and the anticipated spill hazard. According to NFPA 11, spills are considered to

Figure 7.3 If the type of fuel is unknown, make the initial attack with the foam that is available from upwind as shown in this photograph. If the results are satisfactory, apply additional foam until the fire is extinguished.

be less than 1 inch (25 mm) in depth, and most application rates are based on this definition. When depths exceed 1 inch (25 mm), application rates are based on the type of containment area in addition to the types of foam and hazard. For example, spills from rail tank cars or tank vehicles that result in a pool of liquid that is greater than 1 inch (25 mm) require application rates normally required for fixed roof storage tanks.

Required Application Rate

Once the IC has established the size of the fire, the type of fuel and its depth, the application (flow) rate required to extinguish the fire is determined. The flow rate is important when manually applying finished foam through handlines or monitors. Foam application rates through fixed distribution systems are part of the system's design and may not require anyone to make flow-rate calculations during a fire.

The foam concentrate manufacturer along with NFPA 11 and NFPA 11A, *Standard for Medium- and High-Expansion Foam Systems* (1999), provide information on the required application rates for foam concentrates. Each manufacturer usually has a specific application rate for its concentrate, which is based on the type of fuel fire. Whenever possible, fire and emergency services personnel should use the application rate specified by the manufacturer. Consult NFPA 11 for general application rates based on the fuel type involved if fire and emergency services personnel are not certain what application rate the manufacturer specifies for its foam concentrate. It is important to remember that NFPA application rates are listed as minimums. The application rates and times recommended by NFPA 11 for specific types of foam concentrate and fuel hazard type are as follows:

- *Concentrates* — Aqueous film forming foam (AFFF), film forming fluoroprotein (FFFP) foam, alcohol-resistant aqueous film forming foam (AR-AFFF), and alcohol-resistant film forming fluoroprotein (AR-FFFP) foam
  - *Fuel hazard:* Hydrocarbon spills
  - *Application rate and time:* 0.1 gpm per square foot (ft^2) or 4.1 (L/min)/m^2 with a recommended continuous application time of 15 minutes
- *Concentrates* — Protein and fluoroprotein foams
  - *Fuel hazard:* Hydrocarbon spills
  - *Application rate and time:* 0.16 gpm/ft^2 or 6.5 (L/min)/m^2 with a recommended continuous application time of 15 minutes
- *Concentrates* — Alcohol-resistant foams
  - *Fuel hazard:* Polar solvents and other flammable/combustible liquids
  - *Application rate and time:* Foam manufacturer's rate recommendation with a continuous application time of 15 minutes

These application rates are usually sufficient for spill fires less than 1 inch (25 mm) in depth (for example, aircraft rescue and fire fighting incidents), but the fuel depths encountered at incidents involving storage containers, pipelines, tank vehicles, and rail tank cars are much higher. When fuel depths exceed 1 inch (25 mm), the most applicable NFPA recommended application rate is found in NFPA 11, Table 5.2.4.2.2 (Foam Handline and Monitor Protection for Fixed-Roof Storage Tanks Containing Hydrocarbons). This recommended rate is 0.16 gpm/ft^2 or 6.5 (L/min)/m^2 with a minimum discharge time of 65 minutes for crude oil and hydrocarbons with flash points below 100°F (38°C). For hydrocarbons with flash points between 100 and 140°F (38°C and 60°C), the recommended discharge time is 50 minutes. Application rates for polar solvent spills are not included in NFPA 11. Most reference sources, including NFPA, advise the reader to consult the manufacturer for listings on specific products.

Required Foam Concentrate Amount

When the required application rate has been calculated, fire and emergency services personnel can then determine whether enough foam concentrate is on the scene to extinguish a fire or if more is needed. The discharge time, required application flow rate, and proportioning percentage of the foam concentrate are combined to calculate the total amount of concentrate required to extinguish a fire. Use the following formula:

$$\textit{Discharge Time} \times \textit{Flow Rate} \times \textit{Proportioning Rate} = \textit{Foam Concentrate Amount} \qquad (3)$$

Use the following steps to determine if enough foam concentrate is on the scene to extinguish the fire:

Step 1: Calculate the total amount of foam concentrate needed to extinguish the fire.

Step 2: Take an inventory of the foam concentrate on the scene to see if it equals or exceeds the calculated need.

After inventorying the foam concentrate available on the scene, it will become evident whether enough is available to successfully extinguish the fire and prevent reignition. When not enough concentrate is on the scene, employ defensive measures until enough foam concentrate can be delivered or until the fuel is consumed. Additional foam concentrate may be delivered from mutual aid agencies, local fire equipment vendors, or foam concentrate manufacturers (in bulk) (**Figure 7.4**). In the event that it is necessary to contact these sources, emergency phone numbers are available from dispatch/telecommunication centers.

The exact concentrate amount calculated is often inadequate for total extinguishment because of ineffective application techniques, environmental conditions, and other factors. At least twice the calculated amount of foam concentrate needs to be on the scene before beginning a fire attack. However, some experts recommend that three to four times the calculated amount of foam concentrate be available. If the foam concentrate supply is exhausted before a fire is extinguished, the fire will eventually return to its original size and intensity, thus wasting the initial concentrate. After a fire is extinguished, additional foam concentrate is required to keep an effective foam blanket on the fuel to prevent reignition during the incident-termination phase. It is important to have enough concentrate to maintain the vapor-suppressing foam blanket until normal conditions can be restored after a fire is extinguished.

Figure 7.4 Foam concentrate tankers can be used to provide an uninterrupted supply to fire apparatus that are delivering foam to a fire. *Courtesy of Conoco, Inc.*

Required Application Rate Delivery

An adequate amount of foam concentrate on the scene will not sufficiently extinguish a fire if it cannot be applied at the necessary application or flow rate. In an incident involving a ruptured pipe/spill fire, an application rate of 600 gpm (2 255 L/min) of foam concentrate might be required for successful extinguishment. If the required 540 gallons (2 030 L) of 6-percent FFFP foam concentrate were on the scene and the units attacking the fire were equipped with only one or two 95-gpm (360 L/min) foam eductors and nozzles, they would not be capable of flowing the minimum required application rate. The fire attack units need to be able to proportion and discharge the required flow rate. If they cannot, it will be necessary to alter the strategy to a defensive attack.

It is easy to see that any sizable Class B flammable or combustible liquid fire probably requires high-capacity foam proportioning and delivery equipment beyond the capabilities of the equipment carried on most municipal or industrial fire brigade apparatus. This situation is particularly the case when facing a fire in a large (100 foot [30 m] or larger) fuel storage tank. Special proportioning and delivery equipment is available that flows up to 10,000 gpm (37 854 L/min) from a single master stream device (**Figure 7.5**). This equipment is typically brought to the scene and operated by private contractors who specialize in handling fires of this magnitude.

CAUTION

Before beginning a fire attack, ensure that the equipment on the scene can deliver the required application rate and that enough foam concentrate is available to extinguish the fire and maintain the vapor-suppressing foam blanket until normal conditions are restored.

Figure 7.5 High-capacity foam monitors such as this one can apply sufficient foam to control large-sized spill fires.

Adequate Water Supply

The need for an adequate water supply to mix with the foam concentrate for attacking Class B fires is no different from the water supply requirements for standard structure fires. If the water supply available at the scene is less than the calculated application rate for the present fire conditions, an offensive attack will not be successful. If an adequate water supply is not available, a defensive attack must be used until either the fire consumes the fuel or decreases in size to the point where the water supply is sufficient for extinguishing the remaining fire.

Most fixed-site facilities that are classified as Class B hazards such as refineries and tank farms have on-site water supply systems. These water supply systems are capable of flowing adequate amounts of water for major fires that may occur at the facility. Many facilities also have large static water supply sources that can be used if the need arises (**Figure 7.6**). Given an adequate water supply, fire and emergency service personnel must have the necessary equipment and use the appropriate tactics to deliver the water from the supply system to the fire. Typically, large-capacity fire pumps and large-diameter hose are used in these operations. Many industrial fire apparatus have pump capacities as high as 3,000 gpm (11 356 L/min). Some facilities maintain trailer-mounted pumps that can be used to augment standard fire apparatus during large-scale operations (**Figure 7.7**).

Figure 7.6 Most flammable or combustible liquid fixed-site facilities maintain water supplies required for fire suppression and control. In this case, water is being aerated to prevent it from becoming stagnant.

Figure 7.7 Type III portable foam systems may include high-capacity pump-mounted trailers.

Whether a flammable or combustible liquid fire is going to occur in an area where either an adequate municipal water supply system or a large static water source accessible to fire and emergency services personnel is available is unknown. In most cases, it is not realistic to think that the amount of water needed to extinguish these large fires can be sustained by a water-shuttle operation. Therefore, water availability is another factor in making the decision on whether to use a defensive or nonintervention approach to an incident. In defensive situations, use the amount of water that is available to protect exposures and allow the fire to consume the fuel.

Water Distribution Systems

As learned earlier, fires that occur in fixed-site facilities such as refineries or liquid fuel storage facilities usually have either water supply systems or static sources on site that are capable of meeting the demand for major fire-fighting operations. The challenge for fire and emergency services personnel in these situations is to establish distribution systems that move water from the supply sources to the points of attack in sufficient quantity to extinguish the fires. Large-capacity fire pumps and large diameter hoselines can supply adequate amounts of water in these situations.

Fire Pumps

Fire pumps are mounted on fire apparatus or pump trailers or are located in fixed fire pump houses located around fixed-site facilities. Most municipal fire apparatus are equipped with fire pumps in the 1,000- to 1,500-gpm (3 785 L/min to 5 678 L/min) range, although some may have pumps up to 2,000 gpm (7 571 L/min). Industrial fire apparatus may be equipped with pumps with capacities similar to that of municipal apparatus. However, some industrial pumpers have pumps that can deliver up to 3,000 gpm (11 356 L/min). Trailer-mounted pumps driven by diesel motors are commonly found in capacities up to 4,000 gpm (15 142 L/min). Diesel or electrically driven pumps are contained in fixed fire pump houses around fixed-site facilities and have capacities similar to those of trailer-mounted pumps (**Figure 7.8**).

Obviously, fires that require flows in excess of the capacity of the largest single pump at the scene also require multiple pumps. If a fire scene is more than about 1,500 feet (457 m) from a water supply or is uphill from it, additional fire pumps are needed for relay pumping operations. The operation of water supply pumps for the fire and emergency services is detailed in the IFSTA **Pumping Apparatus Driver/Operator Handbook.**

Large Diameter Hoselines

While small (1¾-, 2½-, and 3-inch [45 mm, 65 mm, and 77 mm]) hoselines are used as attack lines for Class B fuel fires, any supply hose less than 4 inches (102 mm) in diameter will probably prove totally inadequate when attacking a large Class B fire. When possible, use only 5-inch (127 mm) or larger supply hose. Some industrial fire departments have custom-made hose that is 10 inches (254 mm) in diameter **(Figure 7.9).** Generally, each 5-inch (127 mm) hoseline can supply about 1,250 gpm (4 732 L/min) up to 1,000 feet (305 m) in distance. Demands above this figure obviously require either relay pumping or the laying of additional hoselines.

Tactical Priorities

[NFPA 472: 5.1.2.1 (2a & b; 3d), 8.3.4.1 (1a), 9.3.2 (3e & f); NFPA 1081: 5.3.1, 6.3.4, 7.3.4]

The prioritization model (regardless of which one is used) must be strictly followed if safe, effective, and efficient outcomes are desired. The *RECEO* model (rescue, exposures, confine, extinguish, and overhaul) applies just as well to ignited Class B flammable and combustible liquids as it does to unignited Class B fuels and Class A incidents. See Chapter 5, Class A Foam Fire Fighting, for more information on the *RECEO* model.

Figure 7.8 Large fixed-site facilities have permanent fire-suppression systems that include water and foam distribution pumps like the one shown. *Courtesy of Conoco, Inc.*

Figure 7.9 Large diameter supply hose, such as this 10-inch (254 mm) diameter hose, must be transported on trailer-mounted reels. Once the hose is in position, it is not moved until the incident is over. *Courtesy of Sam Goldwater.*

Rescue

Most fire and emergency services personnel normally associate the rescue function of their duties to incidents involving structure fires. However, Class B flammable and combustible liquid fuel fires may also require the rescue of trapped workers at industrial sites or passengers involved in aircraft-related incidents, and these incidents involve a level of risk to emergency personnel. Some significant differences exist between performing rescues from a Class A structure fire and those from a Class B fire situation. The main difference is that in a structure fire situation, the position of the victim(s) is not always readily apparent, which is why the function is often referred to as *search and rescue*. Fire and emergency services personnel must go through a methodical process to locate victims and remove them to a place of safety. It is generally assumed that this process takes a considerable amount of time to complete.

In Class B fire situations, the scenario is often quite different. In many cases, victims are in plain view of fire and emergency services personnel. An example would be workers who are trapped with burning liquid between them and the only avenue of escape. This immediate visual contact may create a desire on the part of the fire and emergency services personnel to attempt an immediate rescue. However, without conducting a quick and proper size-up and product identification before taking action, emergency personnel run the risk of also becoming victims.

If a fire is small and controllable, it is always most desirable to knock down or extinguish it in the area where victims are located before attempting a rescue. At the very least, use protective hoselines to push fire away from the rescue area and provide a barrier against excessive heat.

Fire and emergency services personnel who are engaged in fire attacks or rescue on Class B fires must wear appropriate personal protective equipment (PPE) for the type of incident involved. All personnel who are working within the hot zone must wear full PPE, including self-contained breathing apparatus (SCBA). Fire and emergency services personnel who face these types of fires on a frequent basis may be provided with special turnout clothing (proximity clothing) that has an aluminized outer shell that reflects higher levels of radiant heat than normal structural fire-fighting clothing (**Figure 7.10**). Proximity clothing should not be confused with entry-level clothing that is designed to protect against direct flame contact. Limitations of all types of personal protective clothing are always emphasized during training classes.

Figure 7.10 Proximity clothing provides greater protection against high radiant heat than does regular structural protective clothing.

Exposures

Exposure protection is covering any object in the immediate vicinity of a fire with water or other extinguishing agent. The principles used to protect exposures at flammable and combustible liquids fires are not that different from the principles used to protect exposures at typical structural fires. No specific rules exist for determining when to provide protection to an exposure. Exposure prioritization is usually necessary in the early stages of a response when on-scene assets are at a minimum (**Figure 7.11**). During this period of the response, the IC may have to sacrifice low-priority exposures in order to prevent the failure of those where the results would be catastrophic. For example, during the first hours following ignition of an atmospheric flammable liquid storage tank, an office building might need to be sacrificed in order to prevent a BLEVE (boiling liquid expanding vapor explosion) of an LPG (liquefied petroleum gas) vessel that would destroy everything in a square-mile area (3 km^2). All available resources would be directed onto the tank exposure instead of the building exposure. Exposure protection becomes necessary in any of the following situations:

Figure 7.11 Fixed-site master streams can be used to protect exposures (such as the fixed deluge sprinkler system on this tank), while handlines are used to extinguish the seal fire on top of this open roof tank.

- A large fire cannot be immediately extinguished because of a lack of resources (personnel, foam concentrate, water, apparatus, or equipment) on the scene.
- Fire and emergency services personnel are employing a nonintervention strategy and allowing a fire to burn until the fuel is totally consumed.
- Extremely sensitive structures such as other storage tanks, explosives storage, etc. are located in the vicinity of a fire.
- Structures in the vicinity of a fire are beginning to show the effects of radiant heat such as smoking/steaming, discoloration, melting or distortion of vinyl siding, etc.

Direct flame contact is the primary threat to exposures such as piping, foam distribution systems, pumps, and tanks during storage tank and/or dike fires (**Figure 7.12**). Direct the first cooling streams to protect fixed foam application equipment on storage tanks (if they are still serviceable and have not been damaged by fire) and foam distribution piping in the dike. If needed, direct the next stream at any other exposures, including adjacent tanks, manifolds, and piping. A test to determine if these exposures require cooling water is to direct water onto them to see if steam results. If the exposure surface remains wet after the test, then further water appli-

Figure 7.12 Direct flame contact can cause distribution piping to fail and add fuel to the original fire. Direct water streams at the exposed piping to cool it and prevent its failure while directing foam streams at the main body of the fire to extinguish it.

cation is unnecessary. Class A foam concentrates can provide a thick foam blanket for exposure protection on the sides of tanks and structures.

Radiant heat contact occurs from the bombardment of electromagnetic energy waves against the surface of an exposure and is a threat. This contact results in heating of the exposed surface. If left unchecked, the surface materials may eventually heat to their ignition temperature and ignite. At one time, it was thought that this process could be slowed or stopped by placing a curtain of water between the fire and the exposure. However, research has shown that electromagnetic waves that cause radiant heat can travel through a water curtain, glass, or any other opaque material. At best, the water curtain only slightly slows the heating of the exposure; it does not completely stop the heating process.

While some fire service professionals advocate a general application rate formula of 1 gpm (4 L/min) of cooling water for every 10 square feet (1 m^2) of exposure surface area, the most widely accepted method of assessing exposure-protection requirements includes the following procedures:

- ***Size up exposures, including the assessment of the risk to responders*** — In this process, only those hazards that are threatened at the time of assessment (or that will be in the immediate future) are candidates for protection. How much protection is necessary depends on the hazard, its proximity to the fire, and its physical configuration. Examples:
  - — An aboveground flammable liquid storage tank located 120 feet (37 m) from a fire might only require intermittent cooling water applications. Water would be applied using narrow-fog patterns capable of reaching the uppermost areas exposed. This procedure results in the water continuously cooling the tank as it cascades down the shell of the tank.
  - — A compressed gas tank encountering flame impingement from a pressurized liquid fire would require immediate and continuous water application to redirect the flames from the exposure while cooling the tank.
- ***Assess water-delivery capabilities available at the time*** — Consider water that is immediately available from apparatus tanks, fixed water supply systems, or water storage areas. Do not consider what might be available later.
- ***Ensure that assets are sufficient to provide adequate exposure protection and they can be deployed in time*** — At large, prolonged incidents, assets committed to exposure protection may never be available for other tasks.

Closely monitor exposures and cooling water used for exposure protection. Failure to continuously monitor exposures and the need for cooling applications can result in the following problems:

- Overapplication of cooling streams into a diked area will cause the dike to overflow. Because hydrocarbons are lighter than water and float on the surface, they will overflow first and spread the spill or fire.
- Exposure hoselines left unattended can create "swamplike" conditions in an earthen dike, making it difficult for fire and emergency personnel to enter and close valves, etc.
- Water unnecessarily committed to exposure protection can become unavailable when extinguishment is being considered.
- Contaminated water soaking into the soil or flowing into area water supplies can have a negative impact on the environment.

It is an established fact that the most effective way to protect an exposure is to place water or finished foam directly on its surface **(Figure 7.13)**. The water or finished foam absorbs the heat created by the bombardment of the electromagnetic waves. As long as enough water is placed on the exposure to

Figure 7.13 Class A foam can be effectively used to protect exposures.

absorb the heat created by the fire or the foam blanket remains intact, the exposure is protected. Two considerations that must be made by the incident commander regarding exposure protection are the severity of the danger to the exposure and the quantity of water that will be required to protect it.

Danger Severity Factors

How severe the danger to an exposure will be is determined by the following factors:

- ***Amount and intensity of the original fire*** — The larger and more intense the original fire, the larger the number of electromagnetic waves it releases. This larger number results in faster heating of exposed surfaces. Large fires also have serious impacts on fire attack teams' ability to get close enough to set up effective exposure protection. In severe cases, fire and emergency services personnel must advance under the cover of protective fog streams to place exposure protection hoselines and monitors. In these cases, the best way to protect an exposure is to knock down the main body of fire. Even in situations where the ultimate strategy is to allow the fire to consume the fuel, it may be necessary to slightly knock down the main body of fire to reduce the risk to exposures.
- ***Proximity of the exposure to a fire*** — The closer an exposure is to a fire, the greater the level of risk to emergency personnel who must establish the exposure protection. When the IC is determining which exposure to protect first, the exposure that is closest to the fire is usually chosen. In cases where an exposure is movable (such as a tank truck or railcar), the most desirable option might be to move it away from the fire to increase the distance between it and the fire.
- ***Environmental conditions*** — Wind, rain, snow, and extreme temperatures can play a significant role in determining the severity of danger to an exposure. However, do not overestimate weather affects; anything short of a heavy, driving rainstorm most likely is insufficient to provide adequate protection for moderately to severely exposed structures. Fire and emergency services personnel would still need to employ manual exposure protection tactics. Environmental factors:
  - — Radiant heat waves travel against the wind with little resistance. A wind blowing away from an exposure only slightly slows radiant heat impact. However, a wind blowing away from an exposure causes a significant reduction in problems caused by convected heat.
  - — Wind helps spread fire onto exposures in its path **(Figure 7.14)**. It can also affect finished foam distribution and scatter foam streams.
  - — Rain and snow play a beneficial role in the protection of exposures. Moisture added to the surface of exposures by precipitation slows any fire development caused by either radiated or convected heat.
  - — High heat raises the temperature of the exposed material or fuel closer to its ignition temperature making it easier to ignite. Extremely cold temperatures can affect the use of water and foam solution through freezing.
- ***Exposure's exterior composition*** — A building constructed of wood shingles is more hazardous than a tilt-slab concrete building **(Figure 7.15)**. Even if a structure has a generally noncombustible exterior, do not overlook such building features as doors, windowsills, and other items made of wood or similar combustible materials. In addition, metal construction materials can weaken and fail when exposed to high temperatures or direct flame contact.

Water Supply Calculations

Exposure protection usually involves a significant flow of water and is included in water supply calculations performed during pre-incident planning. Failure to consider this factor could result in insufficient water to both make a fire attack and provide adequate exposure protection. Facility specialists can usually provide information to fire and emergency services personnel on how much water is necessary to adequately protect exposures in the facility. Unless otherwise specified, a general guideline for providing adequate water on a spill-fire exposure is 1 gpm for every 10 square feet (or 4 L/min per 1 m^2). Another way to look at it would be 0.1 gpm per square foot (4.1 L/min per m^2) of exposed area. Fire and emergency services personnel can easily calculate the exposure area by multiply-

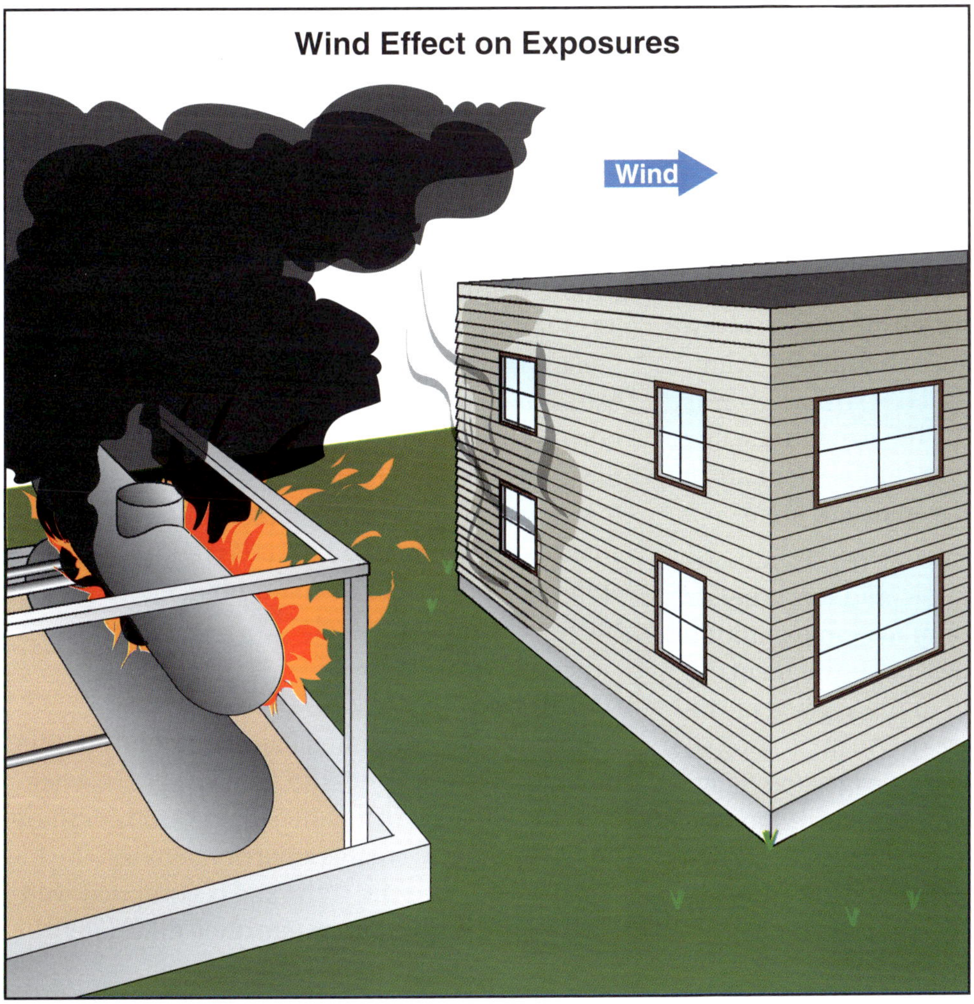

Figure 7.14 Wind can push flames and radiant heat into nearby exposures, heating them to the point of combustion.

Figure 7.15 The exterior materials of structures (such as these shingle-covered apartments) can directly influence the hazard posed to that structure.

ing the height of the exposed structure or tank by its width. Dividing this number by 10 in English units (or multiplying this number by 4 in SI units) gives the flow of water needed to protect the exposure. See the textbox for water supply calculation formulas in both English and SI units.

Water Supply Calculation Formulas

English Units

$$(L \times W)/10 = gpm \qquad (4)$$

where: *L = length in feet*

W = width in feet

Constant = 10

SI Units

$$(L \times W) \times 4 = L/min \qquad (5)$$

where: *L = length in meters*

W = width in meters

Constant = 4

Confine

At flammable and combustible liquid fire incidents, exposure protection and the confine step (or confinement) can be quite similar activities. The process of limiting the size and spread of a fire to the smallest area possible usually serves to protect exposures. Confinement may require redirecting (damming and diking) the flowing liquid fire, then maintaining a foam blanket. The size of the blanket is dictated by the following factors:

- ***Amount of foam available after the higher priorities of rescue and exposure protection have been addressed*** — The prioritization of confinement needs may require eliminating some exposure candidates due to the lack of foam. Identifying some areas as "sacrificial burn zones" may be necessary. Allow the fire to continue unabated in these areas so that confinement areas of higher priority can be covered with foam.

  NOTE: While this discussion is generally not oriented to foam application techniques, it needs to be mentioned that the application technique used must deliver the highest quality foam in the greatest amounts possible to the surface of the burning fuel. In most cases, this foam is applied with aspirating-type nozzles. If water-fog delivery devices are necessary to extinguish the fire, consider the aspirating-type devices for long-term maintenance of the foam blanket covering the extinguished fire.

- ***Ability to deliver foam to an area where it is effective*** — Foam is most effective on static pools of fuel. If the turbulence resulting from flowing liquid fuels results in a foam blanket that does not contribute to the confinement process, discontinue application.

At fixed-site facilities that process, store, or distribute flammable and combustible liquids, NFPA 30, *Flammable and Combustible Liquids Code* (2000), mandates accomplishing confinement by using dikes and containment areas (**Figure 7.16**). If burning fuel breaches a primary storage container, the fuel is contained within the diked area. Finished foam is then applied to blanket this area. Ignited spills that occur in undiked areas require the use of spill-control devices or foam blankets to prevent further spread of the fuel.

Extinguish

Extinguishment of ignited flammable and combustible liquids at a fixed-site facility occurs when a complete blanket of finished foam smothers the ignited vapors and creates a barrier between the fuel and oxygen in the air. If the site is equipped with a fixed foam distribution system, then that system has

Figure 7.16 Dikes are required for the protection of flammable and combustible liquid storage tanks. *Courtesy of Conoco, Inc.*

Figure 7.17 Maintain a complete foam blanket on the surface of the liquid spill fire until the fuel can be removed.

to remain in operation throughout the incident to be effective. If handlines and monitors flowing either water or finished foam are deployed in support of the fixed foam distribution system or as the only extinguishment method, then follow the attack procedures outlined in the Foam Monitors and Handlines section under Basic Foam Application Techniques. When possible, only extinguish a hazardous liquid fire when reignition can be prevented. Therefore, deploy sufficient foam to extinguish the fire and maintain a proper foam blanket until normal conditions can be restored. Once the fire is extinguished, maintain a complete blanket of finished foam to prevent reignition (**Figure 7.17**).

Overhaul

Generally, overhaul of Class B flammable and combustible liquid fires at a fixed-site facility involves the removal of all unburned fuel products, cleanup of the site (following EPA or other mandated guidelines), and return to normal conditions. These tasks are outside the normal responsibilities of fire and emergency services and are performed by the site owner or private contractors. However, fire and emergency services personnel may be required to remain on the scene to protect exposures and maintain the foam blanket while overhaul is undertaken.

The importance of correctly determining the quantity of foam concentrate required for complete control of the incident cannot be overemphasized. Case histories abound where the size of the fire was underestimated and foam concentrate supplies were exhausted before completing the task of extinguishment, not to mention the amount needed for the overhaul task and maintaining the foam blanket.

Many consider the period just following extinguishment to be the most dangerous. Responders have a tendency to relax at this time and, in some cases, have failed to remain focused on the final tasks. No matter how it happens, once vaporization of the fuel resumes, responders are at great risk. When a fire is at its most developed stage, it is consuming most of the flammable vapors present, and its full dimensions are easily identifiable. Following extinguishment, without an adequate and continuously maintained foam blanket, responders may find themselves in a vapor cloud that is undetectable without the use of air-sampling equipment, exposing them to potential injury or death if the fuel reignites.

Application Safety Issues

As mentioned in the Rescue section earlier and in Chapter 5, Class A Foam Fire Fighting, the acceptable risk to fire and emergency services personnel must be kept to a minimum. Therefore, personnel safety must be ensured through the use of the Incident Management System (IMS), use of accountability systems, use of personal protective equipment, use of rapid intervention teams (RITs), and a knowledge of the behavior of liquid fuel pools (**Figure 7.18, p. 220**). Because information on IMS, accountability, PPE, and RITs are available in detail in other IFSTA and Fire Protection Publications manuals, they are not discussed here. Personnel safety is mandatory at all incidents, along with adhering to related policies and procedures.

A key safety axiom that fire and emergency services personnel must remember when fighting Class B fires is that they must avoid walking in or wading through pools of liquid fuels. This axiom also applies to pools of fuel that have been extinguished or covered with foam blankets. Reasons for not walking in or wading through pools of fuels are as follows:

- ***Turnout clothing absorbs liquid fuel with a wicking action*** — Wicking action exposes emergency personnel to any chemical effects that a fuel might pose to the skin. It also exposes them to a serious fire hazard situation if they were to come into contact with an ignition source.

Figure 7.18 One element of personnel safety is the accountability system. This firefighter wears an accountability tag on the back of his helmet.

NOTE: Leather fire-fighting boots and leather trim on personal protective clothing may be damaged by contact with foam concentrate, finished foam, or hazardous materials, including fuels. Decontaminate or dispose of damaged items in accordance with the manufacturer's recommendations or local protocol.

- ***Walking into a pool of fuel disturbs it and the foam blanket*** — Stirring of the fuel increases vapor production and resultant dangers associated with vapor production. This situation increases the chance of igniting the fuel. Even AFFF and FFFP blankets may form gaps when personnel wade through them. These gaps release vapors that the foam blanket is intended to suppress. Ignition of the vapors can result in personnel being severely injured or killed.

In the event that it may be necessary for fire and emergency services personnel to enter a pool of liquid fuel to perform a rescue, they must wear full PPE with SCBA and be protected with foam streams.

Figure 7.19 Personal protective equipment must be thoroughly decontaminated following exposure to both hazardous materials and foam concentrates/solutions. *Courtesy of Ron Jeffers.*

Apply finished foam at a point slightly away from the rescuers' position but not directly at them. This action allows the finished foam to gently flow into position around the rescuers. Moving through the foam blanket very smoothly and slowly minimizes the disturbance.

When rescue personnel exit a foam blanket or pool of fuel, they must remove all of their PPE as soon as possible and go through standard hazardous materials decontamination procedures to remove both the finished foam and the fuel contamination (**Figure 7.19**). PPE must be decontaminated according to manufacturer's instructions so that fuel and finished foam residues are completely removed. If decontamination is not possible, then PPE must be discarded in accordance with the jurisdiction's hazardous waste plan or manufacturer's recommendation.

Basic Foam Application Techniques

[NFPA 472: 5.3.2 (2f), 6.2.4.2 (1), 9.3.2 (3e & f); NFPA 1003: 3.3.4; NFPA 1081: 5.3.1, 5.3.2, 6.3.3, 6.3.4, 7.3.4]

For fire and emergency services personnel to be successful in extinguishing flammable or combustible liquid fires at fixed sites, it is important that they use proper foam application methods. Proper application technique is more important during fire conditions than it is when covering unignited spills because of the destructive effect fire has on foam during application. The extreme temperatures generated by fire break down the finished foam bubbles if the foam is not applied rapidly and properly. Proper application minimizes a fire's effect and allows extinguishment to proceed as quickly as possible. All fire and emergency services personnel involved in foam fire-fighting operations must be familiar with the various considerations for finished foam application and the different application methods.

While the majority of modern fixed-site facilities are protected by Type II (fixed) foam discharge systems, situations can occur where fire and emergency services personnel have to extinguish or confine a fire with Type III (portable) foam discharge systems. As mentioned in Chapter 4, Foam Delivery Systems, these portable discharge systems consist of handlines and apparatus or ground monitors. This section focuses on manual application methods and the portable types of foam discharge systems: foam monitors and handlines using aspirated/nonaspirated foam nozzles.

Foam Monitors and Handlines

Where acceptable to the authority having jurisdiction, small flammable and combustible storage tanks may be protected by portable foam monitors and/or foam handline discharge connections in place of Type II fixed foam discharge systems (**Figure 7.20**). According to NFPA 11, foam monitors may be used for protection of fixed roof tanks up to 60 feet (18 m) in diameter. Foam handlines may be used for the protection of fixed roof tanks not over 30 feet (9 m) in diameter or 20 feet (6 m) in height. In actual practice, these discharge systems are used to protect tanks larger than the sizes specified in NFPA 11. For hydrocarbon storage tanks, the normal foam solution application rate with foam nozzles is usually 0.16 gpm/ft^2 (6.5 [L/min]/m^2) of fuel surface area. The minimum discharge time is 50 minutes for combustible liquids and 65 minutes for flammable liquids and crude oil.

These discharge systems may also be used on storage tanks that contain polar solvent fuels. An alcohol-type (polar solvent) foam concentrate must be used on these installations, and the minimum discharge time is 65 minutes. The foam concentrate manufacturer specifies the minimum application rate, which is based on the type of fuel protected. Refer to the foam manufacturer's instructions for this information.

The primary advantage of these portable systems is that they are less expensive than surface, subsurface, and semisubsurface fixed discharge systems (see Chapter 4, Foam Delivery Systems). Portable systems are also not prone to damage in the initial stages of an incident. Their primary disadvantages are that their methods of foam application are less precise than fixed discharge system application methods and more susceptible to wind and thermal updrafts. Portable systems may also require additional staffing in areas that are exposed to fire. Because of reduced application efficiency, greater application rates and discharge duration times are required. These facts mean more foam concentrate is required to control an incident.

Seal areas on storage tanks less than 250 feet (76 m) in diameter can be protected with handline delivery systems. These tanks require a finished

Figure 7.20 Fixed-site monitors permanently attached to water mains provide fire-suppression systems that can be put into operation quickly.

foam dam like those described in Chapter 4, Foam Delivery Systems. The foam delivery system on these tanks consists of a riser that extends up the outside of the tank to the area near the top of the stairway. A single, fixed foam chamber is required at the top of the stairs (**Figures 7.21 a and b**). This chamber provides automatic finished foam discharge to that area and a safe base of operations for fire and emergency services personnel in that area. The fixed outlet must flow at least 50 gpm (189 L/min). Two 1½-inch (38 mm) discharge outlets may be provided on the tank's topside at the stairs. See the Open Top Floating Roof Tanks section for handline tactics to extinguish a seal fire.

At the beginning of an emergency incident involving Class B flammable and combustible liquid fuels, fire and emergency services personnel must decide how to apply AFFF or FFFP concentrates. They have to make the critical decision of whether to use aspirating or nonaspirating nozzles. Each application method has advantages and limitations that are generally agreed upon by foam fire-fighting professionals. Each fire-fighting situation is evaluated separately to determine which method is most suitable.

Aspirating Foam Nozzles

Aspirating foam nozzles produce aspirated finished foam of the highest quality (see Chapter 3, Foam Proportioning and Delivery Equipment) (**Figure 7.22**). Two advantages of using aspirating foam nozzles are as follows:

- Aspirating foam nozzles produce finished foam that is longer lasting than that produced with fog nozzles.
- Aspirated finished foam has longer drainage times than does nonaspirated finished foam. This factor is a decided advantage when the fuel needs to be blanketed for extended periods.

Limitation considerations apply only when using regular AFFF, alcohol-resistant AFFF, or FFFP concentrates. Aspirating foam nozzles must be used when using protein, fluoroprotein, and some alcohol-resistant foam concentrates. Some of the limitations that may preclude the use of aspirated foam nozzles are as follows:

- Aspirated foam streams generally have a short reach. The reach is reduced because the stream has a high air content, and the energy normally used to propel the stream forward is consumed in generating the finished foam.
- Aspirated foam streams are more subject to adverse effects by wind conditions than are nonaspirated foam streams.
- Aspirated foam streams cannot provide fog patterns to protect fire and emergency services personnel when they make close approaches to fires as good as can nonaspirated foam streams.
- Aspirating nozzles are bulky and only used for applying foam. This factor could add additional equipment costs for those who choose to carry and use them because they would have to purchase both aspirating nozzles and standard water stream nozzles. Also, because aspirating nozzles are not suitable for protection of the operators of the delivery stream, additional hose streams with protection capability (water-fog nozzles set on full-fog patterns) may be required.
- Aspirated foams are less fluid (flowing), which can become a problem when trying to reach all areas of a fire or spill. Aspirated foam blankets take more time to develop because the foam spreads slowly.

Nonaspirating Foam Nozzles

Two advantages of using nonaspirating foam fog nozzles (also see Chapter 3, Foam Proportioning and Delivery Equipment) are as follows:

- Any standard fog nozzle can apply nonaspirated finished foam. A straight stream or fog pattern of 10 degrees or less is the most desirable application pattern. This pattern allows for maximum aspiration of the foam solution. However, fog patterns may be employed if it becomes necessary to provide protective cover for fire and emergency services personnel (**Figure 7.23**).
- Nonaspirating foam nozzles produce foam streams that have greater penetrating power and reach than do aspirated foam streams. Tests conducted by the U.S. Navy indicate that nonaspirated finished foam gives a faster knockdown of fire than does aspirated finished foam.

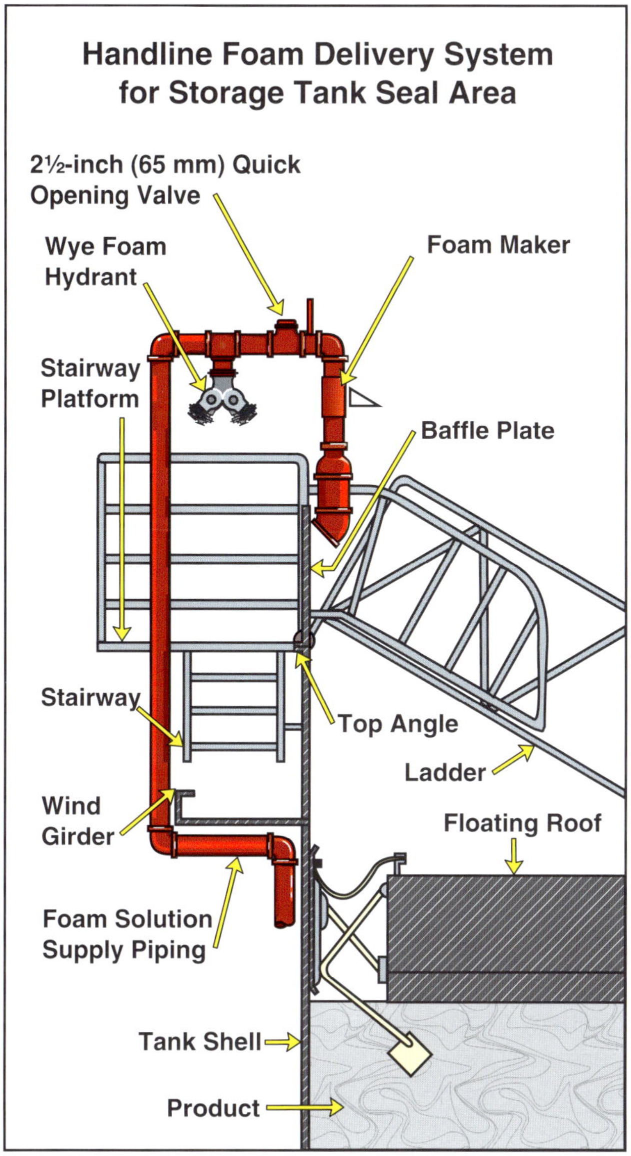

Figure 7.21a Diagram of a connection for foam handlines and a fixed foam discharge chamber on the top side of an external floating roof tank.

Figure 7.21b Photo of handline foam delivery system in use. Firefighters attach handlines and apply foam to fires along the edge of the tank seal. *Courtesy of Shell Oil Co., Deer Park, Texas.*

Figure 7.22 Aspirating foam nozzles produce long lasting finished foam blankets with increased drainage times.

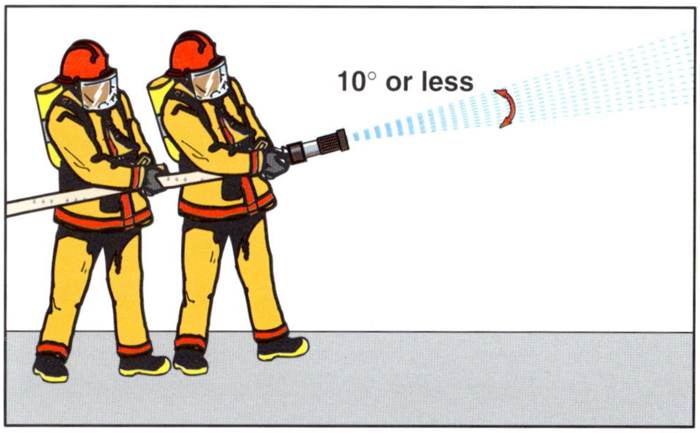

Figure 7.23 Applying nonaspirated foam with a fog nozzle set at less than 10 degrees is the most desirable application pattern.

Two limitations of using nonaspirated foam nozzles are as follows:

- Nonaspirated foam nozzles do not form as thick a foam blanket as do aspirated foam nozzles.
- Nonaspirated finished foam requires more frequent applications to maintain adequate foam blankets over extinguished spills than does aspirated finished foam.

Manual Application Methods

It is important to use the correct techniques when manually applying foam from handline or master stream foam nozzles. If incorrect techniques are used such as plunging the foam into the fuel, the effectiveness of the finished foam is reduced. The techniques for applying foam to a fire are basically the same as for applying foam to unignited flammable and combustible liquid fuel spills. As mentioned in Chapter 6, Class B Foam Use: Unignited Class B Liquid Spills, the methods include the following:

- ***Roll-on (bounce)*** — Direct the foam stream on the ground near the front edge of the liquid pool that is on fire. The finished foam then rolls across the surface of the fuel, creating a complete foam blanket on the surface and extinguishing the fire. This method is used only on undiked pools of liquid fuel on open ground.
- ***Bank-down (deflect)*** — Direct the stream off an object and allow the foam to flow down onto the fuel's surface. It may be employed when an elevated object is near or within the area of the burning pool of liquid. This method is used primarily on dike fires and liquid spill fires around damaged or overturned transport vehicles.
- ***Rain-down (raindrop)*** — Direct the stream into the air above the fire and allow the finished foam to float gently down onto the surface of the burning fuel. It is used when the other two methods are not feasible because either the size of the spill is too large to effectively cover with the roll-on method or no object to bank the finished foam from is available. It is also the primary manual application method used on aboveground storage tank fires. On large fires, it may be more effective to direct foam streams at one location to allow finished foam to take effect and then work its way out from that point.

Fuel Facility Fire Tactics

[NFPA 1081: 5.3.1, 6.3.4, 7.3.4]

Fixed-site facilities that contain large quantities of Class B flammable and combustible liquids include vehicle service/filling stations, transport fuel loading racks, storage tanks, and pipeline pumping stations. Smaller quantities may be found at vehicle maintenance centers, warehouses, lumber yards, and manufacturing facilities (**Figure 7.24**). This section discusses the challenges of fire involving large quantities of flammable and combustible liquid fuels at service/filling stations and fuel loading racks and gives tactics for fire suppression at these sites.

Service/Filling Stations

Besides transportation incidents (discussed in Chapter 8, Class B Foam Fire Fighting: Transportation Incidents), the most common Class B flammable and combustible liquid fuel fires that the majority of municipal fire and emergency services personnel face are those that involve service or filling stations. The most common filling station consists of self-service fuel islands, aboveground or underground storage tanks, and a convenience store. Full-service stations are rare, but those that are in operation have fuel islands, an office, main-

Figure 7.24 Lumberyards contain both large quantities of Class A fuels and smaller quantities of Class B fuels such as paints, mastics, solvents, and cleaning materials.

Figure 7.25a Example of a full-service station.

Figure 7.25b Example of a maintenance garage.

Figure 7.25c Example of a convenience store/filling station.

Figure 7.25d Example of an oil-change facility.

tenance bays, and storage tanks. Oil-change facilities and maintenance garages also exist but rarely have fuel-dispensing or fuel-storage tanks on site (**Figures 7.25 a–d**). The main Class B fire hazards concerning service/filling stations are fires involving fuel-dispensing equipment, tank fill connections and vents, filling pumps, and station interiors.

Fuel-Dispensing Equipment

Fuel dispensed to customers at most service/filling stations is stored in large, underground tanks. Modern underground tanks are made of fiberglass while old steel tanks may still be in service in some areas. Although the steel tanks eventually corrode and leak fuel to the soil, neither the steel nor the fiberglass tanks pose a fire hazard. In some jurisdictions, underground tanks are being replaced with aboveground horizontal storage tanks. See the Storage Tank Fire Tactics section.

Tank Fill Connections and Vents

Fires involving a service/filling station's underground storage tank fill connections and vents are usually minor fires that are a result of either static electricity or other fires in the area. Vent fires can be extinguished with a quick burst from a handheld Class B-rated fire extinguisher or water hoseline (**Figure 7.26**). Fire and emergency services personnel can extinguish tank fill opening fires by simply closing the lid on the opening, thereby shutting off the fuel source. If a fuel spill is on the ground around

Figure 7.26 An underground storage tank vent fire can be easily controlled with a handheld fire extinguisher.

the opening or the fill hose is still connected, extinguish the fire with a portable extinguisher or foam handline and then close the opening. It is virtually impossible for a fire in a fill opening to transmit down into the tank itself.

Filling Pumps

Fires involving the filling pumps at full- and self-service stations are very common. These fires may result from pump or filler hose and nozzle malfunctions, a vehicle striking a pump, a vehicle moving with the pump nozzle still inside the vehicle's fill connection, or a vehicle on fire next to a pump.

Filling pumps are electrically operated, and power to the pumps is turned off at the main electrical panel inside the station. Some stations also have emergency shutoff switches or buttons located either inside or outside the station (**Figure 7.27**). Most service/filling station pumps dispense fuel into the vehicle's tank at a rate of 10 to 15 gpm (38 L/min to 57 L/min). The nozzles are designed to turn off when a vehicle's tank is full. The pumps are also designed to stop the flow of fuel from the tank in the event the hose or nozzle fails.

Figure 7.27 Emergency fuel shutoff switches may be located either inside or outside the building.

Even if all of the preventive equipment fails, a large spill fire from a filling pump incident is rare. It is also rare for more than 50 gallons (189 L) of fuel to discharge on the ground at these incidents. What would more likely cause problems would be ignition of the filling pump, vehicles parked at the pumps, or the service/filling station building.

Fire and emergency services personnel arriving on the scene of a filling pump fire must first ensure that the power to the filling pumps has been stopped. In most cases, the station operator turns off the power before fire and emergency services personnel arrive, but they need to confirm that this step was taken. Pre-incident plans contain information on the location of these controls for each station in a response area. In various jurisdictions, unmanned filling stations are allowed under the local fire code. Pre-incident plans show where the necessary shutoff systems are installed on these types of filling pumps. As mentioned earlier, some jurisdictions now require fuel storage tanks to be installed aboveground. These storage tanks can gravity-feed fuel to the pump area even when the pumps are turned off. Fire and emergency services personnel must know the location of these tanks and turn off the fuel valve at the tank in addition to turning off the power to the pumps (**Figure 7.28**).

The remaining fire tactics include possible rescue situations, exposure protection, water supply requirements, fire confinement, fire extinguishment, and overhaul. These tactics are discussed in the paragraphs that follow:

Rescue situations. Even though a rescue situation is very rare in these types of fires, fire and emergency services personnel must ensure that no such

Figure 7.28 Aboveground storage tanks are equipped with gravity-feed shutoff valves.

situation exists. The possibility is present that an attendant or patron may be trapped inside the station building or victims may be trapped inside nearby vehicles. Make rescues under the cover of protective hoselines. It may be possible to flush burning fuel away from a victim so that a rescue may be made.

Exposure protection. After ensuring that a rescue situation does not exist, consider any exposures that are threatened by the fire. Exposures include other vehicles parked in the area, the service/filling station building, other filling pumps, or the overhead canopy above the filling area. Employ standard exposure protection methods using water streams for this purpose. Fire and emergency service personnel need to be aware of the travel direction of runoff water. Take care not to flush the spill fire into any unwanted areas such as storm drains, creeks, or other bodies of water.

Water supply requirements. Water supply requirements for this type of incident are likely to be greater than the capacity of a fire apparatus water tank. Establish some type of continuous water supply operation to support the operation in the form of large diameter hose or multiple supply hoses, relay pumping from a hydrant, or a tanker (tender) shuttle operation.

Confinement and extinguishment. The main body of fire in these situations can usually be extinguished using one or two foam handlines. Even though a fire may be contained with one handline, deploy and charge a second one to serve as a backup line. Vehicles on fire can be extinguished using the techniques described in Chapter 8, Class B Foam Fire Fighting: Transportation Incidents. Extinguish spill fires by banking finished foam off a vehicle, pump, building, or other object and allow it to roll across the surface of the fire. Create a 4-inch (102 mm) foam blanket over the entire area once the fire is extinguished.

Overhaul. Overhaul is limited to the absorption and removal of the remaining unburned fuel, the securing of fuel supply valves to prevent further spilling of fuel, and generally returning the site to its normal condition. Foam and fuel removal have to meet the requirements of the appropriate hazardous materials waste and environmental regulations. Fire and emergency services personnel may be required to remain on the scene to protect exposures and maintain the foam blanket while overhaul is undertaken.

Station Interiors

Fires inside a full-service station, oil-change facility, or maintenance garage may pose a serious hazard to fire and emergency services personnel due to the large quantities and varieties of Class B fuels that are stored inside the repair area. These fuels include motor oils, transmission fluids, antifreeze, cleaning solvents, paints, and greases that are stored in containers ranging from 1-quart (2 L) containers to 55-gallon (208 L) drums. Few of these containers have relief valves or devices, and they may fail dramatically when exposed to fire. Class A materials such as tires, plastics, vinyl products, and other highly flammable solids may add to the intensity of a fire **(Figure 7.29)**.

Because of this fuel load and the potential for container failure, exercise extreme caution when making an interior attack on fires inside a service station. In reality, it may be safer to concentrate efforts to an aggressive exterior attack using standard structural fire-fighting techniques. Compressed-air foam systems (CAFSs) or Class A foam concentrates may be helpful in the initial attack. Extinguish any Class B fuel fires that are not extinguished by these methods with foam hoselines.

Fire and emergency services personnel must be aware of the additional threat posed by lubrication/oil-change pits found in the garage area of full-

Figure 7.29 Full-service stations, maintenance garages, and oil change facilities are usually equipped with dispensing hoses that carry oil, hydraulic fluid, antifreeze, and solvents from storage tanks to the work area.

Figure 7.30 Flammable vapors may collect in oil-change pits in service bays. Additionally, these pits create a fall hazard to personnel during fire incidents.

Figure 7.31 Tank trucks are filled through inlets located under the tank. Damaged connections or improperly connected supply pipes may result in a fuel spill during the loading operation.

Figure 7.32 A manual *dead-man* control valve — generally has been replaced with electronic controls.

service stations, oil-change facilities, and vehicle maintenance facilities. These pits may be as deep as 6 feet (2 m), and Class B fuels and flammable vapors may collect in them during fires (**Figure 7.30**). Apply a foam blanket to the pit. Overhead vehicle lift racks, which are usually hydraulically activated, may weaken and become unstable during fire conditions. The same would be true of any metal storage racks that usually line the walls of these facilities.

Fuel Loading Racks

Flammable and combustible liquid loading racks are found at bulk fuel handling facilities. Loading racks transfer liquid fuels from pipelines or bulk storage tanks into tank trucks or rail tank cars (**Figure 7.31**). In facilities where safety procedures are strictly followed and equipment is well maintained, fires involving loading rack equipment are rare. However, employee errors or malfunctioning equipment at any facility may create a situation where a fire could occur. Two types of fires associated with loading racks are primary concerns: spill fires and dome/hatch opening fires. However, several safety features designed to prevent spills are built into fuel loading rack systems.

In old fuel loading rack systems, a manually operated *dead-man valve* controls the flow of fuel from the rack to the tank (**Figure 7.32**). This device is a spring-loaded valve designed to turn off immediately when an operator releases the handle. The potential for problems with this type of equipment is obvious: Idle operators may tie the valve open with a rope or prop it open with a stick. They may then leave the area or divert their attention from the operation. Operator inattention can lead to overfilling of the tank and a spill.

New loading racks are designed with numerous safety features to prevent spills and fires. The first safety device is a *plug* that is connected to the truck tank or railcar (**Figure 7.33**). This device assures that the tank is grounded and overfill equipment is working. The loading and vapor return hoses are then connected to the tank. The operation automatically opens the internal valves on the bottom of the truck tank or railcar, which in turn allows the filling of the tank. The next step in the process is for the driver or operator to set the fuel meter to the amount of product to load into each tank compartment. If the meter is set incorrectly, the tank overfill protection systems will still prevent a spill.

Loading racks are usually equipped with secondary valves and remote emergency switches that can stop the flow of fuel if other equipment fails (**Figure 7.34**). Secondary valves are located between the loading racks and bulk storage tanks. In many cases,

Figure 7.33 To prevent static electricity buildup, tank trucks are connected to a grounding plug during loading-rack operations.

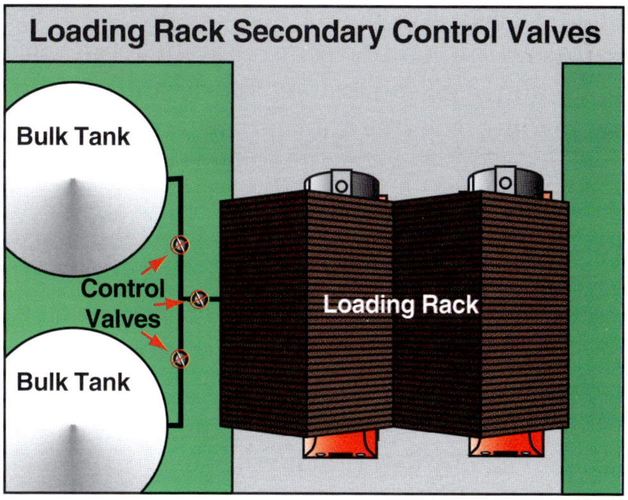

Figure 7.34 Loading rack secondary product control valves can stop the flow of fuel.

the secondary valves are held in the open position by some type of heat-sensitive device such as a fusible link. If a fire occurs, the fusible link melts and closes the valve. Remote emergency switches are located at various locations around the loading rack area. Signs are placed in highly visible locations to denote the position of these emergency switches. The posts or walls with the emergency switches may be painted red to denote the fire protection equipment. Pre-incident surveys of the facility will yield this and other valuable information for fire and emergency services personnel.

Fuel Spill

The most serious type of fire that can occur at a loading rack is a large fuel spill fire. Spills may occur as a result of malfunction of pumping equipment, failure of fill spouts or hoses, or weakened safety equipment that has been tampered with or overridden by personnel. If all equipment is used as designed, spills will seldom occur.

Once the higher priorities of rescue and exposure protection have been addressed at any loading rack fire situation, the next action for fire and emergency services personnel is to ensure that fuel-transfer operations have been turned off. In nearly every case, employees accomplish this step before the arrival of emergency units; however, fire and emergency services personnel must confirm that fuel-transfer operations have been turned off. Emergency shutdown devices (ESDs) are located throughout loading rack areas and are visibly marked (**Figure 7.35**). Pre-incident plans include information concerning the location and operation of all valves and controls and whether the loading rack area is diked. Dikes confine spills and prevent burning fuel from leaving the immediate area, making the extinguishment process a little easier. Pre-incident plans also contain information on the dimensions of a dike (if present) so that fire and emergency services personnel will have an idea of how much liquid it can hold before overflowing. Having this information keeps emergency personnel from adding too much water or finished foam to the area and causing the dike to overflow.

Many fuel loading racks are protected by some type of fixed foam delivery system such as foam/water sprinkler systems, fixed foam monitors, or combination systems (see Chapter 4, Foam Delivery Systems). A combination system is designed so that sprinklers handle the bulk of the surface fire with monitors discharging finished foam to the underside of the truck or railcar (**Figure 7.36**). In most cases, if the foam delivery system operates properly, most of the fire (if not all of it) is extinguished before the arrival of fire and emergency services units. If this is not the case or if the loading rack is not equipped with a fixed foam delivery system, use manual fire-fighting techniques.

Figure 7.35 Loading racks are equipped with emergency shutdown devices to prevent liquid spills. *Courtesy of Conoco, Inc.*

Fire and emergency services personnel should approach loading racks from the side just like they would with any horizontal tank or cylinder (**Figure 7.37**). Never approach from the ends because they are the weakest seals on a tank trailer/truck or railcar and could rupture if exposed to excessive heat. Personnel can quickly determine the approximate size of the fire area and then calculate the foam application rate necessary to extinguish the fire. Apply finished foam by handlines, master stream devices, or a combination, depending on the size of the fire and the required flow rate. Use the previously described finished foam application methods (roll-on, bank-down, or rain-down) to extinguish a fuel loading rack fire. Usually finished foam can be banked off the tank or loading rack equipment and allowed to gently flow down onto the fuel (**Figure 7.38**). When an entire dike area is filled with a foam blanket, the bulk of the fire is probably extinguished.

Figure 7.36 Because overheated brake drums may be ignition sources, loading racks are equipped with foam spray nozzles that are directed at the tank truck wheels and brakes.

Figure 7.37 Because truck and rail car tanks have weak seams on the ends of the tanks, always approach a fire in a loading rack from the side.

Figure 7.38 Foam can be banked off the cargo tank onto a spill fire.

Extinguish all of the fire on the ground and then turn attention to the remainder of the truck and any other objects in the area. Ensure also that any water discharged does not interfere with the foam blanket.

Dome/Hatch Opening

Dome/hatch opening fires occur in tanks that are designed for filling through open hatches on the tops of tanks **(Figure 7.39)**. These fires generally involve only the dome/hatch area and any loading hoses or loading rack components close to the dome. Static electricity is the most frequent cause of dome/hatch opening fires. The process of transferring fuel from the rack to the tank creates a large static buildup. Proper bonding and grounding procedures can eliminate the hazards of static electricity. However, employees who overlook these procedures can create a fire situation. These fires may also be caused by equipment malfunctions or carelessness (such as smoking) on the part of operators during filling operations.

Dome/hatch opening fires are generally easy to control. If the loading rack is equipped with a fixed fire-suppression system, chances are that this system will extinguish the fire before fire and emergency services personnel arrive. If this situation is not the case, handle these fires quickly. Place a protective hoseline in service as soon as possible. This procedure protects fire and emergency services personnel who are working to extinguish the fire and any exposures that may be receiving heat or flame contact. In most cases, the only exposure is loading rack equipment above the hatch area.

Figure 7.39 Dome/hatch opening fires may occur in the loading rack and involve either truck or rail tank cars.

If the loading hose is not connected to the tank, the easiest way to extinguish the fire is to simply close the dome/hatch cover, thereby removing the fuel source **(Figure 7.40)**. A pike pole can usually accomplish this step. If not, fire and emergency services personnel may climb to the top of the tank and close the hatch manually. All fire and emergency services personnel operating from this position must wear full PPE, including SCBA. If the

Figure 7.40 Dome/hatch opening fires may be extinguished by simply closing the hatch and smothering the fire.

intensity of the fire prevents emergency personnel from approaching the hatch area, extinguish the fire by using either a Class B-rated portable fire extinguisher or a foam handline. Once the fire is extinguished, attempt to securely close the hatch. Use caution and be prepared if the fire reignites after the initial extinguishment.

If the loading spout is still in place, stop the flow of fuel into the tank with the shutoff and control valves. With the loading spout still in place, fire and emergency services personnel will not be able to immediately close the hatch. They must leave the spout in place until the fire is extinguished. Attempting to remove the spout while the fire is still burning could spread the fire. Extinguish the fire with a portable extinguisher or a foam handline. After extinguishing the fire, allow the dome area and spout to cool on their own or use a fire stream to cool them. After the threat of reignition is eliminated, remove the spout and close the hatch.

Storage Tank Fire Tactics

[NFPA 1081: 5.3.1, 5.3.2, 6.3.3, 6.3.4, 7.3.4; NFPA 472: 9.3.2 (3e & f)]

Although fairly uncommon, fires involving large fuel storage tanks can be some of the most spectacular and challenging Class B fire incidents fire and emergency services personnel will encounter. In almost every case, extinguishment of well-involved storage tank fires is beyond the capabilities of the local municipal fire department or emergency services organization (**Figure 7.41**). In some cases, an industrial fire department or brigade protecting the facility may have the resources to fight such a fire. When neither the municipal fire department nor the plant fire brigade has the capability to extinguish the fire, outside firms with the necessary expertise and equipment to extinguish such fires are needed on the scene to assist with extinguishment. Fire and emergency services personnel may need foam manufacturers to provide emergency bulk shipments of foam concentrate. Generally an all-out attack on such a fire should only take place when adequate resources are available on the scene. Once any necessary rescues have been accomplished, the exposures protected, and the burning fuel spills confined, choose a defensive mode and wait for sufficient resources to arrive or allow the fire to consume the fuel.

Figure 7.41 Storage tank fires present major challenges to municipal and rural fire departments. *Courtesy of Richard W. Giles.*

In the event that the fire department, emergency services organization, or fire brigade does have the necessary resources to attack the fire, carefully plan the implementation of sound tactics. The physical characteristics of common types of flammable and combustible liquid storage tanks and fixed foam delivery systems provided to protect them were discussed in Chapter 4, Foam Delivery Systems. In the sections that follow, it is assumed that fixed foam delivery systems are either absent or inoperable for the various storage tank types, and manual techniques are necessary to control and extinguish fires.

Cone Roof Tanks

Cone roof tanks are generally used to store refined combustible liquids, crude oil, and in some cases flammable liquids (**Figure 7.42**). A vapor space exists between the fuel surface and the roof of the tank. Cone roof tank roofs over 50 feet (15 m) in diameter are usually equipped with a weak (frangible) seam between the roof and tank shell that is designed to fail very early in a fire incident or after an explosion. This failure allows the roof to separate from the tank and exposes the fuel surface to the atmosphere. It is impossible to predict where the final position of the roof will be. It may fall into the tank, blow away from the tank, or stay partially connected to the tank. This situation may render any fixed foam delivery system useless and necessitate the use of portable extinguishment techniques. The tactics necessary for handling cone roof tank fires involving refined combustible liquids and crude oil are given in the sections that follow.

Figure 7.42 Cone roof storage tank. *Courtesy of Paul Valentine.*

Another version of the cone roof tank is the dome roof tank **(Figure 7.43)**. These tanks are usually less than 50 feet (15 m) in diameter and usually do not have a weak seam between the roof and tank shell. If a dike-area fire in close proximity to one of these tanks is encountered, fire and emergency services personnel must be alert to the possibility of catastrophic failure where the tank is mounted to its base. This situation could actually result in a rocket-like launching of the tank caused by the ignition of the fuel at the base of the tank.

Refined Combustible Liquid Fires

Fires involving refined combustible liquids such as kerosene, diesel fuel, and lubricating oils can be extinguished using Class B foam concentrates. The most desirable tactic for extinguishing these fires is the use of fixed foam delivery systems. When these systems are not available, portable foam application methods are required.

The first priority is for fire and emergency services personnel to determine the need for rescue and perform that function if necessary. The next priority is to address exposures **(Figure 7.44, p. 234)**. Apply cooling water in accordance with recommendations given in the Exposures section earlier in this chapter. Use the application method that consumes the least amount of water necessary to accomplish the task while using the fewest number of personnel. All nonessential personnel must be restricted from the dike area as much as possible. The dike's capacity must always be considered. The confinement area must be able to hold both the contents of the tank and the water/foam used in the fire-suppression operation.

Figure 7.43 Dome roof storage tank, part of a training prop.

In some cases, some portion of the storage tank contents may need to be removed and transferred to another tank in the facility to make room for the foam that is needed to extinguish the fire. A facility operator in cooperation with the incident commander would be responsible for this pumping operation. Transfer of the fuel lessens the time it takes for the fire to consume the fuel. Stop pumping if the fuel being removed shows signs of becoming overheated to the point where ignition is possible.

Fire and emergency services personnel may apply foam via ground-level master streams, elevated master streams, or a combination of both. When using ground-level or elevated master streams, direct the streams to land somewhere near the center of the fuel surface of the tank. Keep streams in this position until sufficient foam blanket buildup is produced to achieve fire knockdown **(Figure 7.45, p. 234)**. Reposition the streams only when the foam stops spreading across the surface of the fire. It is generally less effective for fire and emergency services personnel to sweep finished foam over the entire surface of tank fires by moving the master stream from side to side. This action does not allow the blanket to develop and spread on its own.

When the fire is extinguished, build a 4-inch (102 mm) foam blanket over the surface until normal

Figure 7.44 Adjacent tanks, fixed fire-suppression equipment or other structures may require exposure protection.

Figure 7.45 Streams directed onto burning fuel tanks should remain in position until a sufficient buildup of foam can knock down the fire.

operating conditions are restored. In most cases, reapplication of finished foam is necessary because heat from the wall of the tank or wind starts to break down the foam blanket. Reapplication is required to maintain adequate foam blankets until vapor emissions are no longer concerns.

Crude Oil Fires

Crude and other heavy oils present some unique challenges and hazards to fire and emergency services personnel. When foam streams are initially directed onto the surface of burning crude oil, a frothing (bubbling) of the oil may occur (caused when the finished foam converts into steam). The rapid steam expansion causes a rapid overflow of the oil and water froth from the tank. The overflow hazards at crude oil fires are as follows:

- *Boilover* — Overflow of crude oil from its container; caused when burning crude oil burns the lighter fuel fractions and the heavier ones remain in the solution in the tank. The heavy oil remaining in the tank heats to over 450°F (232°C), creating a heated layer of oil called *heat wave.* The heat wave travels down into the liquid at a rate of up to 3 feet (1 m) per hour **(Figure 7.46).** This heat wave turns the normal layer of water at the bottom of the tank into steam. The steam expands and moves up through the product, creating a violent explosion that ejects burning oil from the top of the tank **(Figure 7.47)**. Prevent boilover by extinguishing the fire before the heat wave reaches the water in the bottom of the tank. If the fire cannot be extinguished by that time, move all personnel a minimum of 500 feet (152 m) in all directions from the tank. Boilover characteristics:

— A fireball may rise hundreds of feet (meters) into the air.

— The ejected burning fuel may fall to the ground hundreds of feet (meters) from the tank in all directions.

- *Slopover* — Results when a water stream is applied to the surface of a burning oil or when the heat wave within the tank contacts a stratified layer of water within the oil product. The water converts into steam and causes the oil to overflow the rim of the tank (often confused with boilover). The resulting slopover may be a slow boil or rapid explosion, depending on the viscosity and temperature of the oil **(Figure 7.48)**.

- *Frothover* — Occurs when hot materials (such as asphalt) being loaded into a tank come in contact with superheated water in the bottom of the tank. The water trapped under the material starts to boil and causes the material to overflow the tank. In some cases the boiling action of the water creates enough pressure to rupture the vessel, spreading the superheated froth over the surrounding area **(Figure 7.49)**. Primary tactical considerations:

— Cool the vessel with water or finished foam.

— Protect exposures.

— Confine the overflow.

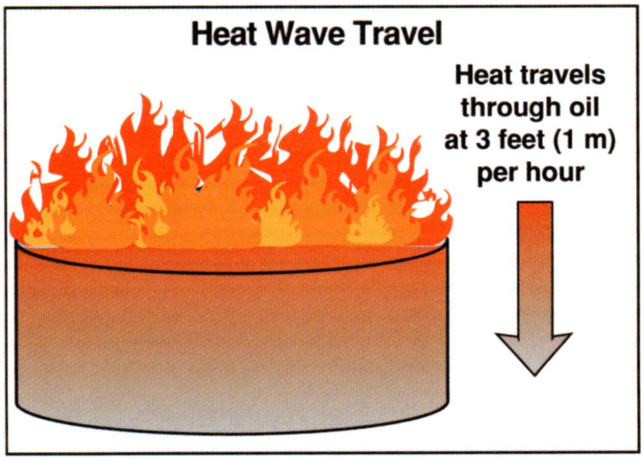

Figure 7.46 Rate of heat wave travel in crude oil tank fires.

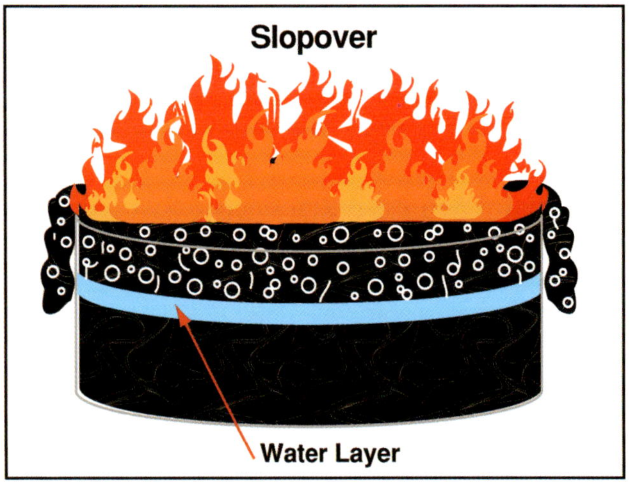

Figure 7.48 A slopover causes oil to overflow the rim of the tank.

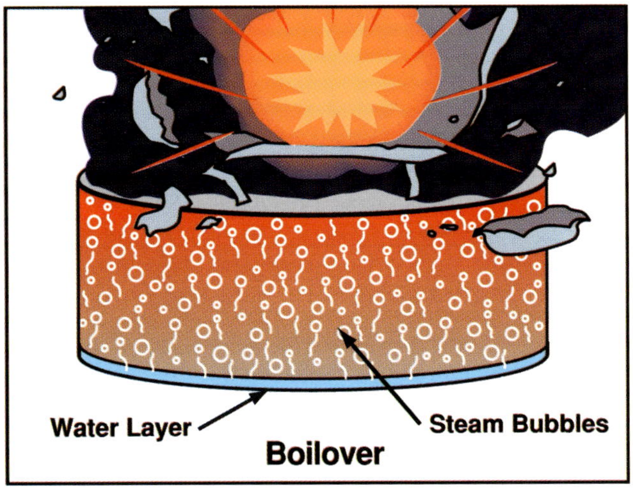

Figure 7.47 A boilover creates an explosion that ejects burning oil from the tank.

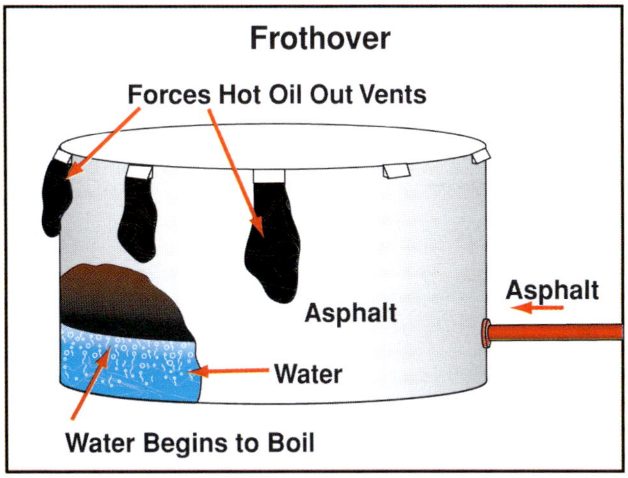

Figure 7.49 A frothover occurs when superheated water in the bottom causes the product to overflow the tank.

Crude oil tank fires may be successfully extinguished only if all of the resources needed to extinguish the fire are on the scene immediately and the attack can begin before a heat wave starts to move through the fuel. The best tactics for large tank fires of this type are to protect exposures, cool the tank on fire, pump as much product out of the tank as quickly as possible, and allow the fire to consume the rest of the fuel. It is important that all personnel stay out of the dike area in case a boilover occurs.

Crude oil tanks may also be found in rural areas as collection points serving operating oil pumps. These tanks may be relatively new or decades old (**Figure 7.50**). In any case, the tanks lack any type of fire-suppression system and are probably not diked. Volunteer and rural fire and emergency services organizations that respond to fires in these crude oil storage tanks may lack the resources and water supplies necessary to handle the incident. Take extreme caution when approaching burning crude oil tanks in rural areas. The fire has probably been underway long enough to create a boilover potential. The best initial decision would be to maintain a defensive mode while additional resources are called.

Covered Internal Floating Roof Tanks

Fire and emergency services personnel can be fairly certain that any fuel inside a covered internal floating roof tank is a finished product such as gasoline, diesel fuel, jet fuel, etc. The tank construction is basically a floor, shell, and fixed roof plus a floating roof that moves up and down with the level of the product in the tank (**Figure 7.51**). The ability of the roof to float directly on the fuel reduces the amount of vapor space between the roof and the product level. If the fixed roof is still intact, fires under it and around the floating roof seal (which connects the

Figure 7.51 A series of vents near the top edge of a tank is an indication that it is an internal floating roof tank. *Courtesy of Conoo, Inc.*

Figure 7.50 Rural fire protection organizations may find old tanks, piping systems, and pump jacks in remote areas.

Figure 7.52 Foam streams can be directed into the open vents in an attempt to force foam into an internal fire. *Courtesy of Frank Bateman.*

solid roof to the walls of the shell) are not easily extinguished from outside the tank. Each individual fire incident of this type requires a different approach, given the fire conditions and the expertise of the fire-suppression personnel on the scene. If the floating roof and fixed roof both fail, employ the tactics discussed in the Refined Combustible Liquid Fires section under Cone Roof Tanks.

One type of fire that is unique to covered internal floating roof tanks is fire emanating from the vents around the top of the outer wall of the tank. In some cases, fuel vapors build inside the tank between the two roofs. These vapors exit through the vents. Occasionally, these vapors are ignited by a lightning strike or other ignition source. These fires have sometimes been controlled using a fixed foam subsurface injection system through the product supply line that is used to fill or empty product from the tank. Applying finished foam through the vents may also be successful **(Figure 7.52).** When a dike fire accompanies a vent fire, extinguish the dike fire first. Do not allow any fire and emergency services personnel to walk on the roof of these tanks. Although subsurface and vent applications have been successful in controlling tank fires, they are a last resort tactic. When to employ these applications is based on the following considerations:

- Type of fuel in the tank
- Intact tank roof (resulting in an oxygen-starved condition)
- Failure of other methods (such as fixed systems)

Open Top Floating Roof Tanks

The most common fire scenario posed by open top floating roof tanks is a fire involving the roof seal that separates the floating roof from the walls of the tank shell. The seal serves as a vapor barrier/cushion between the floating roof and the tank wall. Usually fires occur in these seals as a result of lightning strikes. The fires may either involve only a small portion of the seal or the entire seal. The amount of the seal involved has a direct impact on the fire-extinguishing method used. Another direct impact on choosing an extinguishing method is whether the floating roof is designed to hold the weight of fire and emergency services personnel walking on it.

The volume of fire presented by a roof seal fire is not significant when compared to the other tank fires discussed. Seal fires pose little threat to exposures. The first concern is to protect exposures subjected to direct flame impingement. Cooling water is required only for areas that steam during the stream application test. When steam no longer results from stream application, discontinue application, but periodically reapply water to be sure the exposure is still adequately cool. An indication that cooling spray is required is the evidence of blistering, discoloring, or smoking of the paint on the tank surface.

Do not use ground-level and/or elevated master streams to extinguish seal fires. The use of these streams may result in too much foam being applied to the roof of the tank and create a potential for the roof to sink **(Figure 7.53, p. 238).** Use either portable extinguishers or foam handlines from the top of the tank stairs or tower platform. One method for successful extinguishment of these fires is to advance

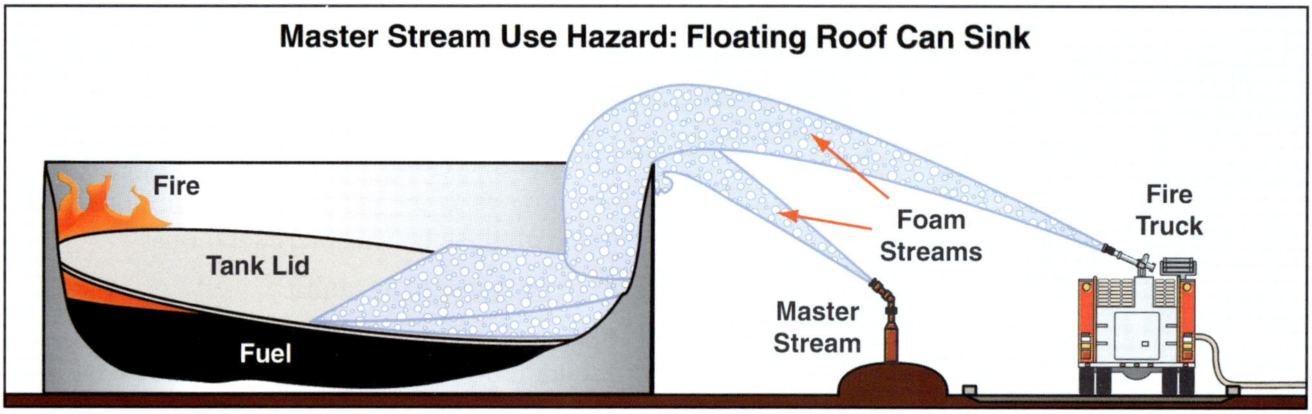

Figure 7.53 Master stream appliances should not be used on open roof floating lid fires. Too much foam and water can accumulate on the roof, causing it to tip and sink and exposing a larger portion of the fuel's surface to fire.

a foam hoseline from the top of the ladder onto the roof. This tactic is only attempted by fire and emergency services personnel who are specifically trained in this technique. A backup foam hoseline team remains at the top of the ladder to provide protection for the crew on the roof. Once the fire is extinguished, ensure that an adequate foam blanket is applied to the seal area and cool any hot spots to prevent reignition. If faced with a situation where the open top floating roof has sunk, handle the fire like a cone roof tank fire.

Horizontal Tanks

A significant exposure problem exists at fixed-site facilities because horizontal storage tanks are commonly grouped together. Horizontal tanks may also be found at service/filling stations as aboveground storage tanks (**Figure 7.54**). See Service/Filling Stations section. Horizontal tanks at fixed-site facilities may hold as much as 50,000 gallons (189 270 L) of liquid. More foam concentrate is required to extinguish this type of fire than with

Figure 7.54 Fuel distribution companies may have horizontal aboveground tanks for filling delivery trucks.

truck or railcar tank fires. Establish effective exposure protection cooling streams on adjacent tanks as soon as possible. Also prepare for the tank that is on fire to fail and spread burning fuel throughout the dike area and around other tanks. Some horizontal tanks are elevated on steel stilts to use the effects of gravity to dispense fuel from the tanks. When an elevated tank is on fire, make sure that cooling streams are applied to the supports to prevent a collapse (**Figure 7.55**).

Three-Dimensional/Pressurized Fuel Fire Tactics

[NFPA 1081: 5.3.1, 6.3.4, 7.3.4]

Three-dimensional/pressurized fuel fires pose an extreme challenge to fire and emergency services personnel. Leaking valves or flanges and damaged aboveground piping are the two most common sources of these fires (**Figure 7.56**). In addition to having a burning pool of fuel on the ground, burning fuel is also being discharged into the pool from some type of elevated, pressurized source. These fires create the following problems for fire and emergency services personnel:

- Wet finished foam works only on relatively flat surfaces. Finished foam usually cannot extinguish fires flowing from a pressurized source.
- Burning fuel discharging onto the ground pool disturbs efforts to control the spill fire by breaking up the foam blanket on the surface of the pool.
- Vapor production and what to do with a flowing unignited product become problems if a

Figure 7.55 Direct water spray at the tank supports to prevent collapse of horizontal tanks.

Figure 7.56 A three-dimensional fire requires foam attacks on both the fire in the pit and the fuel that is feeding it from above.

flowing liquid fire is extinguished but the fuel flow is not stopped (which is similar to pressurized gas fire situations).

The best choice in controlling three-dimensional/pressurized fires is to first stop the flow of fuel to the fire area. This process can usually be accomplished by closing a pipeline valve somewhere upstream of the leak. Consult plant operators or representatives of the company who own the pipeline before any such actions are taken. They are more familiar with the location and operation of such valves and any consequences that closing the valves might present. Never close a valve without first gaining permission from a plant operator or company representative. Closing valves may not immediately solve the problems. Depending on how far the valve is from the leak, a considerable amount of fuel may have to drain from the piping before the fire is halted. If the valve is very close to the fire area, closing it may add the problem of having to protect the closed vessel that is filled with vapors and exposed to fire.

If fire and emergency services personnel cannot stop the flow of fuel immediately, they must first concentrate on protecting exposures and controlling the flow of any burning fuel onto the ground. Construct temporary dikes and dams to control the fuel when the fire is not within a dike area. Attempt to control or extinguish the ground fire with foam. Use either the roll-on foam application method from the edges of the fire or bank-down method using any structures in the fire area. It may not be possible to completely extinguish the ground fire when burning fuel is still falling into the pool, but at least the intensity of the fire is diminished.

If it is determined that the fire flowing from the pressurized source must be extinguished, use either dry-chemical (such as Purple K) or halon-replacement agents. A protective foam fog stream may be required for fire and emergency services personnel to advance close enough to the fire to use the dry-chemical or halon-replacement agents. Multiagent nozzles and equipment may be helpful in these situations and are available from many sources **(Figure 7.57)**.

When a three-dimensional/pressurized fire is extinguished, control any remaining spill fire. Use foam hoselines to cool any closed pipes, exposed vessels, or other heated objects in the spill area until there is no threat of reignition. Use foam hoselines for this purpose because plain water would fall into the spill area and speed the destruction of the foam blanket. Maintain the standard 4-inch (102 mm) foam blanket over the entire spill area. If fire and emergency services personnel extinguish a flowing fuel fire, they must have a good plan for dealing with the fuel that is continuing to add to the spill. This effort has to be coordinated with the plant owner or company representative.

Figure 7.57 A mutiagent attack may prove effective by using foam to suppress part of the fire and using dry chemical on the pressurized source of the fire.

Electrical Transformer Vault Fire Tactics

Most fire and emergency services personnel are familiar with the electrical transformers that are mounted on power poles or within fenced surface enclosures of electrical distribution systems. Many transformers, however, are also located in vaults underground and in the basements of large buildings. Although transformers are electrically charged, (and therefore a Class C fire hazard), they also contain large quantities of cooling oil that make them a Class B fire hazard. An explosion and fire within an electrical distribution system can not only cause major power outages but also spread dioxins (toxic impurities) throughout the atmosphere. These dioxins can have long-lasting, adverse affects on humans if they are inhaled.

Transformers may contain from 100 to 25,000 gallons (379 L to 94 635 L) of cooling oil. Fixed fire-suppression systems used to protect transformers consist of water-spray sprinkler systems for large aboveground installations and automatic sprinklers, water spray systems, foam-water sprinklers, or gaseous total flooding systems for interior transformer vaults. Vaults are also required to have floor drains that are capable of containing a spill from the largest single container plus the discharge of the fixed fire-suppression system for 10 minutes and the discharge of the maximum number of hoselines flowing a minimum of 500 gpm (1 893 L/min) for

10 minutes **(Figure 7.58)**. Vaults are also equipped with ventilation systems that activate in the event of a fire.

Following the *RECEO* model, the rescue of trapped victims and the evacuation of individuals within the affected area take priority. Fire and emergency services personnel must wear full personal protective equipment including respiratory protection within the designated hot zone of the incident. Because NFPA 850, *Recommended Practice for Fire Protection for Electric Generating Plants and High Voltage Direct Current Converter Stations* (2000), requires separation between transformers and exposures, additional exposure protection may not be required. The separation is a cleared space around the transformer group aboveground in addition to water spray nozzles. Interior transformers are separated from exposures by 2- or 3-hour-rated firewalls.

The next step is to have the power company turn off all service to the transformer. The IC must ensure that the transformer is fully de-energized before ordering an attack on the fire. Once the transformer is de-energized, a fire attack can take place. Class B foam applied through hoselines may be used in addition to or in place of the fixed fire-suppression system. High-expansion foam may also be used in a total flooding application to extinguish vault transformer fires.

CAUTION

Transformer oil that has been heated reacts in the same way that crude oil does when water or foam strikes it. Personnel must remain at a safe distance while applying water spray or foam onto any nonenergized oil-cooled transformer.

Overhaul consists of the removal of the oil, foam, and water and is the responsibility of the power company or the owner of the transformer. Keep foam hoselines in place as long as the possibility of reignition exists.

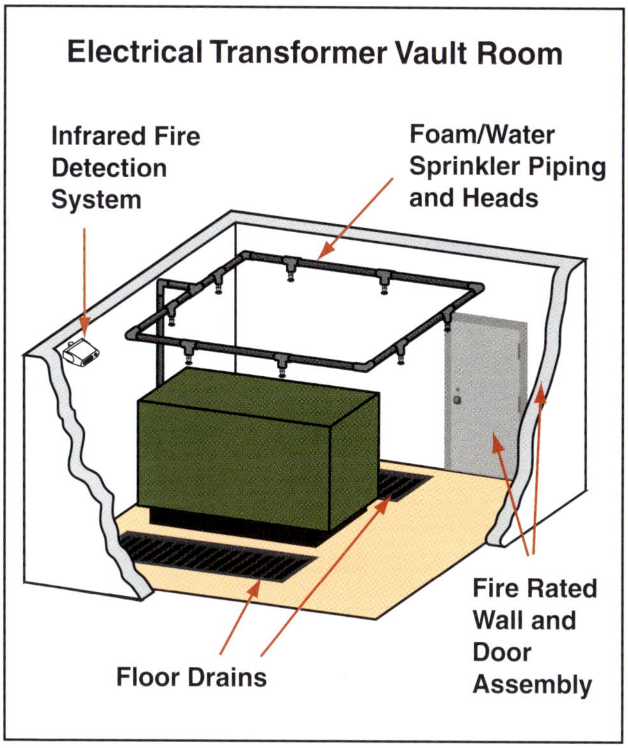

Figure 7.58 Interior electrical transformer vaults are protected by fire-resistant enclosures (walls), fire detection systems, and foam fire-suppression systems.

Summary

Because fire and emergency services personnel, both public and private, are rarely faced with flammable or combustible liquid fires in fixed-site facilities, it is important that they have the opportunity to practice the tactics and skills required for such incidents. Continuous training in classrooms and at simulated Class B fires, target-hazard site visits, pre-incident planning exercises, and review and updating of the emergency organization's standard operating procedures for Class B operations help to maintain an adequate level of experience in case a flammable or combustible liquid fire occurs.

Courtesy of Howard Meile, III

Chapter 8

Class B Foam Fire Fighting: Transportation Incidents

Job Performance Requirements

This chapter provides information that will assist the reader in meeting the following job performance requirements from NFPA 1001, *Standard for Fire Fighter Professional Qualifications*, 2002 edition; NFPA 1002, *Standard for Fire Apparatus Driver/Operator Professional Qualifications,* 1998 edition; NFPA 1081, *Standard for Industrial Fire Brigade Member Professional Qualifications,* 2001 edition; NFPA 1003, *Standard for Airport Fire Fighter Professional Qualifications,* 2000 edition; NFPA 472, *Standard for Professional Competence of Responders to Hazardous Materials Incidents,* 2002 edition; and NFPA 1051, *Standard for Wildland Fire Fighter Professional Qualifications,* 2002 edition. Colored portions of the standard are specifically addressed in this chapter.

NFPA 1001

5.3.7 Attack a passenger vehicle fire operating as a member of a team, given personal protective equipment, attack line, and hand tools, so that hazards are avoided, leaking flammable liquids are identified and controlled, protection from flash fires is maintained, all vehicle compartments are overhauled, and the fire is extinguished.

(A) *Requisite Knowledge:* Principles of fire streams as they relate to fighting automobile fires; **precautions to be followed when advancing hose lines toward an automobile; observable results that a fire stream has been properly applied; identifying alternative fuels and the hazards associated with them; dangerous conditions created during an automobile fire**; common types of accidents or injuries related to fighting automobile fires and how to avoid them; how to access locked passenger, trunk, and engine compartments; and methods for overhauling an automobile.

(B) *Requisite Skills:* The ability to identify automobile fuel type; assess and control fuel leaks; open, close, and adjust the flow and pattern on nozzles; apply water for maximum effectiveness while maintaining flash fire protection; advance 1½-in. (38-mm) or larger diameter attack lines; and expose hidden fires by opening all automobile compartments.

6.3.3 Control a flammable gas cylinder fire operating as a member of a team, given an assignment, a cylinder outside of a structure, an attack line, personal protective equipment, and tools, so that crew integrity is maintained, contents are identified, safe havens are identified prior to advancing, open valves are closed, flames are not extinguished unless the leaking gas is eliminated, the cylinder is cooled, cylinder integrity is evaluated, hazardous conditions are recognized and acted upon, and the cylinder is faced during approach and retreat.

(A) *Requisite Knowledge:* Characteristics of pressurized flammable gases, elements of a gas cylinder, effects of heat and pressure on closed cylinders, boiling liquid expanding vapor explosion (BLEVE) signs and effects, methods for identifying contents, how to identify safe havens before approaching flammable gas cylinder fires, **water stream usage and demands for pressurized cylinder fires,** what to do if the fire is prematurely extinguished, valve types and their operation, alternative actions related to various hazards and when to retreat.

(B) *Requisite Skills:* The ability to execute effective advances and retreats, apply various techniques for water application, assess cylinder integrity and changing cylinder conditions, operate control valves, choose effective procedures when conditions change.

NFPA 1002

2-3.7 Operate all fixed systems and equipment on the vehicle **not specifically addressed elsewhere in this standard, given systems and equipment, manufacturer's specifications and instructions, and departmental policies and procedures for the systems and equipment, so that each system or piece of equipment is operated in accordance with the applicable instructions and policies.**

(a) *Requisite Knowledge:* Manufacturer specifications and operating procedures, policies, and procedures of the jurisdiction.

(b) *Requisite Skills:* The ability to deploy, energize, and monitor the system or equipment and to recognize and correct system problems.

7-1.1 Perform the routine tests, inspections, and servicing functions specified in the following list in addition to those contained in the list in 2-2.1, given an ARFF vehicle and the manufacturer's servicing, testing, and inspection criteria, so that the operational status of the vehicle is verified.

- **Agent dispensing systems**
- **Secondary extinguishing systems**
- Vehicle-mounted breathing air systems

(a) *Requisite Knowledge:* Manufacturer specifications and requirements, policies, and procedures of the jurisdiction.

(b) *Requisite Skills:* The ability to use hand tools, recognize system problems, and correct any deficiency noted according to policies and procedures.

7-2.1 Maneuver and position an ARFF vehicle, given an incident location and description that involves the largest aircraft that routinely uses the airport, so that the vehicle is properly positioned for safe operation at each operational position for the aircraft.

(a) *Requisite Knowledge:* Vehicle positioning for firefighting and rescue operations; **capabilities and**

limitations of turret devices related to reach; and effects of topography, ground, and weather conditions on agent application, distribution rates, and density.

(b) *Requisite Skills:* The ability to determine the appropriate position for the apparatus, maneuver apparatus into proper position, and avoid obstacles to operations.

7-2.2 Produce a fire stream while the vehicle is in both forward and reverse power modulation, given a discharge rate and intended target, so that the pump is safely engaged, the turrets are deployed, the agent is delivered to the intended target at the proper rate, and the apparatus is safely moved and continuously monitored for potential problems.

(a) *Requisite Knowledge:* **Principles of agent management and application, effects of terrain and wind on agent application, turret capabilities and limitations**, tower light signals, airport markings, aircraft recognition, aircraft danger areas, theoretical critical fire area and practical critical fire area, aircraft entry and egress points, and proper apparatus placement.

(b) *Requisite Skills:* The ability to provide power to the pump, determine the appropriate position for the apparatus, maneuver apparatus into proper position, avoid obstacles to operations, apply agent, and determine the length of time an extinguishing agent will be available.

7-2.3 Produce a fire stream, given a rate of discharge and water supplied from the sources specified in the following list, so that the pump is safely engaged, the turrets are deployed, the agent is delivered to the intended target at the proper rate, and the apparatus is continuously monitored for potential problems.

- The internal tank
- Pressurized source
- Static source

(a) *Requisite Knowledge:* **Principles of agent management and application, effects of terrain and wind on agent application, turret capabilities and limitations,** tower light signals, airport markings, aircraft recognition, aircraft danger areas, theoretical critical fire area and practical critical fire area, aircraft entry and egress points, and proper apparatus placement.

(b) *Requisite Skills:* The ability to provide power to the pump, determine the appropriate position for the apparatus, maneuver apparatus into proper position, avoid obstacles to operations, apply agent, and determine the length of time an extinguishing agent will be available.

NFPA 1081

6.3.5 Operating as a member of a team, control a flammable gas fire, given an assignment, an attack line, personal protective equipment, and tools, so that crew integrity is maintained, contents are identified, the flammable gas source is controlled or isolated, hazardous conditions are recognized and acted upon, and team safety is maintained.

(A) Requisite Knowledge. Characteristics of flammable gases, components of flammable gas systems, effects of heat and pressure on closed containers, boiling liquid expanding vapor explosion (BLEVE) signs and effects, methods for identifying contents, water stream usage and demands for pressurized gas fires, what to do if the fire is prematurely extinguished, alternative actions related to various hazards, and when to retreat.

(B) Requisite Skills. The ability to execute effective advances and retreats, apply various techniques for water application, assess gas storage container integrity and changing conditions, operate control valves, and choose effective procedures when conditions change.

NFPA 1003

3-1.1.1 **Fundamental aircraft fire-fighting techniques, including the approach, positioning, initial attack, and selection, application, and management of the extinguishing agents; limitations of various sized hand lines;** use of proximity protective personal equipment (PrPPE); fire behavior; fire-fighting techniques in oxygen-enriched atmospheres; reaction of aircraft materials to heat and flame; critical components and hazards of civil aircraft construction and systems related to ARFF operations; special hazards associated with military aircraft systems; a national defense area and limitations within that area; characteristics of different aircraft fuels; hazardous areas in and around aircraft; aircraft fueling systems (hydrant/vehicle); aircraft egress/ingress (hatches, doors, and evacuation chutes); hazards associated with aircraft cargo, including dangerous goods; hazardous areas, including entry control points, crash scene perimeters, and requirements for operations within the hot, warm, and cold zones; and critical stress management policies and procedures.

3-1.1.2 **General Skills Requirements.** Don PrPPE; operate hatches, doors, and evacuation chutes; approach, position, and initially attack an aircraft fire; **select, apply, and manage extinguishing agents**; shut down aircraft systems, including engine, electrical, hydraulic, and fuel systems; operate aircraft extinguishing systems, including cargo area extinguishing systems.

3-3.2 Extinguish an aircraft fuel spill fire, given PrPPE, an assignment, an ARFF vehicle hand line flowing a minimum of 95 gpm (359 L/min) of AFFF extinguishing agent, and a fire sized to the AFFF gpm flow rate divided by 0.13 (gpm/0.13 = fire square footage) (L/min/0.492 = 0.304 m^2), so that the agent is applied using the proper techniques and the fire is extinguished in 90 seconds.

(a) *Requisite Knowledge:* The fire behavior of aircraft fuels in pools, physical properties and characteristics of aircraft fuel, **agent application rates and densities**.

(b) *Requisite Skills:* Operate fire streams and apply agent.

3-3.3 Extinguish an aircraft fuel spill fire, given PrPPE, an ARFF vehicle turret, and a fire sized to the AFFF flow rate of 0.13 gpm (0.492 L/min) divided by the square feet of fire area, so that the agent is applied using the proper technique and the fire is extinguished in 90 seconds.

(a) *Requisite Knowledge:* **Operation of ARFF vehicle agent delivery systems**, the fire behavior of aircraft fuels in pools, physical properties and characteristics of aircraft fuel, **agent application rates and densities**.

(b) *Requisite Skills:* Apply fire-fighting agents and streams using ARFF vehicle turrets.

3-3.4 Extinguish a three-dimensional aircraft fuel fire, given PrPPE, an assignment, and ARFF vehicle hand line(s) using primary and secondary agents, so that a dual agent attack is used, the agent is applied using the proper technique, the fire is extinguished, and the fuel source is secured.

(a) *Requisite Knowledge:* The fire behavior of aircraft fuels in three-dimensional and atomized states, physical properties and characteristics of aircraft fuel, **agent application rates and densities, and methods of controlling fuel sources.**

(b) *Requisite Skills:* Operate fire streams and apply agents, secure fuel sources.

3-3.5 **A**ttack a fire on the interior of an aircraft while operating as a member of a team, given PrPPE, an assignment, an ARFF vehicle hand line, and appropriate agent, so that team integrity is maintained, the attack line is deployed for advancement, ladders are correctly placed when used, access is gained into the fire area, effective water application practices are used, the fire is approached, attack techniques facilitate suppression given the level of the fire, hidden fires are located and controlled, correct body posture is maintained, hazards are avoided or managed, and the fire is brought under control.

(a) *Requisite Knowledge:* Techniques for accessing the aircraft interior according to the aircraft type, **methods for advancing hand lines from an ARFF vehicle, precautions to be followed when advancing hose lines to a fire**, observable results that a fire stream has been applied, dangerous structural conditions created by fire, principles of exposure protection, potential long-term consequences of exposure to products of combustion, physical states of matter in which fuels are found, common types of accidents or injuries and their causes, the role of the backup team in fire attack situations, attack and control techniques, techniques for exposing hidden fires.

(b) *Requisite Skills:* Deploy ARFF hand line on an interior aircraft fire; gain access to aircraft interior; open, close, and adjust nozzle flow and patterns; apply agent using direct, indirect, and combination attacks; advance charged and uncharged hose lines up ladders and up and down interior and exterior stairways; locate and suppress interior fires.

3-3.7 Attack a wheel assembly fire, given PrPPE, an assignment, an ARFF vehicle hand line and appropriate agent, so that the fire is controlled.

(a) *Requisite Knowledge:* **Agent selection criteria**, special safety considerations, and the characteristics of combustible metals.

(b) *Requisite Skills:* Approach the fire in a safe and effective manner, select and apply agent.

3-3.11 Overhaul the accident scene, given PrPPE, an assignment, hand lines, and property conservation equipment, so that all fires are extinguished and all property is protected from further damage.

(a) *Requisite Knowledge:* **Methods of complete extinguishment and prevention of re-ignition**, purpose for conservation, operating procedures for property conservation equipment.

(b) *Requisite Skills:* Use property conservation equipment.

NFPA 1051

9.5.5 **Hazardous Materials.** Analyze the potential involvement of various hazardous materials, given incident information and resources, so that hazardous conditions are identified and mitigated.

(A) Requisite Knowledge. A working knowledge of the types of hazardous materials that can be involved and the hazards they can pose to the public, fire-fighting personnel, and the environment; NFPA 472, *Standard for Professional Competence of Responders to Hazardous Materials Incidents,* First Responder level.

(B) Requisite Skills. None required.

NFPA 472

5.1.2.1 The first responder at the operational level shall be able to perform the following tasks:

(1) Analyze a hazardous materials incident to determine the magnitude of the problem in terms of outcomes by completing the following tasks:

 (a) Survey the hazardous materials incident to identify the containers and materials involved, determine whether hazardous materials have been released, and evaluate the surrounding conditions.

 (b) Collect hazard and response information from MSDS; CHEMTREC/CANUTEC/SETIQ; local, state, and federal authorities; and shipper/manufacturer contacts.

(c) Predict the likely behavior of a material as well as its container.

(d) Estimate the potential harm at a hazardous materials incident.

(2) **Plan an initial response within the capabilities and competencies of available personnel, personal protective equipment, and control equipment by completing the following tasks**:

(a) Describe the response objectives for hazardous materials incidents.

(b) Describe the defensive options available for a given response objective.

(c) Determine whether the personal protective equipment provided is appropriate for implementing each defensive option.

(d) Identify the emergency decontamination procedures.

(3) Implement the planned response to favorably change the outcomes consistent with the local emergency response plan and the organization's standard operating procedures by completing the following tasks:

(a) Establish and enforce scene control procedures including control zones, emergency decontamination, and communications.

(b) Initiate an incident management system (IMS) for hazardous materials incidents.

(c) Don, work in, and doff personal protective equipment provided by the authority having jurisdiction.

(d) **Perform defensive control functions identified in the plan of action**.

(4) Evaluate the progress of the actions taken to ensure that the response objectives are being met safely, effectively, and efficiently by completing the following tasks:

(a) Evaluate the status of the defensive actions taken in accomplishing the response objectives.

(b) Communicate the status of the planned response.

5.2.1 Surveying the Hazardous Materials Incident. Given examples of both facility and transportation scenarios involving hazardous materials, the first responder at the operational level shall survey the incident to identify the containers and materials involved, determine whether hazardous materials have been released, and evaluate the surrounding conditions and also shall meet the requirements in 5.2.1.1 through 5.2.1.6.

5.2.1.1 Given three examples each of liquid, gas, and solid hazardous materials, including various hazard classes, the first responder at the operational level shall identify the general shapes of containers in which the hazardous materials are typically found.

(A) Given examples of the following tank cars, the first responder at the operational level shall identify each tank car by type as follows:

(1) Cryogenic liquid tank cars

(2) High-pressure tube cars

(3) Nonpressure tank cars

(4) Pneumatically unloaded hopper cars

(5) Pressure tank cars

(B) Given examples of the following intermodal tanks, the first responder at the operational level shall identify each intermodal tank by type and identify at least one material and its hazard class that is typically found in each tank as follows:

(1) Nonpressure intermodal tanks, such as the following:

(a) IM-101 (IMO Type 1 internationally) portable tank

(b) IM-102 (IMO Type 2 internationally) portable tank

(2) Pressure intermodal tanks

(3) Specialized intermodal tanks, such as the following:

(a) Cryogenic intermodal tanks

(b) Tube modules

(C) Given examples of the following cargo tanks, the first responder at the operational level shall identify each cargo tank by type as follows:

(1) Nonpressure liquid tanks

(2) Low pressure chemical tanks

(3) Corrosive liquid tanks

(4) High pressure tanks

(5) Cryogenic liquid tanks

(6) Dry bulk cargo tanks

(7) Compressed gas tube trailers

(D) Given examples of the following tanks, the first responder at the operational level shall identify at least one material, and its hazard, that is typically found in each tank as follows:

(1) Nonpressure tank

(2) Pressure tank

(3) Cryogenic liquid tank

(E) Given examples of the following nonbulk packages, the first responder at the operational level shall identify each package by type as follows:

(1) Bags

(2) Carboys

(3) Cylinders

(4) Drums

(F) Given examples of the following radioactive material containers, the first responder at the operational level shall identify each container/package by type as follows:

(1) Type A

(2) Type B

(3) Industrial

(4) Excepted

(5) Strong, tight containers

5.3.2 Identifying Defensive Options. Given simulated facility and transportation hazardous materials problems, the first responder at the operational level shall identify the defensive options for each response objective and shall meet the following requirements:

(1) **Identify the defensive options to accomplish a given response objective.**

(2) **Identify the purpose for, and the procedures, equipment, and safety precautions used with, each of the following control techniques:**

(a) Absorption

(b) Dike, dam, diversion, retention

(c) Dilution

(d) Remote valve shutoff

(e) Vapor dispersion

(f) **Vapor suppression**

5.4.4 Performing Defensive Control Actions. Given a plan of action for a hazardous materials incident within their capabilities, the first responder at the operational level shall demonstrate defensive control actions set out in the plan and shall meet the following related requirements:

(1) **Using the type of fire-fighting foam or vapor suppressing agent and foam equipment furnished by the authority having jurisdiction, demonstrate the effective application of the fire-fighting foam(s) or vapor suppressing agent(s) on a spill or fire involving hazardous materials.**

(2) Identify the characteristics and applicability of the following foams:

(a) Protein

(b) Fluoroprotein

(c) Special purpose

i. Polar solvent alcohol-resistant concentrates

ii. Hazardous materials concentrates

(d) Aqueous film-forming foam (AFFF)

(e) High expansion

(3) **Given the required tools and equipment, demonstrate how to perform the following defensive control activities:**

(a) Absorption

(b) Damming

(c) Diking

(d) Dilution

(e) Diversion

(f) Retention

(g) Vapor dispersion

(h) **Vapor suppression**

(4) Identify the location and describe the use of the mechanical, hydraulic, and air emergency remote shutoff devices as found on cargo tanks.

(5) Describe the objectives and dangers of search and rescue missions at hazardous materials incidents.

(6) Describe methods for controlling the spread of contamination to limit impacts of radioactive materials.

6.2.1.2 Given three examples of facility and transportation containers, the hazardous materials technician shall identify the approximate capacity of each container.

(A) Using the markings on the container, the hazardous materials technician shall identify the capacity (by weight and/or volume) of the following **examples of transportation vehicles:**

(1) **Cargo tanks**

(2) **Tank cars**

(3) Tank containers

(B) Using the markings on the container and other available resources, the hazardous materials technician shall **identify the capacity (by weight and/or volume) of each** of the following facility containers:

(1) **Nonpressure tank**

(2) **Pressure tank**

(3) **Cryogenic liquid tank**

6.4.3 Performing Control Functions Identified in Plan of Action. Given various simulated hazardous materials incidents involving nonbulk and bulk packaging and facility containers, the hazardous materials technician shall **select the tools, equipment, and materials for the control of hazardous materials incidents** and identify the precautions for controlling releases from those packaging/containers and shall meet the following related requirements:

(9) **Identify the methods and precautions used when controlling a fire involving an MC-306/DOT-406 aluminum shell cargo tank.**

8.3.4.1 Performing Response Options Specified in the Plan of Action. Given an assignment by the incident commander in the employee's individual area of specialization, the private sector specialist employee B shall perform the assigned actions consistent with the organization's emergency response plan and standard operating procedures and shall meet the following related requirements:

(1) **Perform assigned tasks consistent with the organization's emergency response plan and standard operating procedures and the**

available personnel, tools, and equipment (including personal protective equipment), including the following:

(a) **Confinement activities**

(b) **Containment activities**

9.1.2.2 In addition to being competent at the awareness, operational, and technician levels, the hazardous materials branch officer shall be able to perform the following tasks:

(2) **Plan a response within the capabilities and competencies of** available personnel, personal protective equipment, and **control equipment by completing the following tasks:**

(a) Identify the response objectives for hazardous materials incidents.

(b) **Identify the potential action options (defensive, offensive, and nonintervention) available by response objective.**

9.3.2 Developing a Plan of Action. Given simulated facility and transportation hazardous materials incidents, the hazardous materials branch officer shall **develop a plan of action** consistent with the local emergency response plan and the organization's standard operating procedures that is within the capability of the available personnel, personal protective equipment, and control equipment and shall meet the following related requirements:

(3) Given the local emergency response plan or the organization's standard operating procedure, identify procedures to accomplish the following tasks:

(e) **Coordinate with fire suppression services as it relates to hazardous materials incidents.**

(f) **Coordinate hazardous materials branch control, containment, or confinement operations.**

9.5.1 Evaluating Progress of the Plan of Action. Given simulated facility and transportation hazardous materials incidents, the hazardous materials branch officer shall **evaluate the progress of the plan of action** to determine whether the efforts are accomplishing the response objectives and shall meet the following related requirements:

(3) **Determine the effectiveness of the following:**

(a) Hazardous materials response personnel being used

(b) Personal protective equipment

(c) Established control zones

(d) **Control, containment, or confinement operations**

(e) Decontamination process

11.1.2.2 In addition to being competent at the hazardous materials technician level, the technician with a tank car specialty **shall be able to perform the following tasks:**

(2) **Plan a response for an emergency involving tank cars within the capabilities** and competencies **of available** personnel, personal protective equipment, and **control equipment by completing the following tasks:**

(a) **Determine the response options (offensive, defensive, and nonintervention) for a hazardous materials emergency involving tank cars.**

(b) **Ensure that the options are within the capabilities** and competencies **of available** personnel, personal protective equipment, and **control equipment.**

11.4.1 Implementing the Planned Response. Given an analysis of an emergency involving tank cars and the planned response, technicians with a tank car specialty **shall implement or oversee the implementation of the selected response options safely and effectively and shall meet the following related requirements:**

(10) **Given a simulated flammable liquid spill from a tank car, describe the procedures for site safety and fire control during cleanup and removal operations.**

12.2.2 Predicting the Likely Behavior of the Cargo Tank and Its Contents. Technicians with a cargo tank specialty shall predict the likely behavior of the cargo tank and its contents and shall meet the following related requirements:

(1) **Given the following types of cargo tanks (including a tube trailer), describe the likely breach/release mechanisms:**

(a) **MC-306/DOT-406 cargo tanks**

(b) **MC-307/DOT-407 cargo tanks**

(2) **Describe the difference in types of construction materials used in cargo tanks** and their significance in assessing tank damage.

(3) Describe the significance of the jacket on cargo tanks in assessing tank damage.

(4) Describe the significance of each of the following types of damage on different types of cargo tanks in assessing tank damage:

(e) **Flame impingement**

Chapter 8
Class B Foam Fire Fighting: Transportation Incidents

While not all jurisdictions are responsible for protecting fixed sites that manufacture, store, or ship flammable and combustible liquids, all jurisdictions have transportation networks where these Class B materials travel in one form or another. Highways, railways, airports, pipelines, and waterways create shipping routes that cross and intersect North America, creating a variety of opportunities for fire and emergency services personnel to respond to transportation incidents.

Streets, roads, and highways carry thousands of motor vehicles daily that are propelled by gasoline, diesel, or alternative fuels. Those same roadways provide the means for transferring large quantities of Class B liquids across North America. From small quantities in individual containers carried in box van trucks to large quantities in tractor-trailer transport trucks, Class B liquid transports are a common sight on the roadways. Statistics show that nearly one-quarter of all fire department responses involve vehicle fires of some type (**Figure 8.1**).

Figure 8.1 Vehicle fires account for approximately 25 percent of all fire department responses in the United States.

Railroads, which have crisscrossed the landscape for close to two centuries, carry large quantities of flammable and combustible liquids. The vast majority of railroad engines are diesel-powered and carry hundreds of gallons (liters) of fuel in tanks located on the mainframe of the engine. Railroad tank cars in a variety of sizes and designs carry Class B liquids, hazardous materials, and flammable gases. Freight cars carry the same containers of flammable and combustible liquids in generally the same quantities that the box van trucks do on the highway.

Aircraft incidents may occur at any location within a given jurisdiction and not just at the airport that serves that region. Fire and emergency services personnel are often called upon to respond to aircraft crash sites in open fields, rugged terrain, or urban areas (**Figure 8.2**). The fuel that powers aircraft can create intense heat when ignited and

Figure 8.2 Structural fire-fighting units may be the first to arrive at aircraft crash sites that are located away from the airport. Foam extinguishing agents are beneficial in suppressing fires and allowing the safe extrication of trapped victims. *Courtesy of Steven Sabo.*

can be spread over large areas at crash sites. Smaller quantities of other flammable and combustible liquids may also contribute to incidents. These liquids may be part of the aircraft's hydraulic or mechanical system or be part of the cargo carried by the aircraft.

Pipelines that carry crude oil, natural gas, and finished product may be found above- or belowground throughout North America. They connect storage facilities with production facilities and transshipment centers with retail markets **(Figure 8.3)**. While parts of this network are part of the fixed-site facilities discussed in Chapter 6, Class B Foam Use: Unignited Class B Liquid Spills, and Chapter 7, Class B Foam Fire Fighting: Class B Liquids at Fixed Sites, the pipelines themselves and the intermittent valve stations are considered part of a transportation network.

Finally, jurisdictions may have the responsibility for protecting property located on lakes, streams, rivers, manmade waterways, ports, or coastal areas. Recreational powered boats equipped with gasoline-fueled motors are found in all of these areas. Towboats, barges, and shallow-draft cargo carriers may be found on rivers well inland of normal port areas. For example, the Mississippi and Ohio Rivers host large paddlewheel riverboats that are designed as floating hotels and casinos. Tankers that carry hundreds of thousands of gallons (liters) of crude oil and finished petroleum products may be found on inland waterways, in port cities, and close to shore along coastal states. Cargo ships may also carry materials shipped in bulk in small containers, on pallets, or in steel containers the size of truck trailers. In some port cities and islands close to shore, ferries carry motor vehicles and passengers on a daily basis. Ports that act as transshipment points (where cargo is delivered by ship and transferred to other types of transport or vice-versa) have the potential for emergency incidents involving a wide variety and quantity of flammable and combustible liquids, hazardous materials, and other types of fuels. In addition, offshore drilling platforms exist along the shorelines of many coastal states, creating hazards similar to those found at petroleum refining plants.

Figure 8.3 Pipelines are easily damaged by construction equipment. In this photograph, foam is used to control the spread of a natural gas fire in an urban area. *Courtesy of Howard Meile III.*

Foam concentrates are valuable assets in controlling emergencies involving transportation incidents. Regardless of whether the emergency involves unignited or ignited Class B flammable and combustible liquids or other hazardous materials, foam concentrates are more affective than plain water or other types of extinguishing agents. The strategies and tactics for handling transportation incidents generally follow those already discussed for fixed-site unignited and ignited incidents. The primary differences are the locations of the incidents (which may be remote and difficult to access or in the midst of populated areas) and the lack of fixed-site extinguishing systems to control incidents before the arrival of fire and emergency personnel. In this chapter, each of the types of transportation incidents (roadway, railway, airway, pipeline, and maritime) are discussed as they relate to the use of foam concentrates for controlling both unignited and ignited spills at these emergencies.

Motor Vehicles

The greatest potential that fire and emergency services personnel have for transportation incidents involves motor vehicles. All jurisdictions have streets, roads, and highways within their areas of responsibility. An emergency incident may be the result of a collision or a spill caused by overfilling a vehicle's fuel tank or the tank of a transport truck. It may involve the vehicle's fuel system or fuel carried

Figure 8.4 Passenger vehicles and light trucks may carry from 12 to 35 gallons (45 L to 132 L) of fuel that can contribute to the intensity of a fire. *Courtesy of Ron Jeffers.*

Figure 8.5 Many government agencies are purchasing vehicles powered by alternative fuel sources such as compressed natural gas.

as cargo by the vehicle. This section discusses incidents involving passenger vehicles and light-duty trucks, medium- and heavy-duty trucks, and alternative fuel vehicles that result in the spill and ignition of the vehicle's onboard fuel tank(s) used to supply the vehicle's engine (**Figure 8.4**).

Automobiles and Light-Duty Trucks

The most common type of Class B spills and fires that fire and emergency services personnel encounter are emergencies involving the fuel tanks of automobiles and light-duty trucks (such as pickup trucks and vans). These vehicles may be powered by a variety of fuels including diesel fuel, gasoline, a gasoline/alcohol mixture, propane, natural gas, or hybrid gasoline/electric systems (**Figure 8.5**). The procedures for handling spills and fires involving these vehicles is the same, regardless of the fuel types carried by the vehicles.

Passenger vehicles normally carry anywhere from 12 to 35 gallons (45 L to 132 L) of fuel. Fuel tanks in the 14- to 24-gallon (53 L to 91 L) range are the most common sizes. Light-duty trucks and vans may carry up to 40 gallons (151 L) of fuel, unless specially outfitted with extra tanks. Medium- and heavy-duty trucks (which are discussed later) may carry fuel in excess of 100 gallons (379 L) in multiple tanks on the frame rail of the vehicle.

Fuel spills and fires involving these types of vehicles pose no significant difficulties for fire and emergency services personnel. A single engine company and crew can handle most of these incidents easily. Few incidents occur where all of the fuel from the vehicle tank is dumped to the ground. In reality, most vehicle fires are extinguished like a standard Class A fire. Most are successfully controlled with plain water. However, fuel spills and vehicle fires may also be controlled with either Class A or Class B foam concentrates. The authority having jurisdiction should develop a protocol to help determine which method of extinguishment (water or finished foam) to use to control the situation. Factors that must be considered are the location of the incident, *RECEO* (rescue, exposures, confine, extinguish, overhaul) requirements, availability of foam concentrates, quantity of spilled or ignited fuel, the length of time the incident has been underway, and other available resources (such as personnel, water supply, and equipment).

In general, the tactics for controlling a vehicle fuel spill and fire parallel those mentioned earlier in this manual for unignited and ignited flammable and combustible liquid incidents. All fire and emergency services personnel who participate in rescue or attack must wear full personal protective equipment (PPE), including self-contained breathing apparatus (SCBA). The smallest hoseline used for attack is a 1½-inch (38 mm) hoseline with a minimum flow rate of 95 gpm (360 L/min), although a 1¾-inch (45 mm) hose flowing between 95 and 175 gpm (360 L/min and 662 L/min) is becoming more common. A second hoseline is also deployed as backup and to protect the attack line. Although many departments routinely attack these types of spills or fires with 1-inch (25 mm) reel-mounted booster lines, this practice is not recommended. The flow from booster lines (usually less than 30 gpm [114 L/min]) is not sufficient to protect emergency personnel or victims if the spill should ignite quickly or the fire increase in intensity. In addition, booster lines do not have the capacity necessary to produce adequate finished foam.

The initial fire and emergency services personnel to arrive on the scene make a size-up of the incident. Rescue is the primary concern regardless of whether the incident involves an unignited or ignited spill. If a limited supply of foam concentrate is available, it may be used to blanket the unignited spill and provide a safe area to remove trapped victims. Once rescue is accomplished, the foam system may be shut down to conserve the foam concentrate until additional supplies arrive. If the fuel has already ignited, finished foam can be quickly applied to extinguish the fire, provide an entry path, and allow personnel to perform rescues. If the quantity of foam concentrate is limited, water fog may be used to continue extinguishment following the rescues (**Figure 8.6**).

The need for exposure protection is determined during size-up, depending on whether the incident involves an unignited or ignited spill. The limited quantity of fuel involved in automobile or light-duty truck incidents generally does not create a major exposure hazard. The exceptions to this situation are incidents involving vehicles within structures such as residential garages, parking garages, or vehicle repair facilities or in underground roadways such as tunnels. In these cases, the structure then becomes the exposure and must be protected from the potential of an ignited spill. Any fuel that spills in an enclosed garage can result in a buildup of vapors that could reignite with explosive force. The resultant fireball can escape through any open door or window. In rare cases, this fireball has sprayed burning fuel over the fireground, causing serious injury or death to personnel on the scene. For this reason, fire and emergency services personnel should attack these fires slightly to the side of the garage doors rather than straight ahead. When a fire appears to be knocked down, continue to cool the area and establish thorough ventilation to reduce the buildup of vapors inside the structure. Compressed-air foam systems (CAFSs), which have a greater reach than Class A finished foam applied from straight stream or fog nozzles, can also be affective by allowing personnel to put finished foam into the structure from a greater distance. Class A finished foam in its driest concentration may be used to provide a barrier on vertical and horizontal interior surfaces if necessary. Wet, high-expansion finished foam of Class B foam concentrate may also be used for total flooding within a compartment where the vehicle fire has occurred. In either case, foam concentrate can be used to rapidly extinguish the fire or to blanket the spill to protect exposures.

Figure 8.6 If foam concentrate is not available, water fog can be used to extinguish fires in automobiles and light trucks due to the limited quantity of flammable liquids in the vehicles. *Courtesy of Ron Jeffers.*

The containment of an unignited or ignited fuel does not require a great deal of effort due to the limited quantity of fuel involved. Class B finished foam can blanket the fuel and hold it in one area. Absorbent chemicals or dams can contain larger quantities and prevent them from entering stormwater sewers. If necessary, dams can be constructed with dirt to stop the travel of the fuel and absorb it at the same time. However, the dirt must be gathered at the end of the incident and disposed of in accordance with the jurisdiction's hazardous materials protocol.

Extinguishment may take place simultaneously with rescue and exposure protection due to the limited size of the fuel and vehicle involved in an incident. Tactics for extinguishment are the same whether water or finished foam is used as the extinguishing agent. As fire and emergency services personnel approach a vehicle, they should use a narrow (10-degree) fog pattern to blanket the spill or extinguish any ground fire around or beneath the

Figure 8.7 Class A foam can be used for vehicle fires to provide a blanket around and under the vehicle. *Courtesy of Tom Ruane.*

vehicle (**Figure 8.7**). When a spill is covered or the fire around or beneath a vehicle is extinguished, emergency personnel can use a wider fog pattern to extinguish the remainder of the fire in the vehicle. Approach the vehicle from the sides to avoid the hazards of shock-absorbing bumpers (front and rear), which are under pressure, and the fuel tank, which is usually located at the rear of the vehicle.

Explosions or total failures of vehicle fuel tanks are rare. Most fuel tank failures resulting in spills or fires are caused by collision damage. Safety relief valves/plugs or fuel hoses generally melt or burn away during a vehicle fire and allow fuel to slowly escape before a tank actually explodes. If fire and emergency services personnel are concerned with overpressurization of the fuel tank because of heat or impinging flames, cool the tank with a water or finished foam hoseline.

WARNING

Do not remove the fuel filler-line cap to relieve excess pressure from the fuel tank. This removal results in a sudden, large release of heated fuel vapors that can instantaneously ignite and envelop firefighters in a fireball, resulting in serious injury or death.

When using Class B foam concentrate to extinguish a vehicle fire, blanket the entire vehicle and the surrounding area with finished foam 4 inches (102 mm) deep if possible. Cover all remaining fuel puddles with finished foam until they are cleaned up. Standard small-volume (1¾-inch [45 mm]) foam handlines can be used for this purpose. Plain gasoline, which has many additives, blended formulas (see Chapter 6, Class B Foam Use: Unignited Class B Liquid Spills), and gasoline/alcohol mixtures, may react to finished foam in a manner similar to that of polar solvent fuels. Polar-solvent compatible foam concentrates may be needed on these fuels. If fire and emergency services personnel are using a 3 by 6 concentrate, they begin the attack at the 6-percent proportioning setting.

Small vehicle fires such as minor engine compartment fires can be extinguished with dry chemical or aqueous film forming foam (AFFF) portable fire extinguishers. However, when using a portable extinguisher, it is always a good idea to deploy a hoseline as a backup in the event the portable extinguisher attack is inadequate. Regardless of what method is employed, personnel must never turn their backs on a fire once it is extinguished. These fires can reignite at any time. Always be prepared for a reignition.

Medium- and Heavy-Duty Trucks

Medium- and heavy-duty trucks are vehicles larger than a pickup truck or van. These vehicles can range from a medium-sized delivery van up to a large tractor-trailer rig. The vast majority of these vehicles are diesel-powered. In spite of their large sizes, these vehicles do not present any significantly greater hazards than do smaller automobiles. Even the largest over-the-road tractors rarely carry more than approximately 150 gallons (568 L) of fuel (when full), and most are in the 80- to 100-gallon (303 L to 379 L) range. Medium- and heavy-duty trucks have fuel tanks (called *saddle tanks*) that are located on one or both sides of the frame rail near the cab and are usually visible from a distance (**Figure 8.8**). Motor homes and buses, which are built on medium- and heavy-duty truck chassis, have fuel tanks that are located under the body of the vehicle and hidden from sight. Take caution when approaching these types of vehicles. The location of the fuel filler cap or compartment door usually indicates the location of the tank or the existence of a second tank.

The tactics previously described for automobile and light-duty truck fires are also applicable for medium- and heavy-duty truck fires. The only

Figure 8.8 Tractor-drawn trailers carry between 80 and 100 gallons (303 L to 379 L) of fuel in saddle tanks mounted on the chassis. Note the destruction to the tank, which has melted down to the level of the fuel.

difference is that because of the increased vehicle size and fuel capacity, greater flow rates or additional attack lines are required. It is usually wise to put a second 1¾-inch (45 mm) or larger attack line in service. Typically, this procedure means that at least a second engine company and crew are required to provide the backup hoselines, required foam concentrate, and water supplies.

While an automobile or light-duty truck incident can usually be handled with the water supply carried by one engine, incidents involving medium- and heavy-duty trucks may require additional water. Depending on the size of the spill or the intensity of the fire, fire apparatus equipped with water tanks of less than 1,000 gallons (3 785 L) may require additional water to complete extinguishment. This water may be supplied from the second engine company's water tank, a supply line laid to a hydrant, a water tanker/tender, or mobile water supply apparatus.

Just as additional water may be required, additional foam concentrate may also be needed. Apparatus equipped with small quantities of foam concentrate (less than 30 gallons [114 L]) may need additional concentrate to complete extinguishment and to cover any fuel spills with a blanket of 4 inches (102 mm) of finished foam. Identify sources for the additional foam concentrate before any incident and include them in the organization's protocol for foam concentrate use. Available sources include the next-arriving engine company, the organization's logistics section, mutual aid, nearby industrial complexes, and foam concentrate distributors or manufacturers.

Alternative Fuel Vehicles

An increasing number of automobiles, trucks, and busses are powered by liquefied petroleum gas (LPG), liquefied natural gas (LNG), compressed natural gas (CNG), or battery-powered hybrids. A sticker or placard somewhere on the vehicle usually identifies these alternative fuel vehicles. The location of the fuel tank varies depending on the type of vehicle. Some locations are as follows:

- ***Pickup trucks*** — Directly behind the cab in the truck bed **(Figures 8.9 a and b)**
- ***School buses*** — Between the frame rails at the rear of the vehicle
- ***Medium- and heavy-duty trucks*** — Between the frame rails at the rear of the vehicle or on the frame rails near the cab

Figure 8.9a In pickup trucks, compressed natural gas (CNG) is usually carried in tanks that are visible in the bed of the truck. *Courtesy of Conoco, Inc.*

Figure 8.9b All vehicles that operate on CNG are required to display the symbol shown here. *Courtesy of Conoco, Inc.*

As mentioned earlier in this manual, Class A and Class B foam concentrates are not directly effective on gaseous Class B fuels. However, both Class A and Class B foam concentrates can effectively extinguish a vehicle fire, allow a fuel tank to cool, and protect responders while they turn off the fuel source. Finished foam may also be used to provide exposure protection while the fire is consuming the fuel. It may also be used to disperse the gas vapors that are released into the air from an unignited gas leak.

Hybrid automobiles have small gasoline-powered engines that operate the vehicle part of the time and an electric motor that powers it the rest of the time. The fuel tank is small so the hazard is similar to the average automobile or motorcycle. The batteries pose a problem because they may rupture in a collision or explode when exposed to heat. Use finished foam to blanket and cool the fuel tank and battery system.

Tank Vehicles

Fires involving transport tank trucks present a greater challenge to fire and emergency services personnel than do the incidents involving only a vehicle's fuel tank because of the larger amounts of fuel involved. However, most tank vehicles are equipped with safety features that are intended to prevent or reduce the possibility of a spill or fire. These include shutoff valves, pressure relief valves, safety or rupture disks, and combination devices. Descriptions of tank vehicles are given in the sections that follow along with the types of spill and fire hazards these vehicles present and suggested tactics for handling these spills and fires.

Types and Construction

The Vehicle Inventory and Use Survey (VIUS) made in 1997 by the U.S. Census Bureau, indicates that 403,000 trucks of all types are in use to transport hazardous materials in the United States. That number equates to 0.55 percent of the total 72.8 million trucks on the road. In 1997, trucks hauling hazardous materials accumulated over 25 billion miles (40 billion km). Flammable and combustible liquid transport tank vehicles that are part of that hazardous materials fleet are defined in NFPA 385, *Standard for Tank Vehicles for Flammable and Combustible Liquids* (2000), as follows:

- ***Tank Vehicle*** — Any tank truck, tank full-trailer, or tractor and tank semitrailer combination
- ***Tank, Full-Trailer*** — Any vehicle with or without auxiliary motive power, equipped with a cargo tank mounted thereon or built as an integral part thereof, used for the transportation of flammable and combustible liquids or asphalt, and so constructed that practically all of its weight and load rests on its own wheels; that is, a trailer with axles at the front and rear of the frame capable of supporting the entire weight of the tank
- ***Tank, Semitrailer*** — Any vehicle with or without auxiliary motive power, equipped with a cargo tank mounted thereon or built as an integral part thereof, used for the transportation of flammable and combustible liquid or asphalt, and so constructed that, when drawn by a tractor by means of a fifth-wheel connection, some part of its load and weight rests upon the towing vehicle **(Figure 8.10)**
- ***Tank Truck*** — Any single self-propelled motor vehicle equipped with a cargo tank mounted thereon and used for the transportation of flammable and combustible liquids or asphalt; a tank truck with two or three axles on which the tank is permanently affixed (sometimes referred to as a *bobtail* truck)

The cargo tanks of tank vehicles are constructed in accordance with specifications developed by the American Society of Mechanical Engineers (ASME) and requirements of the *Code of Federal Regulations (CFR)*, Title 49 (Transportation) Part 178, *Specifications for Packagings*. Title 49 divides the tank vehicles into five types (see textbox, p.258). While

Figure 8.10 A typical semitrailer tank trailer.

five types of tank vehicles are designed to carry hazardous liquids, only two, U.S. Department of Transportation (DOT) and Transport Canada (TC) DOT/TC 406 and DOT/TC 407, contain flammable or combustible liquids. The most common of the two is the DOT/TC 406 design. It is the typical gasoline carrier found on highways. It has improved safety features and is replacing the MC (multitank cargo) 306 design (which may still be encountered). The safety improvements include lower center of gravity, stronger construction, increased tank shell thickness, and increased dome cover, relief valve, and rollover performance. Over 90 percent of these tanks are constructed of aluminum although some mild steel or stainless steel tanks do exist. The DOT/TC 406 and MC 306 type tank vehicles constitute 60 percent of the tank vehicle fleet in North America.

The DOT/TC 407 or older MC 307 tank vehicle constitutes 20 percent of the fleet. The remaining 20 percent are designs other than 406 and 407 (MC 312, 331, and 338). DOT/TC 407/MC 307 tanks are designed for liquids with low vapor pressures that are not more than 40 psi (276 kPa) at 70°F (21°C). Considered low-pressure containers, they are usually constructed of stainless steel but may also be made of aluminum or mild steel.

Cargo Tank Truck Types and Typical Cargos (Title 49 *CFR* 180.405)

- ***DOT/TC 406/MC 306 atmospheric-pressure tank truck*** — Gasoline, fuel oil, alcohol, flammable/combustible liquids, liquid fuel products, and other liquids
- ***DOT/TC 407/MC 307 low-pressure tank truck*** — Flammable/combustible liquids, acids, caustics, and poisons
- ***MC 312 corrosive-liquid tank truck*** — Acids and corrosive liquids
- ***MC 331 high-pressure tank truck*** — Gases that have been liquefied under pressure such as anhydrous ammonia, propane, butane, and others
- ***MC 338 cryogenic liquid tank truck*** — Gases that have been liquefied by lowering their temperature such as liquid oxygen, liquid hydrogen, liquid nitrogen, and liquid carbon dioxide

While the DOT/TC 406 and 407 cargo tanks mounted on trailers are usually constructed of aluminum, the smaller tank trucks (bobtail trucks) may have steel cargo tanks **(Figure 8.11)**. They may carry from 2,000 to 10,000 gallons (7 570 L to 37 854 L) of product within their single or multicompartment interiors. A minimum of 2 to 3 percent of each compartment must be vapor space to allow for liquid expansion.

The cargo tanks of both DOT/TC 406/MC 306 and DOT/TC 407/MC 307 tank vehicles may consist of a single compartment or multiple compartments separated by a bulkhead. Cargo tank capacities may range from 500 gallons (1 893 L) to as much as 8,000 gallons (30 283 L) of fuel. If the cargo tank is divided into multiple compartments, the fuel carried may be all the same type or a variety of types. Most tank vehicles have a frameless, unibody construction with the cargo tank acting as the frame. Cargo tanks that are mounted on truck bodies are usually mounted on a flatbed or directly onto the truck's frame rails. Access to the top of the tank is by way of a ladder at the rear of the tank and a walkway on the top. Gross vehicle weight (GVW) is comprised of the weight of the tank truck and its cargo. The most common maximum gross vehicle weight is 80,000 pounds (36 288 kg). Most flammable and combustible liquid tank trucks do not exceed 9,200-gallon (34 826 L) capacities or approximately 73,600 pounds (33 385 kg) of liquid. A few states/provinces allow up to 13,000-gallon (49 210 L) capacities or approximately 104,000 pounds (47 174 kg) of liquid.

Loading, unloading, and vapor-recovery piping are usually located on the bottom of full- and semi-trailer tanks while tank trucks have the piping and

Figure 8.11 This airport fuel truck is a small tank truck (bobtail) that has an aluminum tank.

discharge hoses located in rear- or side-mounted compartments. Each cargo tank compartment has a discharge valve located just inside the tank where the piping enters. These spring-loaded valves remain closed when the tank vehicle is in transit or parked. They are only activated during loading and unloading operations. These valves are also equipped with an automatic, heat-activated closure system consisting of a fusible link that operates at temperatures not greater than 250°F (121°C). In addition, the tank piping is designed to shear off in the event of a collision, which prevents damage to the cargo tank shell and the internal valve. All compartments in excess of 2,500 gallons (9 464 L) must have an access hatch. These hatches have rollover protection and a safety device that prevents the hatch from opening fully when excessive internal tank pressure exits.

Fire Strategies/Tactics

Two basic types of fires may involve tank vehicles: spill fires and vent or relief device fires. Each of these types of fires calls for a different control method. In dealing with both types, fire and emergency services personnel must be careful not to make the following common mistakes:

- ***Misuse or overuse of available water*** — Too much water can cause the liquid pool to spread. If the top of the cargo tank has burned, too much water can displace the fuel and cause it to overflow the tank shell.
- ***Poor foam application techniques*** — Direct application of the foam stream can result in plunging the finished foam into the burning liquid, which can spread the liquid out of the tank shell. Improper foam application techniques can also result in inadequate finished foam blankets that are too thin to maintain a complete vapor barrier.
- ***Applying too little finished foam*** — A fire can overpower too little finished foam and waste foam concentrate.
- ***Inappropriate tactics*** — Not adhering to the accepted *RECEO* protocol can cause injuries, deaths, equipment losses and wasted efforts on the part of responders. It may also result in the fire becoming out of control or the contamination of the atmosphere, water, or surrounding soil.
- ***Poor confinement/containment and lack of environmental protection*** — These situations can result in the spread of fuel or fire into uncontaminated areas.
- ***Insufficient training and preplanning*** — Poor training can result in personnel attempting to exceed their capabilities. A lack of preplanning can result in confusion, wasted effort, and increased damage to the site and environment.

Spill Fires

The majority of tank vehicle fires are spill fires. These fires result from physical damage to the tank by a vehicle rollover or a collision with another vehicle. Less common are fires that occur while loading or unloading the tanks. Also possible, but not very common, are fires that begin in the vehicle's tractor or engine compartment and spread to the cargo tank or trailer or fires that occur as a result of the cargo tank being exposed to another fire.

As with the unignited spill incidents discussed in Chapter 6, Class B Fire Use: Unignited Class B Liquid Spills, the most crucial action that must be taken during the initial size-up of the incident is identification of the fuel involved in the fire. The bills of lading or shipping papers that contain information on the material being transported are kept in the truck/tractor cab near the driver. In addition, vehicles transporting flammable and combustible liquids or other hazardous materials are required to be properly labeled using the placards shown in NFPA 704, *Standard System for the Identification of the Hazards of Materials for Emergency Response* (2001). The basic information on identification contained in Chapter 7, Class B Foam Fire Fighting: Class B Liquids at Fixed Sites, also pertains to these fire scenarios. Consult IFSTA's **Hazardous Materials for First Responders** manual for more information on identifying the fuels involved.

Once rescue has been accomplished, exposures protected, and the spill fire contained to a controllable area, direct hoselines on the transport vehicle to cool the container and into the smoke plume to disperse the column into the atmosphere. Keep Class B foam hoselines in place to keep the finished foam blanket on the unignited or controlled spill intact. Details of the various tactics and nonintervention strategy for thse fires are given in the paragraphs that follow.

Rescue tactics. Once the fuel has been identified, fire and emergency services personnel must turn their attention to rescue situations. If safely possible, check the tank vehicle cab or tractor, other vehicles involved in the collision, or any exposures to see if victims are present. Checking for victims may not be immediately possible if the fuel has been identified as having toxic properties. If the fuel has toxic properties, a hazardous materials technician wearing special encapsulating protective clothing is required to perform the search for victims. Either protective fog water streams or foam solution streams may be used to allow fire and emergency services personnel to make a close approach to victims. If possible, use attack hoselines to apply fog streams to flush burning fuel away from victims, reducing the amount of heat, and making the rescue operation safer. If the fire in the victims' area cannot be extinguished quickly, remove victims immediately and take them to a position of greater safety for further evaluation and treatment.

Exposure protection tactics. Most tank vehicle fires occur in the open and present few exposure problems. However, sometimes the incident and fire may be in urban or suburban areas that contain structures and civilians. Obviously, if exposures are involved, they become the next priority after rescue **(Figure 8.12).** Fire and emergency services personnel can use standard structural exposure protection tactics. If feasible, water fog streams may be used to flush burning liquid pools away from exposures. Only do this flushing if the direction of the runoff is controlled and contaminated water is contained. Another possibility is that fire and emergency services personnel can use Class B foam hoselines to control the fire in the area of the exposure. Class B finished foam can be used even if the overall strategy is to allow the bulk of the fuel to be consumed. A dry Class A foam can also blanket and protect the faces of the exposures. If part of the exposure is the tank vehicle tractor or one of the tandem trailers, it may be possible to move the exposure. While providing foam protection, attach a tow truck to the tractor or tandem trailer and pull the vehicle from the spill area.

Figure 8.12 In this photograph, the fire was centralized around the tractor's fuel tanks. The semitrailer tank and the storage tanks became the immediate exposure concern for emergency responders.

During the initial size-up, the incident commander must determine if the quantity of water and foam concentrate on hand is sufficient to properly control the situation. If supplies are limited, then the use of the foam concentrate must be prioritized. The primary use would be for rescue followed by exposure protection.

Spill-confinement tactics. If any unignited or ignited fuel runs from the tank and spreads across the ground, take immediate control measures to dam and dike the spill. The procedures discussed for damming and diking unignited fuel in Chapter 6, Class B Foam Use: Unignited Class B Liquid Spills, generally apply for fire incidents as well. The only exception is not to use containment booms or other materials that are combustible and usually not effective in fire situations. The preferred choices are earthen materials such as sand, pea gravel, or topsoil. While constructing dikes and dams, always pay attention to the surroundings, wear full personal protective equipment, and ensure that burning fuel does not cut off escape routes. Deploy hoselines to protect personnel who are preparing the dam or dike. When possible, keep the fuel as close to the burning tank as possible. This action minimizes the size of the area impacted by the fire and contamination of the soil surface. If it is not possible to confine the fuel to the area of the tank, attempt to divert it to a remote location free of exposures or other hazards.

Nonintervention strategy. Generally, the most desirable strategy for handling fires involving transport tank trucks is to allow the fire to burn until all of the fuel is consumed. This approach is commonly referred to as *nonintervention strategy* (**Figure 8.13**). The reasons to use the nonintervention approach include the following:

- Eliminates much of the work involved with recovering any remaining product (Recovery operations can be time-consuming, costly, and dangerous.)
- Eliminates contaminating groundwater tables or nearby bodies of water with foam, water, or fuel runoff created by the fire attack, which is particularly important if the fuel is toxic
- Requires the use of little or no foam concentrate, which reduces the cost of the operation for the fire department
- Requires less equipment and personnel

Fire-extinguishment tactics. If the incident commander (IC) decides that it is prudent to extinguish the fire, an all-out, well-planned attack is required. The IC estimates, as closely as possible, the surface area of the spill/fire and bases the application rate on that area. Once the IC determines the application rate and the required water supply, the appropriate foam proportioning and application equipment can be selected and the necessary quantity of foam concentrate can be determined. When all the necessary supplies are on the scene and adequate water supply lines are in position, personnel can start a fire attack (**Figure 8.14**). Fire-extinguishment operations are not undertaken until enough foam concentrate is available, which means that foam concentrate must be available in the event of reignition of the flammable or combustible liquid

Figure 8.14 An apparatus-mounted master stream is prepared for operation at this major semitrailer tank truck fire. The foam stream will be applied from the side and upwind of the burning vehicle. *Courtesy of Timothy Rugg.*

Figure 8.13 In this tank truck fire, nonintervention may be the best approach. *Courtesy of Edward J. Prendergast.*

after extinguishment. The position of the tank in the area of the fire typically makes an ideal situation for personnel to use the bank-down application method. If the tank is surrounded by fire, personnel may need to launch this attack from both sides of the tank. Each foam attack hoseline is supported on either side by hoselines providing water fog protection.

As a precaution, fire and emergency services personnel should approach all tank vehicle fires from the sides of the tank and not from the ends — the same tactic that is used for horizontal storage tanks. Apply cooling water fog streams to the entire top of steel cargo tank vehicles to cool the area that would be directly adjacent to any vapor space present in the tank. If the tank vehicle is of aluminum construction, there is no need to cool the tank because it will melt at 1,200°F (649°C). Water streams are not effective for fire extinguishment and will only spread the burning liquid by displacing the liquid in the tank.

Fire extinguishment starts from the bottom up, focusing first on any liquid pool fires on the ground around the tank. Usually three or four hoselines and several hundred gallons (liters) of foam concentrate are required for a typical tank vehicle fire. The rain-down or bank-down methods can be used to apply the finished foam to the fire. Using air-aspirating foam nozzles also ensures a gentle application and thicker finished foam blanket. After the spill fires are extinguished, extinguish any fire left within the cargo tank or tractor.

Sometimes fire will burn from a small breach or hole in a compartment or a compartment hatch. Foam can be bounced or deflected off some type of object so it can gently flow into the opening. A scoop shovel taped to a pike pole may also be used. Never aim the foam stream directly into the opening. This action only plunges the foam stream into the burning liquid and agitates it. A dry-chemical agent can also be used to extinguish these types of fires.

BLEVE situation strategies. Usually, there is little fear of a boiling liquid expanding vapor explosion (BLEVE) at fires involving tank vehicles. Most petroleum transport tank vehicle shells are constructed of aluminum and melt quickly when exposed to fire. Another factor working against the potential for BLEVE is that most fires occur as a result of physical damage to the tank. Often, the tank has a large tear or split that does not allow pressure to build and result in a BLEVE. However, do not immediately discount the possibility of a BLEVE, especially in situations involving steel tanks that have not sustained much physical damage.

Fire exposure to steel tanks can cause the liquid inside to boil, overwhelming the pressure relief system and causing the tank to violently rupture. Two strategy options are possible when steel tanks are involved: offensive or defensive. The offensive approach requires that water streams with a minimum of 500 gpm (1 892 L/min) be directed on the tank to cool it until the fuel is consumed. The second option is defensive and requires that all emergency personnel remain 2,500 feet (762 m) and all civilians remain 5,000 feet (1 524 m) away from the tank and allow the incident to stabilize itself (**Figure 8.15**).

Overhaul tactics. When total extinguishment has occurred, place a 4-inch (102 mm) blanket of finished foam over the entire area. Frequent reapplication of finished foam may be necessary until all debris within the foam blanket has cooled. Smoldering tires, hot metal, and other debris can have an adverse effect on the finished foam blanket, causing it to deteriorate. Remain on the scene to maintain the finished foam blanket until any remaining fuel has been removed. Cleanup is the responsibility of the transport company who usually contracts with a company that specializes in hazardous materials removal or recovery. Foam hoselines remain charged and in place until the recovery operation is complete. Most tank vehicle fires can be successfully controlled and extinguished within an hour, although cleanup or recovery may take much longer.

Vent/Relief Device Fires

Vent/relief device fires may occur when pressure in a tank (caused by excessive heating or overfilling) causes vapors to discharge and find an ignition source. These fires may also be caused by a static electrical discharge such as a lightning strike. Generally, vent/relief device fires are not a major concern. The main concern is determining the reason for the pressure buildup inside the tank that caused the fire to occur. In many cases, it is a result

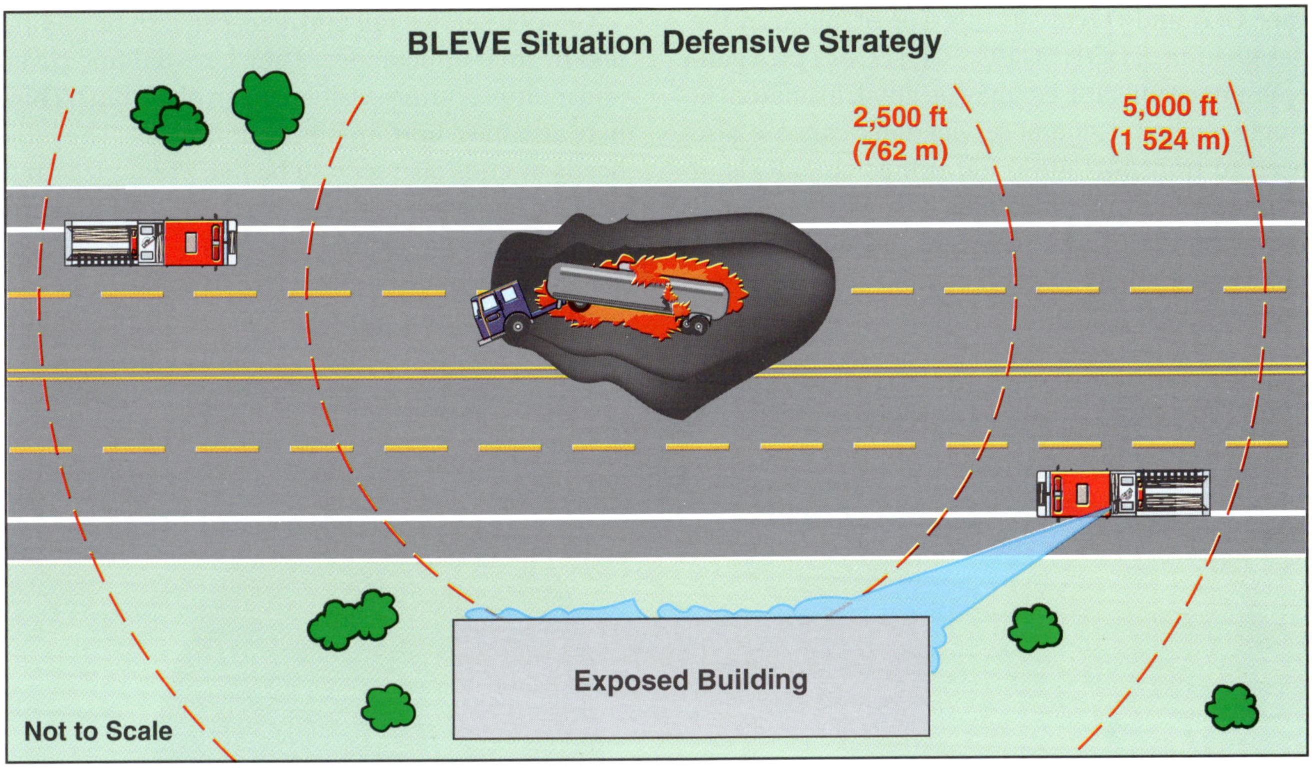

Figure 8.15 Defensive strategy: Minimum distances for fire and emergency responders (2,500 ft [762 m]) and civilians (5,000 ft [1 524 m]) from a potential BLEVE situation.

of the tank being an exposure to another fire. The best corrective measure in thcse situations is for fire and emergency services personnel to cool the tank so that the vapor release discontinues. Once the vent/relief valve resets, the fire usually goes out.

If the fire is not a result of overpressurization or is not completely extinguished, fire and emergency services personnel can manually extinguish the fire using either a short burst from a portable fire extinguisher or by a handline directed at the base of the flames. In most cases, foam concentrate is not required to handle incidents of this type.

Rail Tank Cars

A recent estimate provided by NFPA indicates that more than 200,000 rail tank cars transport 64 percent of the hazardous materials that move by rail in North America. The cargo tanks of rail tank cars are constructed in accordance with specifications developed by ASME and requirements of Title 49 (Transportation) *CFR* 178, *Specifications for Packagings,* in the U.S. and by the specifications and requirements of Transport Canada in the Canadian provinces. Rail tank cars are constructed of steel, stainless steel, or aluminum. The rail tank cars are divided into two categories: (1) nonpressurized (contain the contents at normal atmospheric temperatures) and (2) pressurized (carry the contents under pressure). Like the tank vehicles discussed earlier, rail tank cars can be divided into as many as six compartments, providing a total capacity between 4,000 gallons (15 142 L) and 45,000 gallons (170 343 L). They may contain a single flammable or combustible liquid or a variety of liquids (**Figure 8.16**).

Figure 8.16 Typical railway tank car: May contain one flammable/combustible liquid or a variety.

Types and Construction

Flammable and combustible liquids are transported in nonpressure rail tank cars and are also known as *general service* or *low-pressure* rail tank cars. DOT and TC classify this type of railcar by the designation of DOT/TC 103, 104, 111, or 115. A general service rail tank car may be lined, insulated, or single-shelled. General service tank cars are also used to transport other hazardous chemicals or nonhazardous materials such as fruit juices or tallow. Placarding or product stenciling on the railcar indicate flammable or combustible liquid cargoes **(Figure 8.17)**.

General service rail tank cars can also be identified by their appearance. They have a dome cover and multiple fittings visible on top of the tank. They may also have bottom piping and valves. The tank heads or ends also tend to be flat. Pressurized rail tank cars are easy to recognize because of the protective housing around the utility hole, valves, gauging rod, and sampling well as well as a lack of bottom-unloading piping under the car. Their tank

Figure 8.17 Tanks cars are marked with the contents as shown on this "diesel only" car, and also with DOT-required placards that use the United Nations hazards designation numbers.

Figure 8.18 Double-ended railway tank car has shielding to prevent punctures during a derailment.

heads tend to be more rounded than general service rail tank cars. Pressurized rail tank cars may be covered with a sprayed-on protective coating or have insulation and thermal jacketing, especially if they are carrying flammable gases. Rail tank cars carrying hazardous cargoes also have shielding at the ends to help prevent the couplers from puncturing the tank during a derailment **(Figure 8.18)**.

Fire Strategies/Tactics

Two types of flammable and combustible liquid spills and fires may be encountered with general service rail tank cars: (1) spills and fires that result from a train derailment and (2) those that result from a spill under the tank car or a fire involving the vent or dome on top of the car.

Derailment Spills and Fires

Rail tank car derailments occur when one or more of the cars leave the tracks. Multiple rail tank cars may be involved, impacting each other, jackknifing, rolling over, and stacking on each other. Derailment can cause one or more of the rail tank cars to leak its contents **(Figure 8.19)**. Added to this situation is the possibility that other rail tank cars in the train may contain a variety of hazardous chemicals. Multiple types of hazardous materials may release and mix to form complex and unknown compounds that are difficult and complicated to mitigate. Spilled flammable and combustible liquids may also be on fire and directly impinging on both the leaking and intact rail tank cars.

Responding units should approach from upwind and uphill of the incident if possible. Begin rescue operations with an attempt to account for the location and condition of the train crew and other persons in the area at the time of the derailment. Initial foam resources may be needed to perform a rescue or protect trapped persons. Rescue may not be an option if toxic materials are spilled in the derailment area and the proper chemical protective equipment is not readily available.

To determine the materials in the rail tank cars and, therefore, determine the type of foam to use,

Figure 8.19 Rail tank car derailments may result in spills of flammable or combustible liquids when cars roll onto their sides. *Courtesy of Mike Wieder.*

consult the shipping papers (usually found in the engines or locomotives). Cargo documents include *consists* that list all the rail cars in order as well as the presence of hazardous materials. The hazardous cargo's United Nations (UN) number is usually listed on the consist document. Also, a *waybill* document is made for each hazardous shipment. It has the typical required information found on a hazardous material shipping paper such as the proper shipping name, UN number, hazard class, quantity, emergency phone numbers, and shipper and consignee addresses. A rail tank car may have one or more waybills, depending on the number of different hazardous cargoes carried in the various compartments. Material safety data sheets (MSDSs) or other documents describing the hazardous cargoes being transported may be on the train.

Determine whether there are available resources to extinguish the fire. Offensive extinguishment operations may be an option if the fire is small and it involves only a flammable or combustible liquid. If the fire involves a liquefied or compressed flammable gas, extinguishment is almost never an option. Extinguishment in this situation could allow a flammable gas cloud to spread downwind.

Figure 8.20 A rail tank car that is exposed to fire may rupture in a spectacular and deadly BLEVE.

The most likely strategic option involving rail tank cars on fire is a nonintervention approach. Because rail tank cars are constructed of steel, they are susceptible to BLEVE. Rail tank cars have suffered a catastrophic BLEVE in less than 10 minutes of fire exposure **(Figure 8.20)**. Indications of an impending BLEVE are fire venting from a relief valve, increased venting, increasing relief vent noise, or sounds of the tank metal stressing or pinging. As recommended with most potential BLEVE situations, emergency responders should be at least 2,500 feet (762 m) away in a safe upwind position, and the public should be evacuated to at least 5,000 feet (1 524 m) away. The most dangerous area is opposite the ends of the rail tank car because it is the area most likely to fail during a BLEVE. Entire tank cars have rocketed during BLEVEs; projecting tank shrapnel and wreckage for thousands of feet (meters).

Once the incident has stabilized, firefighters may reenter the immediate incident scene and extinguish residual pool and spot fires. Additional foam may be needed to extinguish spill fires in a railroad right-of-way. Most tracks are located on raised roadbeds and crushed rock. It is often difficult or impossible to establish a foam blanket over burning flammable and combustible liquids because the roadbed is so porous that the fuel soaks into it.

Foam application using a defensive rain-down method from master streams may be the best option **(Figure 8.21)**. Sometimes foam can be applied with the bank-down (deflection) method off the rail

Figure 8.21 The rain-down foam application method is particularly effective on rail tank car fires that are difficult to reach. *Courtesy of Harvey Eisner.*

tank cars from an elevated application device. Both methods require specialized foam apparatus or large flow foam delivery systems and equipment as well as large quantities of foam concentrate.

Spill and Vent Fires

The other types of flammable or combustible liquid rail tank car fires involve a spill fire under the tank car or a vent fire on top of the tank car. In either scenario, the rail tank car is upright on the tracks. The potential for a BLEVE must be taken into consideration during size-up. Follow the procedures for BLEVE incidents discussed in the previous section.

With a spill fire, apply foam using the bank-down or deflection method. Direct the foam stream at the area of the rail tank car that is exposed to direct flame impingement. This action cools the tank and allows the foam to flow off the rail tank car onto the spill fire. Also protect and cool loading piping or valves with either a rain-down or deflection foam application. Use a rain-down application method to extinguish the remainder of the spill fire and any burning liquid runoff. Due to surface conditions, more foam is needed to extinguish and prevent reignition of a spill fire in a railroad right-of-way. Foam hose teams may have to isolate and close open valves in the loading/unloading system. Take measures to contain any liquid or foam runoff.

A fire may occur at the vent/relief valve on top of the rail tank car. Generally, vent and relief device fires are not major concerns. The main concern is determining the reason for the pressure buildup inside the tank that caused the fire to occur. This pressure could be caused by an overfill situation that allowed liquid cargo to escape through the vent. It could also be caused by the contents of the tank being heated by sunlight or exposure to fire, which would cause it to expand and escape through the vent/relief device. First, make sure all loading operations are shut down and valves closed. If there is a chance that flammable vapors may continue to escape and migrate downwind after extinguishment, cool the tank and allow the fire to burn itself out. Vent fires can be extinguished with a narrow fog foam stream, dry-chemical extinguisher, or a combination attack using both foam and dry chemical.

Transportation Fires in Tunnels

In North America, highways and railways pass through hundreds of tunnels carrying traffic under mountains, urban areas, and bodies of water. DOT estimates that there are 300 highway tunnels that are a total of 200 miles (322 km) in length. The majority of tunnels are less than 1 mile (1.6 km) in length. A 1990 survey of railway tunnels longer than 1,000 feet (305 m) in length indicates that there are from 75 to 100 total miles (121 km to 161 km) of tunnels in the U.S. This data is exclusive of subway tunnels. No specific number of how many railway tunnels exist is currently available. Flammable liquid spills and fires in tunnels are considered to be "low-incidence and high-consequence" events. Since 1949, only two major fires have occurred in American highway tunnels. One of those, the Holland Tunnel fire in New York City in 1949 resulted in the banning of any transport of hazardous materials or flammable/combustible liquids through tunnels in most states/provinces. However, this ban is no guarantee that these materials will not find their way into a tunnel. The most recent occurrence of a railway fire took place in the Fort McHenry Tunnel in Baltimore, Maryland, in July, 2001. While no fatalities resulted from this incident, the city infrastructure was severely affected and the cost of repairs was high.

In 1983, the U.S. Federal Highway Administration surveyed the records of highway tunnel incidents to determine the most effective means of prevention and control of fires in tunnels. One of the determinations was that sprinkler systems are not an effective form of fire suppression in a tunnel system due to the fact that fires usually occur in the closed engine compartment of the vehicle. Only three tunnels in the U.S. are currently equipped with sprinkler systems. Those tunnels permit the passage of unescorted hazardous materials cargo carriers. While the conclusions primarily focused on detection systems, the report did note that in those incidents where foam was used, the extinguishment of the fire was more rapid. NFPA 502, *Standard for Road Tunnels, Bridges, and Other Limited Access Highways* (2001), suggests in Appendix D that AFFF foam sprinkler systems be considered in lieu of water sprinklers.

Fire and emergency services organizations that have highway and railway tunnels in their jurisdiction should work with the agency responsible for those tunnels to develop appropriate emergency plans in the event of a spill or fire within the tunnels. The use of foam as an extinguishing agent should be included in the plans.

Aircraft Industry

The objective of this section is to explain the use of foam in aircraft fires and fuel spills. The topic of aircraft rescue and fire fighting (ARFF) is thoroughly covered in the IFSTA **Aircraft Rescue and Fire Fighting** manual as well as other materials available from NFPA, International Civil Aviation Organization (ICAO), U.S. Federal Aviation Administration (FAA), U.S. Air Force and Navy, and other training organizations.

Large quantities of flammable or combustible fuels accompany aircraft fires; therefore, Class B fire-fighting foam concentrate is commonly used to control these fires. The use of foam concentrate is only one part in an overall process of handling ARFF incidents. In this section, general information is provided as background for the use of foam concentrates and includes the following topics:

- Common categories of aircraft
- Types of fuel carried by aircraft and basics of fuel systems
- Basics of hydraulic systems, compressed air systems, and structural materials
- Types of emergency incidents in which aircraft are involved
- Aircraft incidents response procedures and tactics
- Aircraft foam concentrates/application systems

Aircraft Categories

Aircraft may be classified in a variety of ways. The following ways are the most common:

- ***Design*** — Aircraft may be lighter than air (blimps, balloons, dirigibles) or heavier than air (fixed wing or rotary wing).
- ***Propulsion*** — Another category is the method that propels or moves aircraft through the air. Some (gliders and sailplanes) depend on a powered aircraft to tow them aloft, and they then remain airborne by using the prevailing air currents. Other aircraft are powered by piston or reciprocating engines that power one or more propellers and are powered by aviation gasoline (AVGAS). A third group is referred to as jet aircraft and is powered by jet engines, turbojets, or turboprops. The latter two engines are gas turbines that drive either a jet stream or propeller. Both lighter-than-air and heavier-than-air design aircraft may depend on any one of these propulsion systems.
- ***Function*** — The FAA classifies aircraft according to its intended use or function. The FAA lists nine classes of aircraft based on intended use, number of engines, type of propulsion, and range (distance of travel for intended purpose).

In general, however, the primary categories of aircraft are commercial transport, commuter/regional, cargo, general aviation, business/corporate, and military. Some aircraft are included in more than one category such as passenger aircraft that are also used for military purposes. These categories are described as follows:

- ***Commercial transport aviation*** — Aircraft for the commercial transport of passengers comprise the fleets of both national and international carriers or airlines. Commercial airliners generally have a passenger seating capacity of more than 20 and as many as 900 passengers. The fuel capacity of commercial transport aircraft may range from 13,000 gallons (49 210 L) to 58,000 gallons (219 553 L) of Jet A or Jet A-1 fuel (see Aircraft Fuels/Fuel Systems section for descriptions of these types of fuel).
- ***Commuter/regional aviation*** — Commuter aircraft usually carry between 19 and 100 passengers and are operated by commuter airlines on short flights between cities with small airports and the major airports (referred to as *hubs*) in large cities. Aircraft that carry fewer than 30 passengers may not have a flight attendant onboard to assist in an emergency. Aircraft designs may be fixed wing or rotary, and the propulsion systems may consist of piston or jet engines.
- ***Cargo aviation*** — Freight-hauling aircraft, similar to passenger aircraft (but without passenger seating) are used to move cargo quickly around the country and the world and are usually operated by freight companies or cargo airlines **(Figure 8.22)**. Some cargo carriers may also be fitted with seats for limited passenger capabilities. Generally, the aircraft will be the same type as those used for commercial transport.

- ***General aviation*** — General aviation aircraft are used primarily for pleasure or training. These aircraft usually hold from 1 to 10 passengers and are small, lightweight, and nonpressurized. One or two reciprocating engines or turbine engines that drive propellers or two small jet engines typically power general aviation aircraft (**Figure 8.23**). Fuel capacities of AVGAS or jet fuel may range from 90 gallons (341 L) to 500 gallons (1 893 L). The general aviation category includes amateur (home) built, antique, and experimental types of aircraft. Aircraft may be fixed-wing or rotary-wing designs. The FAA also includes aircraft used for fire fighting, medical evacuation, aerobatics, and skydiving in this category.
- ***Business/Corporate aviation*** — These aircraft are primarily used for business-related transport and can range in size from small, lightweight, nonpressurized aircraft to large commercial type jets. Generally, they accommodate from 6 to 19 passengers. They are usually fixed-wing design and powered by piston or jet engines. Fuel types and capacities are the same as those given for commercial and general aviation.
- ***Military aviation*** — Military aircraft include cargo, fighter, bomber, trainer, rotary-wing, and special-mission aircraft. Aircraft types ranging from single-engine fighters to large, multiengine transports and bombers are used by the military services for a variety of functions (**Figure 8.24**). Jet-powered military aircraft usually carry a limited number of crew and may have armaments and explosive ejection devices.

Figure 8.22 Commercial cargo aircraft.

Figure 8.23 General aviation aircraft. *Courtesy of Jim Rackl.*

Figure 8.24 Military aircraft.

Aircraft Fuels/Fuel Systems

Aircraft fuels are designed specifically to power aircraft engines. These fuels are found primarily at aviation facilities such as commercial airports, military airfields, and onboard vessels that carry operating aircraft or helicopters. They may also be encountered at hospitals or office buildings that operate heliports; in petroleum processing, storage, and terminal facilities; and in tank trucks, rail tank cars, and pipelines that carry finished petroleum products. Because aircraft fuels may be found in so many locations, it is important for fire and emergency services personnel to be able to recognize the various types and their properties.

An understanding of the basic components of aircraft fuel delivery systems is also important. Knowledge of these systems, including aircraft fuel tanks, control valves, pumps, and piping system, permits the emergency responder to determine the appropriate tactics required for aircraft that are leaking unignited fuel or have been involved in a crash.

Fuels

Three basic types of fuels are used in aircraft: aviation gasoline, kerosene, and blends of gasoline and kerosene. The last two types are jet fuels and cover a broad range in the hydrocarbon series. Jet fuels are divided into two grades: Jet A (including A-1) and Jet B. Jet fuel is used much more extensively and in greater quantities than aviation gasoline. Jet B was originally developed for the U.S. Air Force following World War II. It was known then as JP (jet propulsion)-2 (obsolete), JP-3 (obsolete), and JP-4 (still in use in some parts of Canada and Alaska). Jet B had the advantage of greater availability than gasoline or kerosene alone. The disadvantages, though, included higher volatility, greater loss at high altitudes due to evaporation, increased risk of fire during handling on the ground, and lower victim survivability rates in crashes that resulted in fires. By 1998, the U.S. Air Force had changed from Jet B to JP-8 and JP-10, both kerosene-based fuels that are less volatile and similar to Jet A and A-1. The minimum flash point temperature for JP-8 is 100°F (38°C) and for JP-10 is 140°F (60°C). The U.S. Navy has used JP-5 (also kerosene-based) since 1952 due to safety issues involving fuel storage and handling on board ships. The minimum flash point temperature for JP-5 is 140°F (60°C). NFPA includes JP-5, JP-8, and JP-10 with Jet A and A-1 fuels as kerosene-based fuels. In addition, some military aircraft use hydrazine to power auxiliary engines on the aircraft.

Aviation gasoline. Aviation gasoline is the same as the gasoline used in automobiles except that it has a higher octane rating than automotive fuel. While automotive gasoline have ratings up to about 92 octane, ratings for AVGAS range from 100 to 145 octane. The flash point of aviation gasoline is approximately -50°F (-46°C); thus AVGAS can release sufficient vapor to form an ignitable mixture with air in virtually any weather. Flammability limits of AVGAS are from 1 to 7 percent, and autoignition temperatures range from 825 to 960°F (440°C to 515°C). Once AVGAS ignites, flame spread is from 700 to 800 feet (210 m to 240 m) per minute or about 12 feet (4 m) per second. Four grades of AVGAS are available, and each is dyed a different color. Fire and emergency responders will most likely encounter grade 100LL (low lead) that is dyed blue. Antique aircraft normally use grade 80, which is dyed orange or red. The other two grades are 80/87, which is dyed red, and 120, which is dyed green. Aircraft gasoline still has tetraethyl lead as an additive. Oxygenates or polar solvents are not added to AVGAS as they are with automobile gasoline. The fire-fighting characteristics found in the MSDS for each grade of fuel are the same, no matter what the grade.

Jet fuel: Jet A and A-1. Jet A and A-1 are kerosene-grade fuels that have flash points between 95 and 145°F (35°C and 63°C), depending upon the particular fuel mixture. Jet A and A-1 fuels mix with air above their flash points and become flammable when the fuel-to-air mixture is just less than 1 percent. The upper flammability limit is just over 5 percent. Autoignition temperatures range from 475 to 500°F (246°C to 260°C) for Jet A and 440 to 475°F (227°C to 246°C) for Jet A-1 with a flame spread rate of less than 100 feet (30 m) per minute for both. The rate of flame spread in Jet A and A-1 fuels is substantially slower than that in AVGAS. Jet A fuel is commonly used by U.S. commercial air carriers, while the rest of the international community uses Jet A-1. The primary difference between the two is that Jet A-1 has a lower maximum freezing point (-52.6°F [-47°C]) than Jet A (-40°F [-40°C]), which makes Jet A-1 more suitable for long distance, international flights.

Jet fuel: Jet B. Jet B fuel is a blend of gasoline and kerosene and has a very low flash point of -45°F (-43°C). Flammability limits are from 1 to just over 7 percent (a slightly wider range than that of Jet A and A-1). Autoignition temperatures range from 470 to 480°F (243°C to 249°C). The flame spread rate of Jet B fuel is from 700 to 800 feet (210 m to 240 m) per minute or 12 feet (4 m) per second. Flames spread just as rapidly in Jet B fuel as in any type of gasoline including AVGAS. At higher temperatures, the rate of flame spread across any fuel is increased.

Hydrazine. The U.S. National Aeronautics and Space Administration (NASA) and the U.S. Air Force use hydrazine (a hypergolic fuel) as the fuel supply for spacecraft, rockets, and the auxiliary power unit on some aircraft. *Hypergolics* are substances (such as H-70 that is 70 percent hydrazine and 30 percent water) that ignite spontaneously on contact with each other (for example, hydrazine with an oxidizer). Hydrazine is clear, has an odor similar to ammonia, is toxic in both liquid and vapor form, and may explode. It is a strong reducing agent and is hypergolic with some oxidizers such as nitrogen tetroxide and the metal oxides of iron, copper, and lead. Autoignition may occur if hydrazine is absorbed in rags, cotton wastes, or similar materials. Those aircraft that use hydrazine carry a minimum of 7 gallons (26 L). The need for a highly reliable and quickly responsive way of obtaining emergency electrical and hydraulic power aboard aircraft is likely to increase the use of hydrazine.

WARNING

Always wear full personal protective equipment including respiratory protection when dealing with hydrazine emergencies because it may be absorbed through the skin. Even short exposures may have serious effects on the nervous and respiratory systems.

Fuel Systems

The largest system of any type in an aircraft is the fuel system. This system both stores and distributes the fuel. The components of the fuel system (tanks, lines, control valves, pumps, etc.) are located throughout the aircraft (**Figure 8.25**). Therefore, because of the generally hazardous nature of aviation fuels, the fuel system presents the greatest hazard in an aircraft accident. The fuel system consists of two major parts: the tanks and distribution system. In order to effectively use foam on aircraft fires, fire and emergency services personnel need to understand the types of fuel and fuel systems.

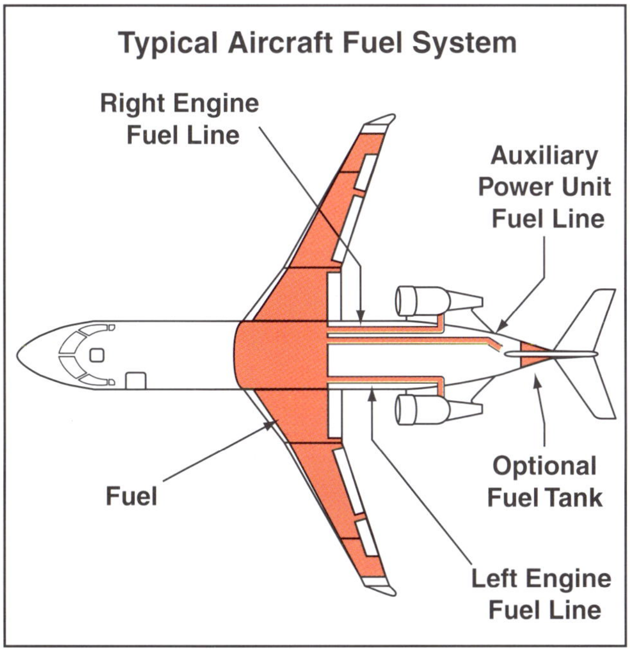

Figure 8.25 Typical aircraft fuel tanks and distribution system.

Fuel tanks. Aircraft fuel tanks are constructed either as separate units or as an integral part of the fuselage, wings, or tail (horizontal or vertical stabilizers). Some aircraft have both types of tanks. Often, to extend their range, military aircraft have auxiliary tanks mounted externally. Fuel tank capacity may range from 30 gallons (114 L) in auxiliary tanks to 58,000 gallons (219 553 L) in the main tanks of large airliners and military aircraft. Changes in temperature cause fuel in the tanks to expand and contract, so aircraft fuel tanks are equipped with vents that release vapors and reduce the pressure caused by expansion. Under normal conditions, the minute amounts of fuel that do escape evaporate quickly, so such venting is usually not hazardous.

Fuel distribution system. Fuel is distributed from the aircraft's tanks to the engines through fuel lines, control valves, and pumps located throughout the aircraft. Fuel lines vary in sizes up to 4 inches (102 mm) in diameter and are constructed of metal tubing or flexible piping. Pumps capable of producing pressures from 4 to 40 psi (28 kPa to 276 kPa) control the fuel flow within these lines.

Other Aircraft Components

In addition to the aircraft fuel system, other components of the aircraft may also constitute hazards by contributing to a fire. Those components include hydraulic systems, compressed air systems, and aircraft structural materials.

Hydraulic Systems

Although they are generally not thought of as being as much of a hazard as the fuel system, fire and emergency services personnel should also be familiar with the hazards associated with aircraft hydraulic systems. Nearly all aircraft have a hydraulic system of some sort. The larger the aircraft, the larger the hydraulic system will be.

The hydraulic system of an aircraft consists of a hydraulic fluid reservoir, electric or engine-driven pumps, appliances, various hydraulic accumulators, and tubing that interconnects the system. Hydraulic fluid is supplied to a pressure pump that moves the fluid throughout the hydraulic system and to accumulators where some of the fluid is stored under pressure. This stored fluid may then be used to supply hydraulic pressure to critical aircraft systems. The accumulator may store this fluid under pressure even after the engines have stopped.

Hydraulic fluid under pressure is used to operate equipment such as landing gear, nose-gear steering, brakes, and wing flaps. Most modern aircraft hydraulic systems operate at a pressure of 3,000 psi (20 684 kPa) or higher. During rescue or fire-suppression operations, extreme caution must be exercised to avoid cutting a pressurized hydraulic line and releasing the fluid into a fine mist that is both toxic and flammable. If sprayed on hot brakes or hot engine components, the fluid may ignite. A hydraulic fire produces a torch effect, or the hydraulic fuel vapors may explode if they are confined.

Synthetic hydrocarbon hydraulic fluids (such as Skydrol™) that are also toxic and flammable are widely used. However, these fluids present a significantly reduced flammability hazard because the flash point is twice that of nonsynthetic fluid. In addition, the flame-spread rate is slower.

Almost all of these fluids are combustible liquids and may give off toxic products of combustion when ignited. In addition, some are corrosive and can cause damage to protective equipment as well as irritate skin, nasal passages, and eyes. Full personal protective clothing and respiratory protection must be worn when working around these systems. Clothing and equipment have to be decontaminated according to the manufacturer's recommendations.

Compressed Air Systems

All aircraft that have pressurized cabins carry compressed oxygen cylinders. These cylinders can rupture during a crash and feed the fire or explode if exposed to heat. Cylinders are painted green for easy recognition, although this sign may not be the case with old aircraft (**Figure 8.26**).

Structural Materials

Aircraft may have composite materials in the construction of its structure (**Figure 8.27**). Some of the carbon and boron composite fibers can be a skin and respiratory irritant when released into the atmosphere. Magnesium and titanium, both combustible metals, are also common aircraft construction materials. Both are difficult to extinguish, present a constant fuel spill reignition problem, and give off very toxic products of combustion. Solid streams of water or foam can extinguish combustible metal fires.

Emergency Incidents

Aircraft emergency incidents on or off the airfield usually occur without any warning. These emergency incidents are caused by a variety of problems that can develop while the aircraft is in flight or on the ground. Types of incidents, response procedures, and fire and emergency tactics are discussed in the sections that follow. The general types of aircraft emergency incidents with which fire and emergency services personnel are confronted include the following:

- ***In-flight emergencies*** — Include fires as well as other problems that may cause or contribute to fire such as hydraulic failure, engine failure, landing-gear malfunction, or other system malfunctions. While in flight, aircraft frequently develop minor difficulties that may or may not be cause for alarm. If the pilot declares an emergency, then control tower personnel notify the airport fire department of the emergency condition and which runway will be used. ARFF resources are then dispatched to predesignated standby positions.
- ***On-ground emergency incidents*** — Include fuel spills or leaks that occur during fueling, defueling, or maintenance procedures. These incidents may occur during landing or takeoff or when the aircraft taxis or parks. A fire may not have occurred by the time fire and emergency services person-

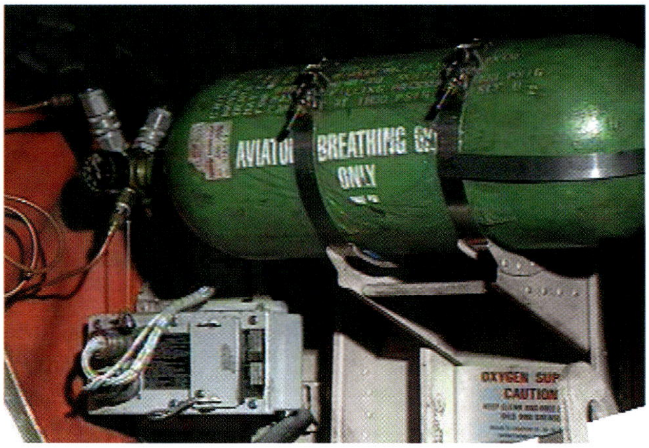

Figure 8.26 Aircraft oxygen tanks are usually painted green for easy recognition. *Courtesy of William D. Stewart.*

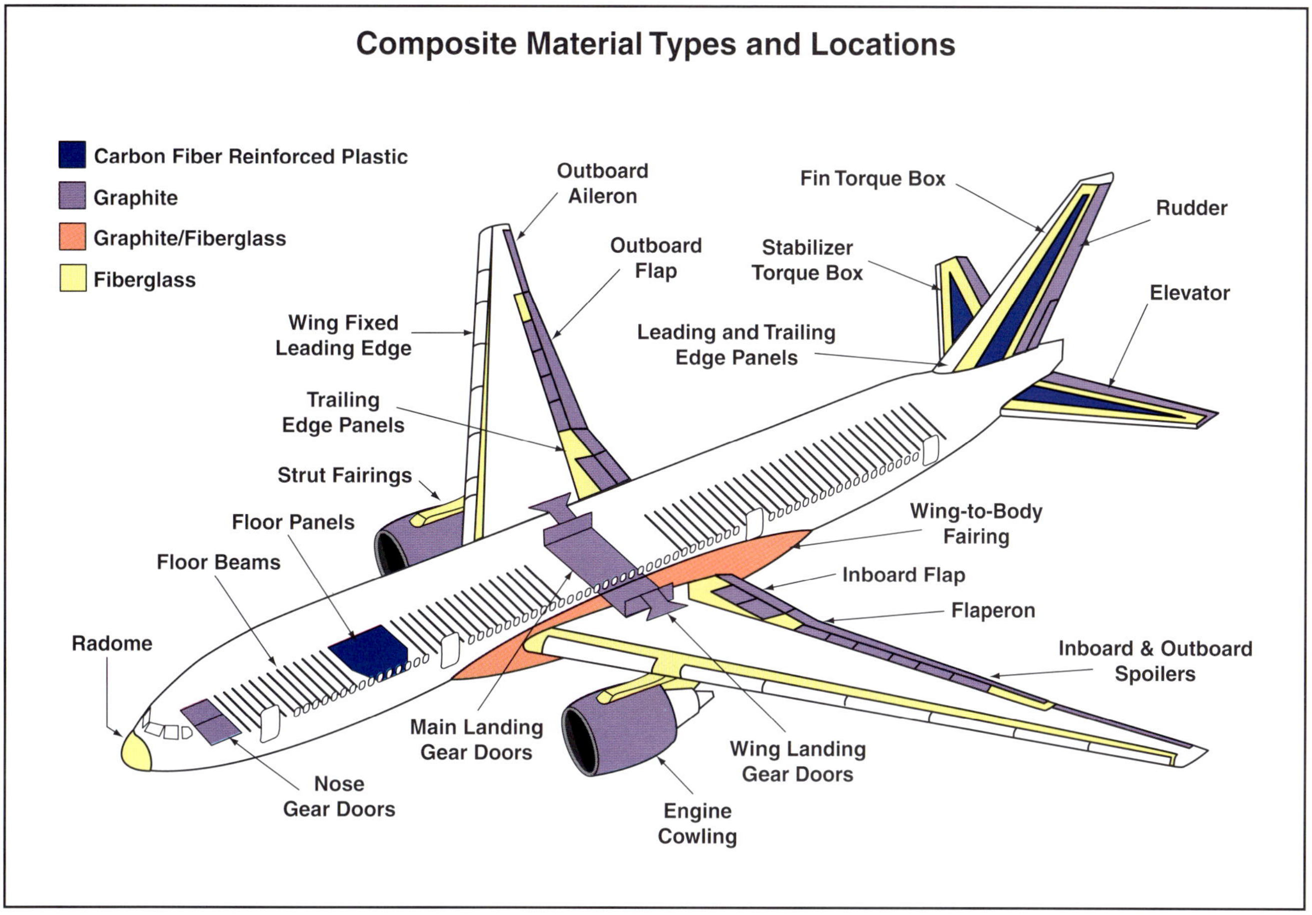

Figure 8.27 Typical aircraft construction materials and their locations.

nel arrive at the scene of an on-ground emergency incident. However, there is a possibility of fire occurring at any instant if some part of the fuel system has ruptured.

In-flight emergencies may result in the aircraft making an uncontrolled landing or crashing. The various types of crashes to which fire and emergency services personnel may respond are described as follows:

- ***Low-impact crashes*** — Do not severely damage or break the fuselage and are likely to have a large percentage of survivors (for example, a wheels-up or "belly" landing) (**Figure 8.28**). These types of incidents may involve fuel fires, although nonfire incidents are common. Occupants are often able to extricate themselves and walk away from low-impact crashes. However, rescue operations may have to be performed in conjunction with fire-suppression efforts if occupants are trapped and/or seriously injured. In addition, handline teams should back up rescue personnel for protection from a flash fire. Rescue personnel may discharge finished foam over areas of obvious fuel spills to aid with vapor suppression.

Figure 8.28 Collapsed or malfunctioning landing gear can result in a low-impact crash such as this one. *Courtesy of Air Line Pilots Association.*

- ***High-impact crashes*** — Cause severe damage to the fuselage and have a significantly reduced likelihood of occupant survival. In many cases, only small portions of the aircraft remain intact, and debris is scattered over a wide area (**Figure 8.29**). Large fuel spills and fires are commonly associated with incidents of this magnitude.

- ***Crashes involving structures*** — Aircraft may crash into one or more buildings, either on or off the airport facility (**Figure 8.30**). The aircraft may break open upon impact, and flying debris may damage surrounding properties. Damage to the roofs and upper stories of buildings may occur, floors and walls may collapse, and people both inside and outside the affected buildings may be injured. Fire-suppression operations begin as soon as rescue operations have been completed. Spread Class B finished foam on the burning fuel while using Class A finished foam and water fog to extinguish the structural fires and provide exposure protection. These fires may be widely separated and may spread rapidly because of scattered fuel, severed gas lines, and damage to domestic electrical systems.

Figure 8.29 High-impact crashes result in the spread of debris over a large area. *Courtesy of Air Line Pilots Association.*

Figure 8.30 This photograph illustrates the difficulty of reaching some aircraft crashes that involve structures. In this case, minor damage was done to the structure and no fire resulted. *Courtesy of Maryland Fire and Rescue Institute.*

- ***Water Crashes*** — May result when an aircraft skids off a runway, lands short, aborts a takeoff, "ditches," or crashes into the water. The surface of the water may be covered with fuel (which may or may not be ignited). If practical, apply a blanket of finished foam to the entire area. Constant application of finished foam may be required to maintain an effective finished foam blanket because wind, heat, and movement can deteriorate it. If the aircraft is partially submerged and has not ignited, fire and emergency services personnel must remain alert while performing rescue operations because fuel rising to the surface may come in contact with heated engine parts and ignite.

- ***Rotary-wing aircraft crashes*** — Because of relatively light construction, helicopters do not withstand the violent forces encountered in vertical impact. The undercarriage, rotors, and tail units usually break apart, leaving the wrecked interior of the fuselage as the main debris (**Figure 8.31**). The main wreckage usually contains the engine and fuel tank. In a low-impact crash, the engine may still be running and the rotors turning. Even after the engine is turned off, the main rotor continues to turn for a few minutes, propelled by its own inertia. Approach the aircraft from the front in a crouching position in order to avoid being hit by the turning main rotor blade or the tail rotor.

Figure 8.31 Low-impact helicopter crashes may cause heavy structural damage and potential fuel spills. *Courtesy of Metro Washington Airports Authority.*

- *Military aircraft crashes* — Aircraft may be armed with live ordinance in the form of ammunition, rockets, flares, missiles, or bombs, and ordinance may spread over the crash site or be exposed to fire. Approach the aircraft from the rear, staying clear of the forward firing weapons systems. Locate ordinance and mark for ordinance disposal teams to disarm and remove. Mark and use safe approach lanes until explosives are removed. Crew ejection seats are either propelled by rocket, cartridge, or compressed gas. In addition, an explosive charge can jettison the canopy. Take extreme care when disarming both the canopy and ejection seats (**Figure 8.32**).

Response Procedures

Personnel that may respond to aircraft incidents should be familiar with their agency's standard emergency response procedures. Most fire and emergency services emergency response agencies have standard operating procedures for both standard emergency and unannounced emergency responses.

Standard emergency. The aircraft crew is aware that there is a problem and notifies the airport authorities of the emergency before they prepare to land. Runway standby positions for fire and emergency services vehicles in anticipation of an emergency are predetermined in the standard operating procedures and include predetermined response routes to use unless unforeseen conditions dictate otherwise (**Figure 8.33**). Promptness and safety are equally important response considerations, but responders must take weather, visibility, terrain, and traffic into consideration.

Unannounced emergency. Fire and emergency services units are dispatched immediately to the scene of the emergency incident by the most direct route, and the control tower continues to keep responding units informed while they are en route. Approach the scene of the accident with caution to avoid hitting persons who may have been thrown clear of the aircraft or who have escaped from the aircraft. In darkness or if vision is obscured, have a person on foot precede the apparatus to make sure that the way is clear. Respond in a way that avoids damaging the responding apparatus and equipment, and avoid running over aircraft debris scattered throughout the accident scene.

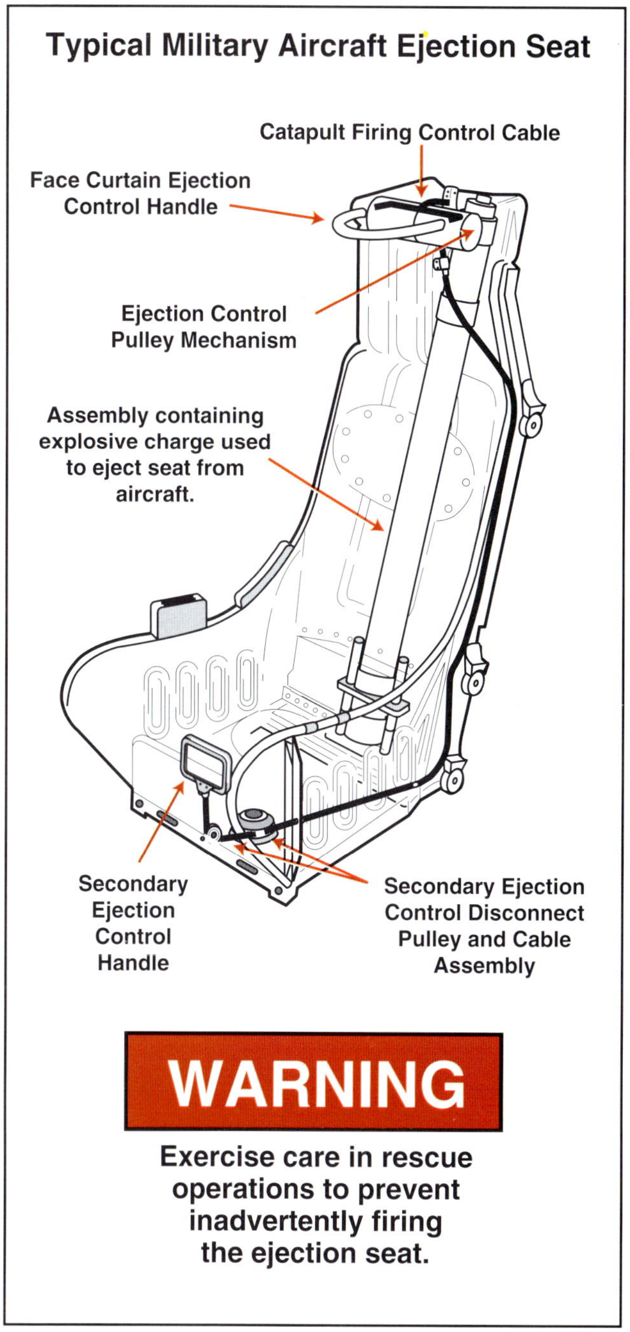

Figure 8.32 A typical military aircraft ejection seat.

Fire and Emergency Tactics

The tactics used for aircraft emergency incidents are dictated by the severity of the incident. Tactics are given for the following aircraft emergency incidents:

- *Fuel spills on airport property* — At large airport facilities, ARFF apparatus and crews will be the first and probably only unit to arrive. At small, unprotected airports, local structural fire-fighting personnel are assigned to respond to these situations. Handle aircraft fuel spills in the same way that the unignited fuel spills in Chapter 6,

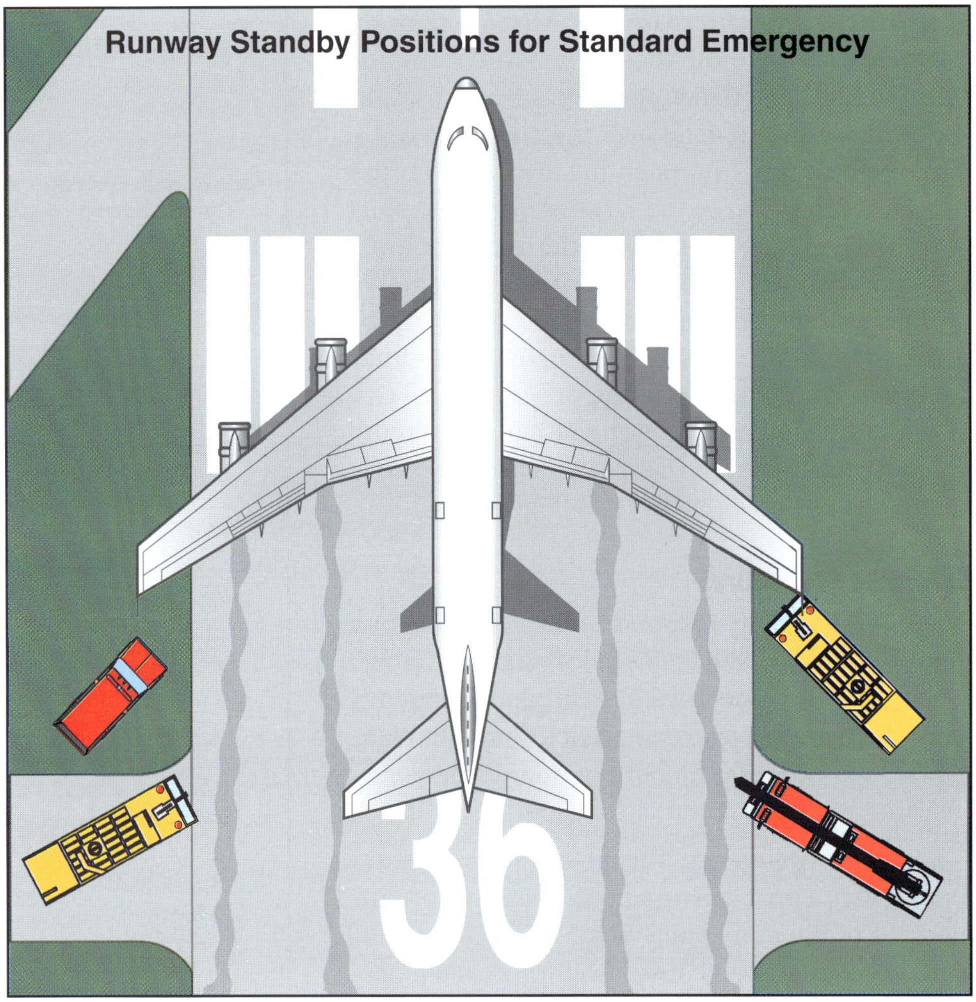

Figure 8.33 Standard emergency positions for fire and emergency services vehicles.

Class B Foam Use: Unignited Class B Liquid Spills, were handled. Apply a 4-inch (102 mm) deep blanket of finished foam as a vapor barrier over the spill and use booms or dikes to contain the fuel until it can be removed. Once applied, do not disturb the finished foam blanket.

- ***Standard emergency on airport property*** — Airport fire-fighting apparatus and/or responding structural fire-fighting units must be positioned correctly if rescue and fire-fighting operations are to be successful. Because ARFF apparatus often respond single file, the first-arriving apparatus may establish a responding route onto the airport property for other vehicles and dictate their fire-fighting positions (**Figure 8.34**). In positioning apparatus for a standard emergency, use the following guidelines:
  - — Consider the slope of the ground and wind direction.
  - — Do not place vehicles so that they block the entry or exit from the accident site.
  - — Do not place vehicles in a hazardous position downhill or downwind from a fuel spill.
  - — Place vehicles so that they may operate effectively in the event of fire.
  - — Place vehicles so that they can help in the egress or rescue of persons from the aircraft.

Figure 8.34 ARFF apparatus are designed to put large quantities of foam onto a fire quickly and with minimal personnel. *Courtesy of Michael T. Defina, Jr., Metro Washington Airports Authority Fire Department.*

— Place vehicles so that they can reposition as easily as possible.

— Place vehicles so that turrets and handlines may maintain the route of egress if necessary.

Once the aircraft is on the ground and has come to a complete halt, use the following guidelines (**Figure 8.35, p. 278**):

— Position handlines and apparatus nozzles to protect the primary escape route for passengers and crews.

— Surround the aircraft with apparatus and personnel because passenger emergency exits are located on both sides and sometimes under the rear of the aircraft.

— Apply a foam blanket to prevent vapor ignition if fuel is leaking.

— Suppress the fire if the fuel has ignited, and push it away from the routes of escape.

— Proceed with fire extinguishment from the bottom up once rescues are complete.

- ***Aircraft interior fires*** — May start on passenger and cargo aircraft both while in-flight and while parked on airport ramp areas. Deploy initial fire attack hoselines to protect the evacuation of passengers and crew. A good initial attack position is through the aircraft's over-wing exit doors (**Figure 8.36**). The wing surface provides a platform to work from and fire spreading from the front or back of the aircraft can be stopped from this location. Use Class A or B foam for the interior attack. Start an aggressive interior attack once rescue has been accomplished. A mass application of extinguishing agent will quickly overwhelm the heat energy of the fire and cool the interior. Use fog streams and forced air ventilation to remove smoke, heat, and steam.
- ***Unannounced emergency on/off airport property*** — May be either a low-impact or high-impact crash. Response time is critical to initiating an effective rescue effort. The aluminum construction may provide passenger protection for 3 minutes, but complete burn through may occur in as little as 60 seconds (**Figure 8.37**). Maintenance of the wreckage is a major concern. FAA crash site investigators must be able to collect and tag each piece of debris in order to determine the cause of a crash. Therefore, fire and emergency services personnel must not move or remove any debris unless it is necessary for rescue, body recovery, or fire extinguishment.
- ***Low-impact crash on or off airport property*** — Place handlines into service to protect the rescue operations. Water fog may be used if foam concentrate is limited or if it will take too long to place the foam into operation. Deploy handlines to provide an escape route or a path for rescuers to reach the passenger compartment. Disperse fuel vapors with water fog. However, do not use power extrication equipment until vapors have dispersed or provide a vapor barrier with a finished foam blanket.
- ***High-impact crash*** — Determine the possibility that survivors might be within the wreckage or thrown free of it. If an adequate supply of foam concentrate is available to the first responding units, then it will be valuable for rapid extinguishment in the areas that need to be searched for survivors. Provide exposure protection, especially in urban areas, with the application of water fog Class A finished foam. Because of the intensity of fires involving aircraft fuels and the vaporizing of those fuels during a crash, it is likely that containment will not be a factor in a high-impact crash. Fuel that is covered with a finished foam blanket will probably absorb into the soil at the site.

Foam Concentrate/Application Systems

Foam concentrates have proven very effective for aircraft crash fire fighting. All of the low-expansion foam concentrates discussed in Chapter 2, Foam Concentrate Technology, may be used for aircraft emergency operations. However, the two film-forming concentrates, AFFF and film forming fluoroprotein (FFFP) foam, are most commonly used. AFFF was originally developed by the U.S. Navy to more effectively control aircraft fires that occurred on flight and hanger decks of aircraft carriers. AFFF and FFFP remain popular today because of their flexibility in application methods, fast extinguishment characteristics, vapor-suppression abilities, and compatibility with secondary (dry-chemical) extinguishing agents.

The application of finished foam at aircraft incidents may be performed by specially trained aircraft rescue and fire-fighting personnel operating ARFF vehicles or by structural firefighters operating foam-

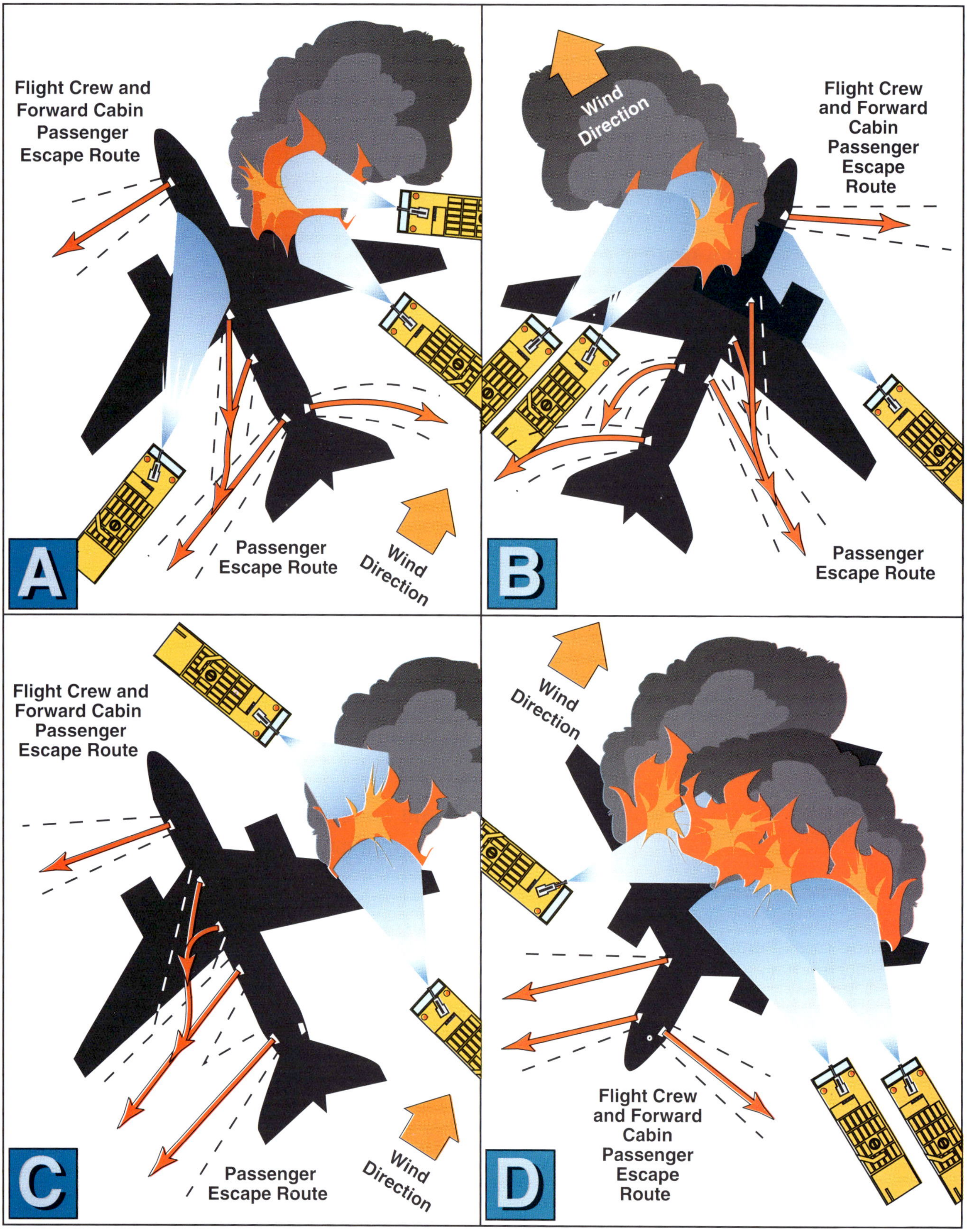

Figure 8.35 Examples of proper ARFF positioning for aircraft fire attack.

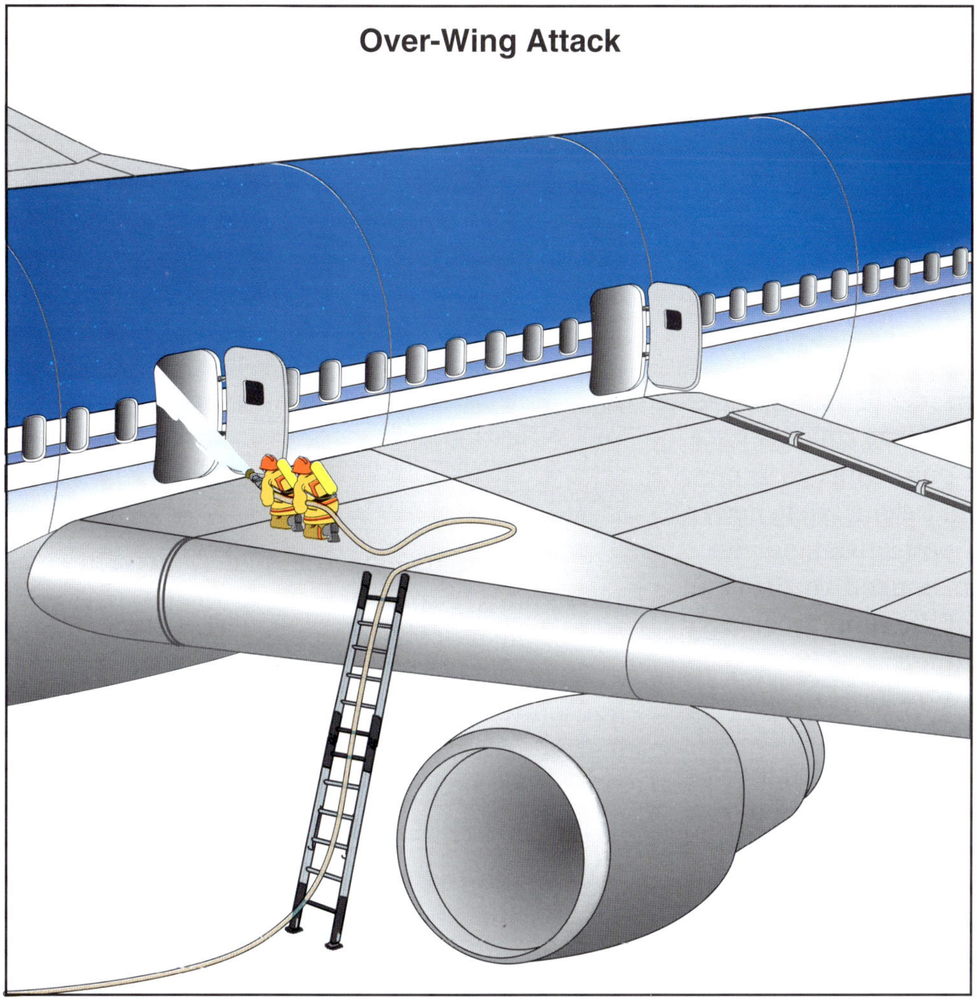

Figure 8.36 When possible, use the over-wing exit doors for fuselage access to fight interior fires.

Figure 8.37 Aircraft fuselages can burn through in as little as 60 seconds. *Courtesy of Air Line Pilots Association.*

equipped engines. The ARFF vehicles are capable of applying various extinguishing agents from turrets, handlines, ground-sweep and under-truck nozzles, or secondary agent devices. The structural fire-fighting apparatus may provide finished foam through handlines or master stream appliances such as deck guns or monitors. See Chapter 3, Foam Proportioning and Delivery Equipment, for discussions of these types of apparatus and mobile application systems.

Pipelines

In the U.S., the network of more than 160,000 miles (257 488 km) of pipelines (lengths of steel pipe welded together) is the safest method for transporting petroleum products according to the National Transportation Safety Board of the U.S. DOT. Pipelines are efficient, are low maintenance, and can move large quantities of products including crude oil, natural gas, liquefied petroleum gas, jet fuel, and hazardous chemicals. They transport 67 percent of all crude and refined petroleum products in the United States. This compares to 28 percent carried on ships, 3 percent carried on tank vehicles, and 2 percent carried by rail.

The DOT's Office of Pipeline Safety, U.S. Federal Energy Regulatory Commission, U.S. Environmental Protection Agency (EPA), National Energy Board in Canada, and various state/provincial commissions or departments regulate pipelines within their jurisdictions. The Office of Pipeline Safety, National Transportation Safety Board, and Association of Oil Pipe Lines (AOPL) compile statistics on pipeline safety.

The diameter of the pipe in pipelines may range from 6 to 40 inches (152 mm to 1 016 mm). Pipelines built in the 1940s and 1950s are still in service today. They have 1-inch (25 mm) Somastic® protective coatings (that also add weight) while newer pipelines are wrapped with protective tape. Electrical cathodic protection is used to prevent electrolysis and corrosion of the pipe. Operating pressures generally range from 85 to 350 psi (586 kPa to 2 413 kPa) although pressures may reach 1,000 psi (6 895 kPa). Generally the pipelines are buried approximately 3 feet (1 m) underground, although they may also be located aboveground or underwater. Pumping stations that push the product through the pipe are located between 20 and 100 miles (32 km to 161 km) apart **(Figure 8.38)**. Pipeline locations are generally marked with DOT-required markers along the pipeline right-of-way. The markers (usually round metal signs or plastic posts) give the name of the pipeline company, an emergency phone number, and the type of material transported through the pipeline **(Figure 8.39)**. Because the majority of pipelines are buried underground, they are a low-risk target for terrorists or third-party vandals.

Figure 8.38 Pipeline shutoff and control valves are usually located on highway or railroad right-of-ways and are secured within locked fences.

Figure 8.39 Warning labels are required for all pipelines in North America. Information provided on the label includes the type of product, the pipeline owner, and an emergency phone number.

Pipeline spills have dropped 40 percent since 1969 from 318 spills per year to 197 per year average for the past six-year period. Spills amount to 1 gallon per million barrel-miles (a 42-gallon barrel transported 1 mile equals a 1-barrel mile) [a 159-liter barrel transported 1 km equals a 1-barrel kilometer]. Spills are usually the result of the following events:

- Third-party vandalism: 33 percent
- Corrosion of the pipe, valves, or connections: 30 percent
- Mechanical failure: 25 percent
- Operational failure: 7 percent
- Natural hazards (weather): 4 percent

These spills generally contaminate the soil by seeping into it. Fires generally occur when a pipeline is severed by a piece of construction equipment that becomes an ignition source. The fuel not only soaks into the soil, but can pool and ignite. The strategies and tactics used for both unignited and ignited pipeline leaks are based on the *RECEO* model and other basic factors.

Pipeline Incident Statistics (1986 to 2001)

- Liquid Product Pipelines:
  - —3,035 incidents
  - —36 fatalities
  - —2,811,000 barrels of product lost
  - —900 barrels average of product lost per incident
- Natural Gas Pipelines
  - —3,447 incidents
  - —340 fatalities

Fire and Spill Strategies/Tactics

Fire and spill tactics vary depending on the need for rescue, exposure protection, and confinement/extinguishment. The type of fuel and available resources are also important factors to consider. The first objective, which is the responsibility of the pipeline company, is to turn off the flow of fuel to the incident site. This procedure is accomplished while the emergency responders are en route to the incident. Central pipeline control computers can detect leaks and suspend the flow of product by closing valves on the supply side (upstream) of the leak. Valves downstream of the leak remain open to allow the material to drain from the line and reduce pressure at the leak site **(Figure 8.40)**. Communications between the responders and the pipeline company are essential during the initial response. The basic decision to use an offensive or defensive strategy may be based on whether the flow of fuel has been controlled by closing valves upstream of the leak or not. If the incident involves the malfunction of one of the control valves, then a larger quantity of product will be available to fuel the incident because of the increased distance to the next upstream control valve. The tactical priorities are summarized as follows:

- ***Rescue*** — If rescue is necessary, consider the use of foam or water fog spray to provide a safe area for the extrication or removal of victims. Tactics follow those described for tank vehicle spills or fires.
- ***Exposures*** — Exposure protection may determine whether intervention (offensive) or nonintervention (defensive) is the strategy of choice. If the incident is surrounded by potential exposures, then use foam to coat those exposures if sufficient foam concentrate is available for both exposure protection and extinguishment. Generally, a pipeline incident occurs in rural areas or in open spaces. Allowing the fire to consume all the fuel (nonintervention) may be the best strategy. Deploy foam handlines or master stream appliances if it becomes necessary to apply foam should there be a change in the situation.

Figure 8.40 The closest control valve up stream of a fire or leak is turned off, while the valves downstream are left open to drain the remaining product from the pipeline.

- ***Confine*** — Confinement can be accomplished by following the procedures listed in Chapter 7, Class B Foam Fire Fighting: Class B Liquids at Fixed Sites, and the Fire Strategies/Tactics section under Tank Vehicles in this chapter. If fuel has saturated the soil, a potential catalytic fire (one where chemicals in the soil can add to the fire when they mix with the fuel in the soil) may occur. Apply a foam blanket over the surface of the soil to prevent the release of flammable vapors. Give consideration to the confinement of all foam and water used to control the incident. Water runoff can contaminate water tables, sewers, streams, and other bodies of water.
- ***Extinguish*** — Pipeline fires are usually extinguished by turning off the flow of the fuel. Foam is used to extinguish the spill fire that has formed around the leak. Tactics for extinguishing this type of fire are described in Chapter 7, Class B. Foam Fire Fighting: Class B Liquids at Fixed Sites.
- ***Overhaul*** — Overhaul is the responsibility of the pipeline company. Overhaul requires the removal of all spilled product and the decontamination of all soil and exposures in the area of the spill. Fire and emergency services personnel may be required to remain at the incident to provide foam protection in the event of the ignition of the remaining product.

Basic Considerations

Besides the elements of the *RECEO* model that help establish priorities and tactics involved in pipeline incidents, other factors must be taken into consideration. They include the type of fuel, topography, wind direction, relative humidity, available water supply, available foam supply, delivery systems, and personnel availability. Refer to the appropriate sections of this manual for discussions of each of these factors as they apply to unignited and ignited fuel spills.

Maritime Transportation Industry

Jurisdictions that border rivers, lakes, canals, or coastal areas have the potential for fires onboard a variety of vessels. These vessels range from small powered recreation boats to large tank vessels that carry thousands of gallons (liters) of flammable or combustible liquids. Foam extinguishing agents provide the fire and emergency services personnel who respond to maritime emergencies with an advantage over the use of water alone. Personnel responsible for fire extinguishment may be members of the vessel's crew who are trained in fire fighting, land-based personnel who respond to incidents, or a combination of both. Because shipboard fire fighting by vessel crew members is so detailed, those personnel should refer to the IFSTA **Marine Fire Fighting** manual. The land-based fire and emergency responder should have a knowledge of the general protocol for dealing with maritime incidents, the types of vessels and hazards within the jurisdiction, types of onboard fire detection and suppression systems, and strategies and tactics for using foam concentrates to extinguish fires on board vessel types. A general overview is provided in this section, while a detailed explanation is found in the IFSTA **Marine Fire Fighting for Land-Based Firefighters** manual.

General Protocols

Commercial cargo vessels, passenger vessels, and military vessels are the responsibility of the master or captain of the particular ship. During fire-fighting operations onboard a vessel, the master or captain retains full command of all fire-suppression operations. Clear and concise communications are essential during an incident. The Incident Management System (IMS) helps maintain the correct level of communication and cooperation among all responding parties. It is also important for land-based personnel to have a good working relationship with the harbor or port authorities who help coordinate activities between fire and emergency services personnel and the ship's crew (**Figure 8.41**).

Fire and emergency services that are responsible for port or harbor facilities should prepare preplans for the types of vessels that frequent the facility. Attention should also be paid to marinas, docks, warehouses, pipelines and storage tanks, and dry docks or maintenance facilities. Not only is each of these items a hazard in its own right, but it also exists as an exposure hazard to fires on vessels that are nearby.

Figure 8.41 Harbor officials and emergency services authorities should meet regularly to review preplanned responses to the harbor and port facilities. *Courtesy of Captain John F. Lewis.*

Vessels Types and Hazards

Types of vessels that may be involved in an emergency incident include the following:

- Recreational
  - — Houseboats
  - — Sailboats
  - — Inboard, outboard, or combination-powered boats
- Passenger vessels (small and large)
  - — Ferries (both passenger and vehicle)
  - — Cruise ships
  - — Passenger liners
- Tank vessels
  - — Petroleum carriers
  - — Liquefied flammable gas/chemical carriers
- Cargo vessels
  - — Bulk cargo (dry)
  - — Break bulk cargo
  - — Container
  - — Roll-on/roll-off vessels
  - — Car carriers
  - — Barges
- Tugboats, offshore workboats, and commercial fishing vessels
- Mobile offshore drilling units and drillships
- Military vessels
  - — Warships
  - — Cargo vessels
  - — Troop/aircraft carriers

Although each type of vessel is different in construction, size, capacity, and purpose, they all have one thing in common. They all carry some quantity of flammable or combustible liquid for use in the vessel's propulsion system or in the auxiliary power or cooking systems (such as in the case of sailboats). Obviously, the larger the vessel, the larger the quantity of fuel carried on board.

In addition to the ship's fuel supply, cargo vessels carry a wide variety of flammable materials, some of them in large steel containers that may not be properly marked for the material carried within. Shipping containers may range from cartons of 1-gallon (4 L) containers to bulk water-reactive dry chemicals carried in the ship's cargo holds. Tankers may carry crude oil, finished petroleum products, hazardous chemicals, or liquefied flammable gases. Military vessels also contain explosives for ship's weapons or as cargo. Aircraft carriers are particularly hazardous due to the large quantities of ordinance and aircraft fuels that are stored and distributed throughout the vessel.

All seagoing vessels larger than recreational craft must have fire detection and suppression systems to protect the engine room and cargo spaces. Land-based fire and emergency services personnel must be familiar with these systems, their effectiveness, their limitations, and how they can be supported from shore.

Several areas exist on a vessel where a flammable or combustible liquid fire may occur. The most common area and location of many vessel fires are machinery spaces. Vessel engine rooms can be extremely large, both in square feet (square meters) and depth. They also are very congested with piping and other systems. Propulsion systems can involve boilers and steam turbines in old vessels or diesel and gas turbines or nuclear reactors in new vessels. Flammable and combustible liquid fires may be caused by spills, as well as fuel piping, hydraulic system, and lubrication leaks. Water and other liquids may accumulate in the *bilge* (lowest inner part of a vessel's hull) **(Figure 8.42).** Flammable or combustible liquids spilled in an engine room accumulate in this area, as does foam discharged into the engine room.

Fuel tanks on vessels are located adjacent to engine rooms. Fuel is usually marine diesel or the much heavier No. 6 fuel oil (also known as bunker C). The heavier fuels may be heated to temperatures well over 100°F (38°C) to facilitate flow through the fuel system. Heated fuels tend to ignite easier when released in a spray or spill. Fuel storage tanks may be termed *deep, double bottom,* or *day* tanks, depending on the phase of the use of the fuel.

Engine room fires may cause fuel to expand in fuel tanks and discharge into the engine room or onto the decks through vents and sounding tubes. Overfilling the fuel tanks during fueling operations may also cause a spill on a deck. Most vessel decks are sloped or cambered so water can drain to the sides and stern (back) of the vessel. Spilled flammable and combustible liquids on a deck also flow in the same manner. Flammable and combustible liquid cargo may also be stored in intermodal containers or tanks above or below deck on container ships or in drums and other smaller containers on break bulk vessels. Fire and emergency services personnel may encounter a Class B liquid fire inside a cargo hold. Class A fires in cargo holds and vessel living spaces may require the use of a high-expansion or Class A foam operation.

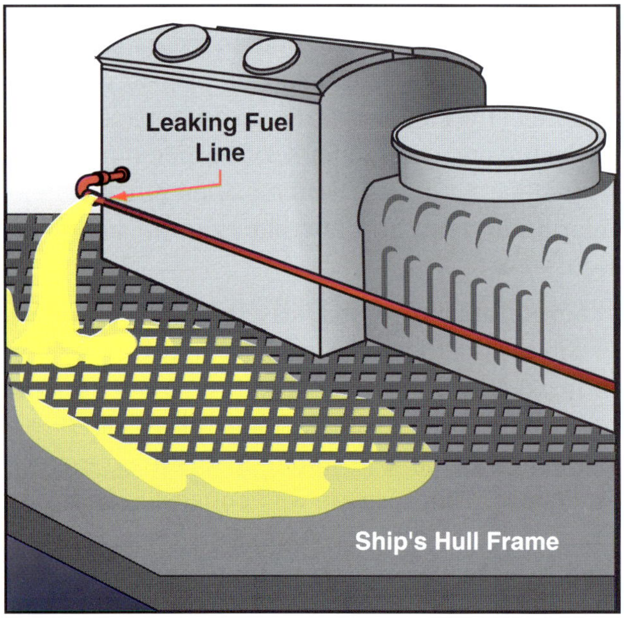

Figure 8.42 Water and other spilled liquids can accumulate in the bilge and engine room areas of a vessel.

Onboard Fire Detection and Suppression Systems

Fire detection and suppression systems on vessels are similar to those found in land-based warehouses, manufacturing facilities, fuel storage tanks, and high-rise type structures. The degrees to which the systems are installed or maintained vary based on the country in which the vessel is licensed. U.S. and Canadian vessels probably have the highest level of fire-fighting equipment maintenance and training for their crews. Vessels that are licensed in smaller foreign nations may not adhere to the same standards. Therefore, land-based fire and emergency services personnel must not assume that a vessel's fire-suppression system is 100 percent operable. Shipboard fire hoses may be worn or have damaged couplings, valves may be tight and difficult to open, and foam systems may be clogged with dried foam solution and may not have been tested recently. In addition, the onboard foam fire-suppression system or the ship's crew using handlines may have expended the shipboard supply of foam concentrate. Finally, the international shore connection (ISC) on the vessel may be damaged or have bolt threads that are not compatible with those on the shore-based fire service organization's fitting. The ISC is a standard flange that is fitted to the vessel's fire main. Once the flange from the fire service organization is bolted to the ISC, fire pumpers can supply water to the vessel's fire main; however it is the responsibility of the fire service organization to have the matching flange with threads matching its supply hose **(Figures 8.43 a and b)**. A potential problem with the ISC is that it is limited in size to 2½ inches (65 mm) and 200 psi (1 379 kPa) in operating pressure. Therefore, the best policy is for land-based fire and emergency services personnel to provide their own equipment and foam concentrates when possible.

Using foam agents is advantageous for all types of shipboard fire fighting because foam can quickly extinguish a fire and blanket any liquid fuels without adding a great deal of weight to the vessel. The addition of thousands of gallons (liters) of water to

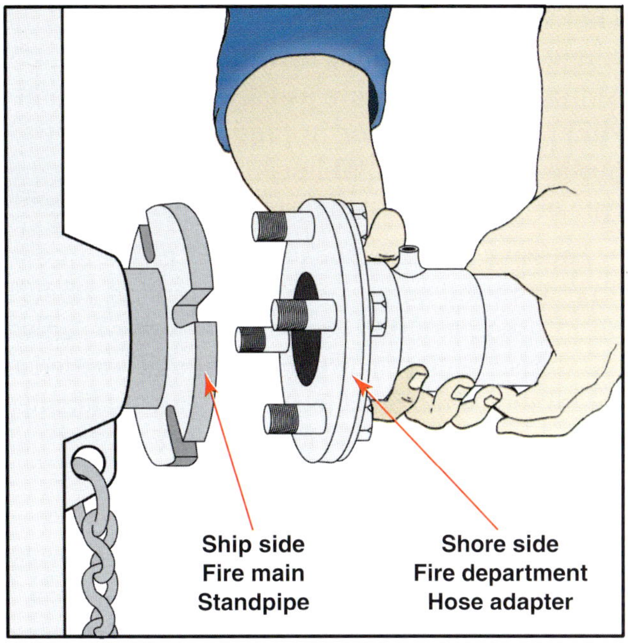

Figure 8.43a When the shore-side connection is attached to the international shore connection (ISC), hose can be secured to the shipboard fitting. *Reprinted with permission from NFPA 1405,* Guide for Land-Based Fire Fighters Who Respond to Marine Vessel Fires, *copyright © 1996, National Fire Protection Association, Quincy, MA 00269. This printed material is not the complete and official position of the National Fire Protection Association on the referenced subject, which is represented only by the standard in its entirety.*

Figure 8.43b Typical shipboard ISC fitting. *Courtesy of Captain John F. Lewis.*

a vessel, regardless of size, can cause it to become unstable and increases the potential for the vessel to list (lean) and capsize. Because less water is required when foam concentrates are used, the potential for instability is lessened.

Fire Strategies/Tactics

The strategies and tactics used should always follow the *RECEO* pattern described earlier, regardless of the type of vessel involved. However, how those strategies and tactics are applied varies based on

Figure 8.44 Devastating fires can occur onboard ships either at sea or while in port. Master streams and large quantities of foam are required to control fires of this magnitude. *Courtesy of Chris Mickal.*

the type of vessel involved, the location of the vessel, and the exposures that it endangers. Normally a vessel at sea extinguishes its own fire (if possible) because it is usually too far from port for other ships to reach in time. Crew members will attempt to reach a point within range of fire-fighting vessels when a fire is beyond their control. Vessels that are close to shore or moored within a harbor area will require a fire attack by fireboats. If a vessel is tied to a dock or pier, a fire attack may be made from the shore as well as from fireboats. If a two-pronged attack is undertaken, it is important to coordinate the attack so the opposing foam streams do not disrupt each other's spray pattern or coverage **(Figure 8.44)**.

Recreational Vessels

Recreational vessels or pleasure craft may contain a quantity of fuel (usually automotive grade gasoline) to power the engine or an auxiliary motor. Quantities range from 5 to 50 gallons (19 L to 189 L) in portable or permanently fixed fuel tanks. Cooking fuel usually consists of 5-gallon (19 L) containers of propane or butane. Due to the limited quantities of fuel, Class A or Class B foam can be used to extinguish a fire on board the craft.

Passenger Vessels

When a fire occurs on a passenger vessel while it is at sea or underway, the crew activates the onboard fire-suppression system, gathers the passengers at predesignated locations, and stands by to abandon ship if necessary. If the vessel is close to land or in port, a land-based fire and emergency services organization may be called upon to assist in evacuation and fire-extinguishment activities.

While high-expansion foam fire-suppression systems are usually found in the machinery and fuel-handling system spaces, the majority of the vessel is protected by a water sprinkler system. Land-based personnel may provide additional water or foam concentrate to support the ship's fire-suppression system. The international shore connection on the vessel permits land-based fire and emergency services personnel to supply the vessel's fire main. Either water or foam solution may be pumped through this system by a fire service pumper.

Cargo Carriers (Bulk and Dry)

Cargo carriers (either powered or nonpowered) are found in both coastal regions and inland waterways. They transport lumber, paper, flour, grain, metal scrap, and other bulk materials (**Figure 8.45**). Nonpowered barges are usually grouped together either side-by-side or in tandem and pulled or pushed by a towboat. While cargo vessels have some type of fire-suppression system, barges do not.

Fires may occur in machinery spaces, living quarters, or in the cargo spaces or holds of vessels. Because some of the cargo may be water-reactive, the ship's manifest (a list of all materials carried as cargo) should be consulted before water or finished foam is directed into the involved space.

Tank Vessels (Liquids/Chemicals)

Tank vessels (or tankers), like their land-based road and rail counterparts, present the greatest hazard for fire and emergency services personnel. Tankers may carry from 15,000 to 3,680,000 barrels of petroleum products. Spills not only pollute the environment, they create a massive potential for fire in the vicinity of the vessel.

Shipboard fire-suppression systems not only protect the machinery spaces but also provide a deck foam system. This system is designed to provide a finished foam blanket onto the ship's deck through fixed monitors, much like the Type II systems mentioned in Chapter 7, Class B Foam Fire Fighting: Class B Liquids at Fixed Sites. Between 20 and 30 minutes of foam generation capacity is provided by the system.

Figure 8.45 A typical dry bulk cargo carrier. *Courtesy of Howard Chatterton.*

If the vessel is discharging unignited fuel onto the ocean or other body of water, containing the spill is the first priority once the need for rescue has been addressed. Floating booms are usually used for this purpose and installed by personnel who are specially trained for this task. If the spill has ignited, then Class B finished foam is applied in the bank-down method, using the ship's hull to disperse the foam. If the fire is being fed by fuel from above, treat the fire as a three-dimensional fire as outlined in Chapter 7, Class B Foam Fire Fighting: Class B Liquids at Fixed Sites.

If the vessel is fully involved in fire, rescue has been completed, and exposures are not an issue, it may be necessary to allow the fire to burn itself out. Water fog can be used to cool the hull and superstructure, and foam can be used to blanket the water's surface.

Tugboats, Commercial Fishing Vessels, Workboats

Rivers and coastlines are the homes of small tugboats, towboats, workboats, and commercial fishing vessels. Flammable and combustible liquids found on these vessels are usually contained in the vessel's fuel tanks. The vessels may range in length from 60 to 200 feet (18 m to 61 m) and carry crews from 6 to 25 people. They may have cooking and sleeping quarters onboard for the crews.

Onboard fire-fighting equipment consists of a fire pump, a water supply, and standpipes equipped with hoses and nozzles. The condition of this equipment may be questionable, so land-based personnel should never rely on the vessel's equipment. No foam capability is found on these small vessels.

Class A or Class B finished foam may be applied to extinguish fires from dockside or from specially equipped fireboats or tugboats (**Figure 8.46**). Interior fire fighting is handled with the same precautions as interior structural fire fighting. Care must be taken to maintain the vessel's stability and prevent capsizing.

Offshore Drilling Platforms

Offshore drilling platforms are a common sight along the coast of most states/provinces in North America (**Figure 8.47**). Some are permanently

Figure 8.46 Fires on small workboats or commercial fishing vessels may have to be fought from vessels alongside. *Courtesy of U. S. Coast Guard.*

Figure 8.47 A typical offshore drilling platform. *Courtesy of Steven Stokely, Sr.*

anchored to the ocean floor while others float in place. Due to the potential for crude oil or natural gas fires on drilling platforms, water spray and foam fire-suppression systems are built into the facilities.

It is unlikely that land-based fire and emergency services personnel will be required to extinguish fire on a drilling platform offshore. The more likely incident requiring fire service response will be for an incident on a drilling platform in a shipyard. If the offshore situation arises, however, it is approached as a three-dimensional fire, using finished foam to blanket the fire on the water's surface first and then directing foam where appropriate on the structure. Fireboats, resupplied with foam concentrate from barges or nurse vessels, are usually the primary foam delivery systems, using high-volume deck guns or monitors.

Military Vessels

Military vessels belong to a country's navy, coast guard, or other branch of its armed forces. They may be combatants (warships) or noncombatants (cargo) vessels contracted for use by the military. The largest combatant military vessels are aircraft carriers, while the smallest are rescue and patrol crafts used by a country's coast guard. All carry fuel for the propulsion systems and cooking facilities. The majority carry ordinance for weapons systems or aircraft that are assigned to the vessel. All vessels have onboard fire-suppression systems similar to the ones found on commercial vessels. However, the systems on military vessels have a much greater potential for operating correctly due to the fact that required inspections and testing are more strictly enforced. In addition, military vessels have larger crews, which means that fires are detected earlier and more people are available to take immediate fire-suppression actions.

Fires have occurred on military vessels both in times of war (as a result of enemy action) and peace (as a result of a collision or accident). The fires may also occur at sea or while in port. Land-based fire and emergency services personnel who respond to military port facilities should have a good knowledge of the types of vessels that are in port at any given time and the fire-suppression systems onboard them. It is essential that the local fire and emergency service organization know the base commander, base staff, and officer who are responsible for port operations.

In the case of military vessels in port, the land-based fire and emergency services personnel generally only provide exposure protection and limited support to the ship's crew. Class A finished foam can be used to blanket adjacent exposures and cover any fuel that has leaked onto the surface of the water. Military support vessels are often operated by civilian crews and are more likely to need and welcome shoreside fire service assistance.

Summary

Fire and emergency services personnel can benefit from the use of foam concentrates to control and extinguish all types of transportation incident fires. While additional training should be pursued for aircraft and maritime fire incidents if the hazard warrants, all fire and emergency services should provide basic foam training for tank vehicle, rail tank car, and pipeline flammable and combustible liquids incidents. Continual training and preplanning prove invaluable for successful foam operations.

Chapter 9

Foam Concentrate Use Training

Job Performance Requirements

This chapter provides information that will assist the reader in meeting the following job performance requirements from NFPA 1002, *Standard for Fire Apparatus Driver/Operator Professional Qualifications,* 1998 edition; NFPA 1081, *Standard for Industrial Fire Brigade Member Professional Qualifications,* 2001 edition; and NFPA 1051, *Standard for Wildland Fire Fighter Professional Qualifications,* 2002 edition. Colored portions of the standard are specifically addressed in this chapter.

NFPA 1002

3-1.1 **Perform the specified routine tests, inspections, and servicing functions** specified in the following list in addition to those contained in the list in 2-2.1, given a fire department pumper and its manufacturer's specifications, so that the operational status of the pumper is verified.

- Water tank and other extinguishing agent levels (if applicable)
- Pumping systems
- **Foam systems**

(a) *Requisite Knowledge:* Manufacturer specifications and requirements, policies, and procedures of the jurisdiction.

(b) *Requisite Skills:* The ability to use hand tools, recognize system problems, and correct any deficiency noted according to policies and procedures.

6-1.1 **Perform the specified routine tests, inspections, and servicing functions** specified in the following list, in addition to those contained in 2-2.1, given a wildland fire apparatus and its manufacturer's specifications, so that the operational status is verified.

- Water tank and/or other extinguishing agent levels (if applicable)
- Pumping systems
- **Foam systems**

(a) *Requisite Knowledge:* Manufacturer specifications and requirements, policies, and procedures of the jurisdiction.

(b) *Requisite Skills:* The ability to use hand tools, recognize system problems, and correct any deficiency noted according to policies and procedures.

NFPA 1081

5.1.2.4 Return equipment to service, given an assignment, policies, and procedures, so that the equipment is inspected, damage is noted, the equipment is clean, and the equipment is placed in a ready state for service or is reported otherwise.

(A) **Requisite Knowledge.** **Types of cleaning methods for various equipment, correct use of cleaning materials,** and manufacturer's or facility guidelines for returning equipment to service.

(B) **Requisite Skills.** The ability to clean, inspect, and maintain equipment and to complete recording and reporting procedures.

NFPA 1051

5.3.3 Maintain assigned suppression hand tools and equipment, given tools and equipment and agency maintenance specifications so that assigned equipment is safely maintained and serviceable and defects are recognized and reported to the supervisor.

(A) **Requisite Knowledge.** **Inspection of tools and assigned suppression equipment, the recognition of unserviceable items, and required maintenance techniques.**

(B) **Requisite Skills.** Sharpening and other maintenance techniques for assigned suppression equipment, and use of required maintenance equipment.

Chapter 9
Foam Concentrate Use Training

Fire and emergency services personnel must be prepared for any emergency situation that may confront them in their daily schedules. Thorough and ongoing training programs help emergency personnel to develop skills and keep them fresh. One of the topics that is essential for fire and emergency services personnel to learn is foam concentrate use, including the techniques involved with the selection, proportioning, and application of Class A and Class B foam concentrates and the care and maintenance of proportioning equipment. The primary reason for Class B foam concentrate use training is that incidents involving the use of Class B foam concentrates occur infrequently and most fire and emergency services personnel have little opportunity to gain practical or on-the-job experience with Class B foam proportioning equipment and application techniques. The main reason for Class A foam concentrate use training is to educate emergency personnel on the benefits of this type of extinguishing agent and make them proficient with it. Fire and emergency services organizations that possess compressed-air foam systems (CAFSs) and multiagent nozzles must provide this specialized training to gain the greatest benefit from using the equipment and justify the cost of acquiring it.

This chapter discusses the training necessary for basic and advanced knowledge of Class A and Class B foam concentrate use and equipment care and maintenance. It also discusses foam training facilities, foam training practical evolutions (both interior and exterior), training foam concentrates, and foam training without foam concentrates.

Foam Use Training Course Contents

Fire and emergency services organizations should develop foam concentrate training programs that realistically cover the duties that fire and emergency services personnel are expected to perform. However, realism must take a backseat to safety in all training exercises. Basic knowledge of foam operations can be provided in the entry-level training program, including information on the types of hazards that foam concentrates are intended to control, types of foam concentrates, foam terminology, foam concentrate proportioning equipment, foam application techniques, and the care and maintenance of foam proportioning equipment. Advanced courses contain information required for personnel assigned to specialized units such as aircraft fire fighting teams, hazardous materials teams, wildland firefighter crews, and industrial fire brigades that protect plants and storage facilities from hazardous liquids and chemicals.

NOTE: Simulated flammable liquids training fires will never duplicate the intensity, radiant heat, or noise that is encountered in actual flammable liquids incidents.

Basic Foam Use Training

Entry-level municipal firefighters and industrial fire brigade members should receive fundamental training in foam concentrate use basics. The course consists of classroom learning, practical field exercises, and continuing education programs.

Classroom Topics

Specific information related to the use of foam concentrates is conveyed to all fire and emergency services personnel before they participate in practical exercises (**Figure 9.1**). The following list contains some of the important topics that are included in the classroom lecture portion of the course:

- ***Flammable and combustible liquid hazards*** — Covers the hazards of hydrocarbon, polar solvent, and blended fuels
- ***Class A hazards*** — Covers the types of Class A fuels and incidents that foam concentrates can be used on effectively
- ***Foam terminology*** — Explains terms such as foam solution, foam concentrate, finished foam, etc.
- ***Foam concentrates*** — Describes the types of concentrates and the uses, advantages, and disadvantages of each
- ***Foam-proportioning equipment*** — Addresses the proper operation, care, cleaning, and maintenance of foam proportioning and application equipment (**Figure 9.2**)
- ***Aspirated versus nonaspirated foam*** — Defines each type and describes the advantages and limitations of each
- ***Application techniques*** — Reviews the different foam application techniques and discusses when each technique is the desired method

Practical Field Exercises

The second portion of the basic foam fire fighting training course provides practical evolutions that allow fire and emergency services personnel to use foam proportioning and application equipment to extinguish fires. Fire and emergency services personnel are required to perform the following tasks:

- Assemble the proportioning and application equipment necessary to produce and sustain a foam stream (**Figure 9.3**).
- Demonstrate the ability to operate the equipment within its design specifications.
- Demonstrate the different techniques for applying foam streams (**Figure 9.4**).

Figure 9.1 Include information on hazard types, foam terminology, foam theory, the types of proportioning and distribution systems, and application techniques in classroom presentations

Figure 9.2 Demonstrate the operation, care, cleaning, and maintenance of foam proportioning equipment before the actual use of the equipment.

Figure 9.3 Students are required to demonstrate the classroom concepts by assembling a foam proportioning and delivery system.

- Demonstrate the proper methods for cleaning equipment used in the application of finished foam, including nozzles, hoses, and proportioning devices.

Figure 9.4 Train students in foam extinguishment techniques in situations that are as close to reality as possible while still maintaining a safe environment. Safety always takes precedence over reality on the training ground.

Continuing Education

To remain proficient with the skills learned during entry-level training, fire and emergency personnel are required to attend continuing training programs throughout their careers. These programs are basically refresher courses that include both classroom and practical field exercises. Emergency personnel who advance in rank to driver/operators need additional coursework in the operation of apparatus-mounted foam proportioning devices, pump operations for foam distribution, and maintenance of onboard systems (**Figure 9.5**). If emergency personnel are assigned to operate CAFS apparatus, aircraft rapid-response apparatus, or fireboats that have foam distribution systems, they must then be provided with training on that specific piece of equipment.

Figure 9.5 Personnel who are assigned to operate foam-equipped apparatus require additional training in the operation of onboard systems.

Advanced Foam Use Training

Fire and emergency services personnel such as airport firefighters, hazardous materials technicians or team members, and petrochemical industrial fire brigade members who are specifically assigned to respond to flammable and combustible liquids fire-fighting situations need advanced training in foam fire-fighting operations. Advanced training is a combination of theory and practical evolutions. In addition, supervisory personnel such as company officers and chief officers are provided with similar information to give them the knowledge to make appropriate decisions in the use of foam concentrates.

Classroom Topics

The classroom portion of the advanced course includes everything contained in the basic-level classes with additional material to cover the following topics:

- *Application rates* — Provides information on how to determine appropriate rates based on the fuels involved and the foam concentrate being used; also determines the total amount of foam concentrate required to control an incident
- *Proportioning systems* — Covers all major foam proportioning systems and portable proportioning equipment
- *Fixed-site foam discharge systems* — Describes the various types of foam concentrate discharge systems at fixed-site facilities within the jurisdiction
- *Fixed-site facilities* — Describes the various types of flammable and combustible liquids, hazardous materials, and Class A high fuel load hazards at fixed-site facilities within the jurisdiction and the types of storage vessels and containers used at those sites
- *Foam concentrate storage* — Discusses the various methods of storing foam concentrates; explains that proper storage methods keep foam concentrates effective and provide easy access when they are needed
- *Foam concentrate compatibility* — Addresses foam concentrate compatibility with other foam agents; ensures that personnel do not use agents that work against each other **(Figure 9.6)**
- *Foam concentrate environmental concerns* — Highlights the impact of foam concentrates, foam solutions, and finished foam on the environment

Practical Field Exercises

When the classroom portion of the training program is complete, extensive field training exercises are performed. Fire and emergency services personnel who are involved in foam operations are trained under realistic conditions within the training capabilities of the organization. The practical portion of the advanced class include performing the following tasks:

Figure 9.6 Include information during training about the results of mixing incompatible types of foam together. As this photograph shows, it can be very easy to accidentally mix foams during an incident. Mixing concentrates could clog the proportioning system.

- Select the appropriate foam concentration, application rate, water source, and proportioning and application equipment for a specified hazard.
- Assemble the proportioning and application equipment necessary to produce and sustain a foam stream.
- Demonstrate the ability to operate the equipment within its design specifications.
- Demonstrate the different techniques for applying foam streams from all types of distribution systems.
- Demonstrate the proper method for cleaning equipment used in the application of finished foam, including nozzles, hoses, proportioning devices, and apparatus-mounted systems.

Use large-scale training fires for advanced foam use training. The fires may also be handled like simulated incidents using a command structure, which gives students a realistic idea of everything needed to control a large fire. The students size up the incident, determine the necessary concentrate type and application rate, set up the delivery system, and apply the finished foam. These types of training exercises also give the organization an opportunity to reevaluate the effectiveness of the existing operational procedures.

Because the majority of fire and emergency service organizations do not have the facilities, training personnel, or funds to provide this level of training,

Figure 9.7 Fire and emergency services organizations that do not have facilities for large Class B training fires may want to contact the state training agents for hands-on training at their facilities.

it may be necessary to use an outside contractor. An outside contractor may also be necessary if local clean air and water ordinances prohibit such training fires. State/provincial universities have degree programs in fire protection, and state/provincial fire service training programs have facilities that can be used for advanced-level training (**Figure 9.7**). In addition, large fire departments may also work with small organizations to include them in scheduled training sessions. This cooperation is especially important in areas where mutual aid agreements exist between the organizations. Finally, foam concentrate manufacturers and nozzle and proportioner manufacturers also provide courses in the use of their products (**Figure 9.8**).

Additional Specialized Foam Use Training

Any fire or emergency services organization that is going to undertake foam fire-fighting operations must also be prepared to support those operations. Therefore, maintenance technicians must be trained to repair, clean, and test foam concentrate

Figure 9.8 Foam concentrate and foam delivery system manufacturers provide training courses throughout the country on a regular basis.

Figure 9.9 Training aids, such as this cut-away of a foam proportioner, can be provided by the manufacturer. Such devices are essential in illustrating operation, cleaning, and maintenance of the units. *Courtesy of Tom Ruane.*

proportioners, eductors, CAFSs, nozzles, and monitors. The equipment manufacturer usually provides this type of training when the equipment is purchased (**Figure 9.9**). If the fire or emergency services organization does not have the personnel or resources to provide this type of support, an outside contractor or the manufacturer can maintain the equipment for the organization. However, basic care and cleaning are still the responsibilities of emergency services personnel, and they require training to that level.

Foam Equipment Care and Maintenance Training

[NFPA 1002: 3-1.1, 6-1.1; NFPA 1051: 5.3.3; NFPA 1081: 5.1.2.4]

Fire and emergency services personnel must be trained in the basic skills necessary for the care and maintenance of foam proportioning and delivery systems. Proper care and maintenance ensure that equipment is ready when needed and reduce the cost of repairs due to neglect or negligence.

Equipment Care

Care of the equipment involves inspecting it at the beginning of each work shift or period, cleaning it following each use, and storing it so that it will not be damaged. These activities follow the same pattern that fire and emergency services personnel use in handling all tools and equipment assigned to them. Of particular importance, however, is the postincident cleaning of foam-delivery equipment. Nozzles, proportioners, eductors, and hoses must be thoroughly flushed with clean water (**Figure 9.10**). Use a minimum of 5 gallons (19 L) of water to flush the foam equipment. This flushing prevents foam concentrate buildup and caking that may prevent the equipment from working properly. It also helps to ensure that incompatible types of foam concentrate are not mixed together, creating a further obstruction in the delivery system. Hoses must be flushed to prevent foam solution from deteriorating the lining while the hose is in storage on the apparatus. Because Class A foam and emulsifiers contain degreasers, it is necessary to apply lubricants to the moving internal parts of handline nozzles, master stream nozzles, proportioners, and eductors after using Class A foam. Lubricants are not necessary if Class B foam has been used with this equipment.

Equipment Maintenance

Although equipment technicians are usually responsible for the repair and testing of foam concentrate proportioners, nozzles, and eductors, field personnel can perform limited maintenance. Replacement of worn parts, gaskets, and pickup hoses may take place during the inspections performed at the beginning of the work shift or period and during postincident cleaning. Manufacturers usually establish the level of maintenance that can be performed in the field and the required amount of training necessary to perform this maintenance. Field personnel should never attempt repairs that are beyond their training and absolutely never modify a nozzle, proportioner, or eductor without the authority of the manufacturer. Such modifications can lead to equipment malfunction, injury, or loss of warranty coverage on the equipment.

Figure 9.10 Follow the foam manufacturer's recommendations on care and cleaning methods. Foams are mild corrosives and should always be flushed out of hoses, proportioners, pumps, and nozzles. *Courtesy of Tom Ruane.*

Foam Use Training Facilities

Most fire and emergency services organizations have access to some type of training facility. Some organizations own training centers with elaborate burn buildings, exterior props, and support facilities. Other organizations may only have a small classroom building and an area large enough to provide basic entry-level training (**Figure 9.11**). Some organizations may only have access to training facilities that are provided by nearby universities, state/provincial training agencies, or metropolitan departments. In any case, the training facility used for foam concentrate training should have structures and props that provide a variety of scenarios and the opportunity for live-fire exercises. The following sections give an overview of training structures that can be used for foam concentrate use training as provided in NFPA 1402, *Guide to Building Fire Service Training Centers* (2002). The

Figure 9.11 Train maintenance technicians in the repair of proportioners based on the equipment manufacturer's recommendations.

U.S. Federal Aviation Administration (FAA) also provides information in its publications on the design of aircraft rescue and fire-fighting training simulators. Information on infrastructure requirements is also discussed.

Class A Foam Structures and Props

Structures and props that are used for nonfoam firefighting training can also be used for foam concentrate use training. Training center burn buildings (mentioned earlier), concentrations of Class A materials, and wildland training areas are included in this category.

Training Center Burn Buildings

Structures that are specifically designed for use in live-burn exercises must have characteristics that provide a safe yet realistic training experience for students **(Figure 9.12)**. The following recommended characteristics are based on NFPA 1403, *Standard on Live Fire Training Evolutions* (2002):

- Construction materials that are resistant to fire and high temperatures
- Multiple interior room arrangements
- Direct exit or means of egress from each room
- No underground rooms or spaces
- Fire-resistant steel doors and window shutters
- Makeup or intake air supply to ensure complete combustion
- Ventilation systems
- Emergency fuel shutoff valves or switches **(Figure 9.13)**
- Instrumentation to monitor temperatures to ensure that a fire is maintained within safe parameters
- Adequate access from all sides
- No exposure hazards

Class A Materials Concentrations

Most training exercises that use concentrations of Class A materials only require an open space and a large pile of wooden pallets. This situation may be adequate for water stream training, but it is not suitable for foam concentrate use training. The reason is that the finished foam and contaminated water must be captured and the water cleaned before it enters the water runoff system. The following characteristics provide environmental safeguards at Class A materials burn areas:

Figure 9.12 Permanent training center burn buildings are constructed of nonflammable materials such as concrete and are equipped with ventilation louvers, steel shutters on windows, and standpipe and sprinkler connections.

Figure 9.13 All live fire props must be equipped with fuel shutoff valves that are constantly monitored during training operations.

- Nonporous base such as concrete to prevent soil contamination with a slight (6- to 12-inch [150 mm to 300 mm]) retaining wall around the perimeter **(Figure 9.14)**
- Large enough space to contain the burning material and room to operate attack and backup hoses
- Adequate drainage to carry contaminated water to the decontamination area (See Infrastructure Requirements section.)
- Located away from exposures and populated areas that might be in the path of smoke

Figure 9.14 Class A burn pits are designed with curbs to retain the water/foam runoff and drains to collect and channel it to retention ponds.

Figure 9.15 Wildland fire training may be combined with controlled burns. In this photograph, personnel spread a foam blanket as a barrier for a burn. *Courtesy of National Interagency Fire Center.*

Wildland Training Areas

Fire and emergency services organizations that have the primary responsibility for protecting rural, suburban, and urban-interface areas from wildland fires should develop training areas specific to their needs. NFPA 1402 does not address this type of facility, but a few suggestions may be appropriate.

Areas that could be dedicated to this type of training may be available on unused military bases, national and state/provincial parks, industrial parks, or other large open spaces. Live-fire exercises could be included with controlled burns (with proper precautions). Training foam that remains visible after application may be used to simulate the application (both by hand and from apparatus) of finished foam as defensive control lines (see Training Foam Concentrates section) **(Figure 9.15)**.

Airdrop training can use both training foam concentrates and raw water. A supply source such as a lake or reservoir is necessary to train helicopter pilots on bucket pickups. Airdrops over all types of terrain can be practiced to determine the appropriate tactics for each type of terrain (See Chapter 5, Class A Foam Fire Fighting, Wildland Fires section).

Class B Foam Props

Flammable and combustible liquid and gas training props may take the form of a variety of installations. In general, they must be located in open areas away from exposure hazards and populations. Each prop must be located in a pit designed to contain the maximum quantity of contaminated

Figure 9.16 Flammable/combustible liquids fire extinguishment training can involve props such as this diked storage tank.

water that could be used in the training exercise **(Figure 9.16)**. Each pit must be provided with an adequate water supply, a fuel source with safety shutoff valves, and a drainage system connected to the water decontamination area. Hazardous materials training and transportation training areas have similar requirements.

Flammable and Combustible Liquid and Gas Training Areas

Props designated for flammable and combustible liquids and gases include the following:

- Pits to simulate both unignited and ignited fuel spills
- Aboveground storage tanks; types:
  - — Vertical
  - — Horizontal
  - — Cone top

— Internal floating roof
— External floating roof

- Overhead flanges
- "Christmas trees" (piping networks that resembles Christmas trees; they are equipped with shutoff valves and used to simulate gas pressure fires) **(Figure 9.17)**
- Liquefied petroleum gas (LPG) facilities **(Figure 9.18)**
- Loading docks
- Pump islands
- Pump stations
- Chemical plant processing facilities

Figure 9.17 The "Christmas tree" prop is a simple training aid that allows students to advance on the burning prop, apply a protective foam or fog stream, and turn off the flow of gas at the valve.

Figure 9.18 LPG storage tanks and piping systems are types of props that can replicate facilities found at most refineries.

The types of props selected for a training facility that conducts Class B foam training depends on the types of hazards located within the jurisdiction. It is recognized that not all of these props are required at most training facilities. In some cases, petroleum and chemical manufacturers within the jurisdiction may be able to provide out-of-service facilities for training purposes. Like purpose-built facilities, these out-of-service units must meet the requirements of NFPA guidelines in order to be acceptable for training.

Hazardous Materials Training Areas

Props used for hazardous materials training may include some of those described in the previous section. They may also include the following items:

- Chemical processing facilities
- Chemical storage tanks **(Figure 9.19)**
- Chemical stage areas (small container and drum)
- Chemical spill pits

Figure 9.19 Chemical plant props may include both vertical and horizontal tanks.

Although only a simulated hazardous material is involved in the training and training foam concentrate is used in the exercise, the runoff water still has to be contained and decontaminated. All training areas must be within diked areas.

Transportation Training Areas

Because transportation and vehicle accidents are the most common incidents for fire and emergency services personnel to encounter, the majority of training facilities have some type of prop representing these types of incidents. For foam training, the props consist of the following items:

- Pits to simulate both unignited and ignited fuel spills
- Flammable and combustible liquids transport truck and trailer **(Figure 9.20)**
- Simulated vehicle collision (may also be used for Class A foam training) **(Figure 9.21)**
- Railroad tank car **(Figure 9.22)**
- Simulated oil tanker (ship)
- Aircraft fuselage or simulated aircraft

Most airport fire-fighting organizations have a burn pit with a simulated aircraft made of heavy gauge steel **(Figure 9.23)**. Major universities and seaboard jurisdictions have access to simulators for training in maritime fire fighting. The remaining types of transportation props are found at most municipal and state/provincial fire service training facilities. All transportation props require the same safety and environmental requirements of the other exterior foam training props.

Figure 9.21 Training props can include simulated collisions between tanker trucks and automobiles that result in unignited or ignited spills.

Figure 9.20 Tanker truck props should be placed on concrete pads that have retention curbs and drainage facilities like the Class A pits.

Figure 9.22 Railcar props can be used for spills, spill fires, vent fires, and loading rack fires. Obsolete railcars may be acquired from railroads or tank car companies.

Figure 9.23 Aircraft training props are constructed from steel. This image shows, left to right, a large commercial fuselage and wing, an engine nacelle, and a small commuter aircraft fuselage.

Infrastructure Requirements

To properly support these training props, a training facility has to provide an infrastructure that consists of the following items:

- ***Adequate water supply*** — Required by NFPA 1142, *Standard on Water Supplies for Suburban and Rural Fire Fighting* (2001); considered to be the volume and pressure required to support training operations based on the following factors:
  - — Number of attack lines and backup lines used
  - — Need for potable (suitable for drinking) water
  - — Need to supply sprinkler and water spray systems with necessary volume and pressure
  - — Need to supply water for other types of exercises that may take place at the site
- ***Fuel source*** — Piped to the prop from a main supply; may use flammable or combustible liquids, liquefied petroleum gas, or natural gas (**Figure 9.24**)
- ***Breathing air supply*** — May include the following items:
  - — Spare replacement breathing air cylinders
  - — Portable breathing air compressor
  - — Piped breathing air distribution system from a centrally located compressor (**Figure 9.25**)
  - — Fixed or portable cascade system
  - — Supplied-air system
- ***Equipment decontamination area*** — Designated for the washing and cleaning of personal protective clothing, hoses, nozzles, and other equipment; pipe contaminated waste water to the water decontamination system (**Figure 9.26**)

Figure 9.24 Large training facilities require flammable liquids fuel storage for the props. Natural gas may also be used.

Figure 9.25 Breathing air supplies may be provided at each prop by remote fill stations connected to a central breathing air compressor.

Figure 9.26 Decontaminate personnel and equipment following each exercise. In this instance a shower and hose are used to rinse foam off training participants.

- ***Apparatus staging, approach, and operational area*** — Parking for units not involved in the training; factors considered:
  - — Short travel routes to the designated training prop
  - — Space to park the apparatus as though it is at an emergency incident
  - — Concrete surfaces capable of supporting the weight of apparatus
- ***Communications system*** — Radio frequency dedicated for the training function or two-way communication devices with limited range; factors considered:
  - — All personnel engaged in the training exercise must have contact with one another
  - — Include the incident safety officer and the individual assigned to the fuel shutoff valve control in the communication loop **(Figure 9.27)**
- ***Water decontamination system*** — Accomplishes separation and removal of contaminates by one or more of the methods listed and returns decontaminated water to the system for use in further training exercises **(Figure 9.28)**
  - — Using oil separators
  - — Containing contaminated water in a pond that separates the oil from the water through natural processes (ponding)
  - — Adding hydrocarbon-eating bacteria into the water to destroy the oil (bacterial breakdown)
- ***Weather (wind direction and air speed) monitoring equipment*** — Installed at the highest point of the facility; allows personnel to determine the wind affect on the burning material and foam solution streams and plan attack tactics to take advantage of it
- ***Environmental controls*** — Besides the water decontamination system, other less obvious environmental controls include the following:
  - — Use of natural gas or environmentally friendly fuels
  - — Use of nonporous concrete surfaces to prevent soil contamination
  - — Protocols to determine the impact of weather (temperature extremes, wind direction, etc.) on training exercises
  - — Noise pollution controls such as mandatory hearing protection
- ***Security*** — Secured from public access; prevents vandalism and injury to the public; can use the following items:
  - — Fencing
  - — Controlled access gates, guards, and lighting
  - — Evacuation signaling system
  - — Automatic fire detection and alarm system

Figure 9.27 During all training exercises, a designated safety officer must be present and in radio communication with training instructors and fuel shutoff monitors.

Figure 9.28 To protect the environment and to reclaim water used to cool the props and extinguish the fires, some facilities have water retention and decontamination facilities to remove the fuel and foam such as this one.

Foam Use Training Practical Evolutions

Although events that occur at real fire incidents are sometimes beyond the control of the personnel on the scene, this situation should never be the case with live-fire training evolutions. Personnel who are responsible for the design, planning, and delivery of live-fire training evolutions must present a course that is realistic, yet as safe as possible for all participants and observers. Achieving the balance between realism and safety can be very difficult, but it can be accomplished with experience and guidance from the appropriate NFPA training safety standards. The following sections highlight some of the more important safety considerations that instructors must follow when conducting both structural and flammable/combustible liquids live-fire training evolutions in both interior and exterior training evolutions.

CAUTION

If there is a choice between realism and safety, safety always takes precedence.

Interior Structural Live-Fire Training Evolutions

Interior structural live-fire training evolutions are conducted to train personnel in the techniques associated with using Class A foam concentrates to extinguish fires inside structures. Requirements for conducting these exercises are contained in NFPA 1403. The standard covers fires in both specially designed burn buildings and acquired structures.

Student Preparation

Just as personnel in the field cannot be allowed to perform duties for which they are not trained, students are not allowed to participate in live-burn training evolutions until they have received some prerequisite training. NFPA 1403 requires students who plan to participate in live structural fire evolutions to have met the objectives for Fire Fighter I on the following topics in NFPA 1001, *Standard for Fire Fighter Professional Qualifications* (2002):

Figure 9.29 Before each live-fire exercise, participants must be given a briefing on the prop, the exercise, and what to expect during the training.

- Safety
- Fire behavior
- Portable extinguishers
- Personal protective equipment
- Ladders
- Fire hose, appliances, and streams
- Overhaul
- Water supply
- Ventilation
- Forcible entry

Before the start of the actual live-fire evolution, instructors should perform the following tasks:

- Brief all participants on the exercise.
- Walk participants through the building to familiarize them with the layout.
- Tell each individual what his or her role in the operation will be (**Figure 9.29**).
- Make a trial run through the process under nonfire conditions for inexperienced firefighters.

Before the start of the actual live-fire evolution, brief all participants on the exercise. Inspect personal protective equipment, including coat, trousers, boots, gloves, protective hood, helmet, respiratory protection, and personal alert safety system (PASS) device for condition and operation. Replace defective equipment before starting the evolution.

Training Center Burn Buildings

As previously mentioned, live structural fire training evolutions may be conducted in either specially designed burn buildings or acquired structures.

Figure 9.30 Permanent live-fire training structures may use solid fuels such as straw or natural gas piping to simulate a structure fire.

Training center burn buildings that are specifically designed for the purpose of simulating live-fire conditions must meet the design requirements found in NFPA 1403. NFPA 1402 also provides detailed information for the design and construction of all types of burn structures. Training center burn buildings may be gas-fired using natural gas systems or nongas-fired systems using Class A fuels (**Figure 9.30**). Flammable and combustible liquids must not be used to simulate or ignite interior structural fires. (An exception is made by NFPA for the use of combustible liquids with a flash point above 100°F [38°C] in limited quantities and in structures designed to accommodate this fuel.)

Preparation for training center burn buildings requires checking for any obvious structural damage and making certain that the following building components are in proper working condition:

- Window shutters and doors
- Standpipe and sprinkler systems
- Ventilation hatches
- Products of combustion monitors
- Evacuation alarms
- Exits (properly marked and accessible)
- Safety equipment (including fuel shutoff valves)

Acquired Structures

Fire and emergency services organizations that do not have access to burn buildings commonly use structures that are acquired for the sole purpose of burning them. These are typically old structures that landowners want removed from their properties or have been acquired by the jurisdiction for right-of-way, urban renewal, or some other use. Usually, the fire department sets and extinguishes several training fires in the structure before allowing it to become fully involved and burn to the ground (**Figure 9.31**).

Perform the following tasks before beginning a training evolution in an acquired structure:

- Ensure that all the legalities are in order. Obtain a written document confirming the ownership of the building and giving the fire department permission to burn it.
- Obtain documentation that proves there is no insurance on the building.
- Notify any local or state/provincial environmental agencies that require permits for burning as required by local laws.
- Remove and properly dispose of any hazardous materials inside or near the building including asbestos in any form.
- Remove all tanks or vessels such as heating oil tanks and water heaters from the building.

Perform the following list of preparation tasks before starting the fire:

- Make all stairs and railings sturdy and safe.
- Patch any holes in the floor to prevent trainees from falling through during vision-obscured conditions.
- Patch any holes in the walls or ceilings to prevent fire from spreading through hidden spaces.
- Remove large, unstable objects inside the building and potentially unstable chimneys outside the building.
- Disconnect all utility service to the building.
- Remove any large insect nests or hives inside the building.
- Create adequate ventilation holes in the roof. Cover these holes with flaps that can be quickly removed when necessary during burn evolutions.
- Clear all natural cover and brush away from the building.
- Cordon off the area to vehicular or pedestrian traffic.

Figure 9.31 Live-fires exercises in acquired structures are extremely dangerous and require a high level of control to prevent injuries to participants.

Pay attention to any other buildings that may become exposures to the fire building. If exposures are so close that they cannot be completely protected from damage, do not burn the acquired structure. Do not burn an acquired structure if buildings are downwind from it; these buildings may be subject to smoke damage.

Training Fuel Materials

Only use materials that burn in a predictable manner. Remove any objects found in an acquired structure that have uncertain burning characteristics. Do not use pressure-treated wood, rubber, or plastics in any amounts. The most commonly used materials are hay, straw, and wood pallets. Regardless of what materials are used, only use them in amounts that simulate the normal fuel load that would be expected in a room or building of that size. Ignite flammable materials with a propane torch or pole with an ignited rag on the end of it. Do not use flammable or combustible liquids in any amount for interior structural fire-fighting training sessions. Furthermore, do not use them as accelerants to ignite the Class A materials.

Training Fire Fuels	
Use	**Do Not Use**
Hay	Pressure-Treated Woods
Straw	Rubber
Wood Pallets	Plastic
	Class B Flammable/Combustible Liquids

Training center burn buildings may be constructed with natural gas fueled simulators. These systems have the advantages of being safer than other fuel sources, are economical to operate, have built-in safety controls, do not require the cleanup that other fuels do, and are environmentally friendly. They can be designed to simulate a variety of fire conditions including single ignition points, multiple ignition points, and flashovers. Liquefied versions of gases are not used as fuel for these types of systems.

Training Safety

A substantial portion of the NFPA 1403 standard is dedicated to tactical requirements designed to ensure that evolutions run safely. The major tactical requirements for safe training evolutions pertain to both training center burn buildings and acquired structures. These requirements are as follows:

- Establish an adequate water supply before starting the evolution. Calculate the required water supply and fire flow demands by using the requirements in NFPA 1142, *Standard on Water Supplies for Suburban and Rural Fire Fighting* (2001).
- Ensure that additional water in the amount of 50 percent of the fire flow demand required by NFPA 1142 is available to handle exposure protection and any other unexpected problems that might occur. Use separate sources for both the attack and backup hoselines. This procedure eliminates the possibility of losing water to both lines at the same time (**Figure 9.32, p. 306**).
- Ensure that all hoselines used for training evolutions are capable of flowing at least 95 gpm (360 L/min). Determine the minimum number of both attack and backup lines needed.
- Park or stage all emergency vehicles used in the evolution in a location that allows them to be quickly put into service when needed. This requirement means vehicles have the option of being put into service at the training evolution or to make emergency responses if necessary (**Figure 9.33, p. 306**).
- Ensure that ambulances or emergency medical services (EMS) responder units standing by at the training evolution have quick access to the

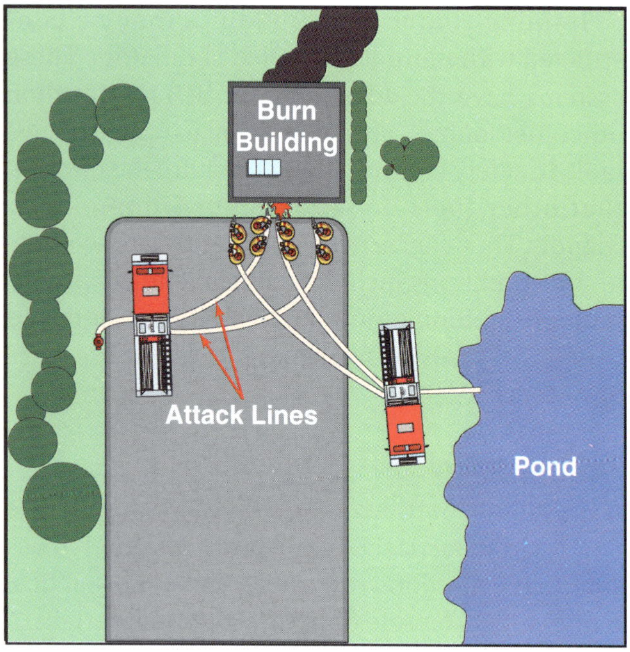

Figure 9.32 Separate water supplies must be used for attack and backup hoselines during live-fire exercises.

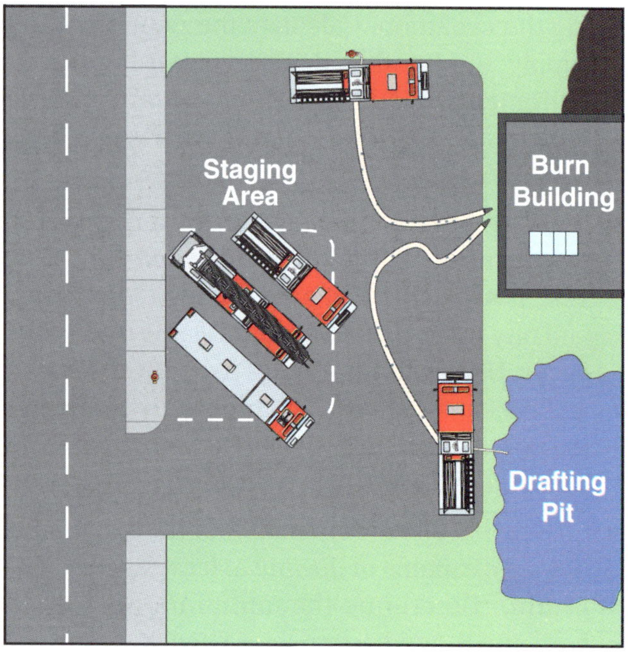

Figure 9.33 Stage vehicles at practical locations at the training facility.

fire building and easy access off the site in the event that they must transport an injured participant.

- Designate an incident scene safety officer (ISO) for the live-fire training evolution just like an actual fire incident. Do not assign the incident scene safety officer any duties other than monitoring the evolution for any unsafe acts or conditions. The ISO must have the authority to correct any safety recognized hazards immediately, which may mean stopping the evolution. Place qualified instructors throughout the interior of the burn structure to monitor conditions and student progress. They must also keep the ISO apprised of any important information. Qualifications for incident scene safety officer can be found in NFPA 1521, *Standard for Fire Department Safety Officer* (2002), and IFSTA's **Fire Department Safety Officer** manual.
- Assign one instructor to each functional crew and ensure that the student-to-instructor ratio does not exceed 5 to 1. The number of instructors needed to conduct structural fire training evolutions depends on the number of students and the complexity of the fire evolution. This student-to-instructor ratio includes backup crews.
- Designate one instructor as the ignition officer **(Figures 9.34 a and b).** The ignition officer must wear full personal protective equipment, including SCBA. Keep a charged line in place to protect this person while igniting the fire. Do not light more than one fire at any given time in acquired structures. Burn buildings that are designed for concurrent multiple fires may have fires at more than one location at the same time. In these buildings, the ignition officer ignites the fires starting at the point farthest from the exit and continues setting fires on the way to the exit.

Exterior Live-Fire Training Evolutions

Exterior live-fire training evolutions are used to train firefighters to attack the following types of fires:

- Piles or stacks of ordinary combustibles
- Vehicle fires including autos and tank trucks
- Pan or pool fires of combustible liquids **(Figure 9.35)**
- LPG fires
- Flammable or combustible liquids production or storage tanks
- Aircraft
- Maritime facilities (including vessels and cargo)

NFPA 1403 sets the requirements for conducting these exercises. In general, the requirements for conducting outside live-fire training evolutions fol-

Figure 9.34a A designated ignition officer is the only person permitted to start the training fire.

Figure 9.35 Small gas-fired burn pans are available for portable foam-extinguisher training.

low those previously discussed for live-fire training evolutions in buildings. The following sections discuss the more important aspects of NFPA 1403 and any deviations from the requirements discussed in previous sections.

Student Preparation

NFPA 1403 requires students planning to participate in outside live-fire training evolutions to have met the objectives for Fire Fighter I on the following topics in NFPA 1001:

- Safety
- Fire behavior
- Portable extinguishers
- Personal protective equipment
- Ladders

Figure 9.34b A handheld propane tank is used as the igniter.

- Fire hose, appliances, and streams
- Overhaul
- Water supply
- Ventilation
- Forcible entry

Before the start of the actual live-fire evolution, brief all participants on the exercise. Inspect personal protective equipment, including coat, trousers, boots, gloves, protective hood, helmet, respiratory protection, and PASS device, for condition and operation (**Figure 9.36, p. 308**). Replace defective equipment before starting the evolution.

Site Preparation

Some of the requirements for preparing outside live-fire training sites are different or more critical than those for preparing structural fire sites. Ground cover and topography concerns are particularly important, especially when conducting fires using combustible liquids. Clear all vegetation and other combustible materials away from the area. The surface on which the fuel is placed must be impervious to keep it from sinking into the ground and contaminating it or the groundwater. The topography must also not allow fuel or extinguishing agents to flow away from the scene in an uncontrolled manner. The topography must direct the fuel and runoff to a collection area where separating equipment separates the water from the leftover fuel.

Equip any props that have fuel supplied to them under pressure (such as those that simulate three-dimensional or pressurized liquid fuel fires) with

Figure 9.36 Full personal protective clothing must be worn during all live-fire training.

remote fuel shutoffs (**Figure 9.37**). The remote shutoff must be far enough away from the prop so that the operator is not within the safety perimeter of the exercise but is still close enough to have an unobstructed view of the entire evolution. Station a trained operator at the remote shutoff at all times during live-fire evolutions. Shut down the prop immediately if it malfunctions, the fire gets out of control, or students become endangered for any reason.

Some outside live-fire evolutions, particularly those that involve combustible liquids, generate large volumes of smoke. Notify owners of any properties that might be affected by the smoke in advance that an exercise will be conducted. Survey any streets or highways that might be affected also. If the wind is going to obscure the view of motorists, close the road or postpone the evolution until this hazard does not occur.

Figure 9.37 Besides the fuel shutoff for the individual props, a master fuel-line shutoff should also be provided.

Training Fuel Materials

Fuels used for outside Class A fire training evolutions must follow the same criteria as those previously discussed for structural fire evolutions. Flammable or combustible fuels may not be used for igniting these fires. Vehicles that are used to simulate automobile fires must be made safe before being ignited. If fuel tanks cannot be easily removed, drain the fuel from them and fill with an inert material such as water or sand; leave the fill cap off. Take the following vehicle-safety measures:

- Drain all fluid reservoirs and tanks.
- Remove all shock absorbers and drive shafts.
- Remove any other components that have gas-filled closed containers.

Use only fuels that have known burning characteristics and are as controllable as possible for Class B live-fire evolutions. The fuel must be "clean" and free of any toxic contaminants or unknown materials. Use only enough fuel to make the scenario realistic. Avoid excessively large fires unless the training simulator is specifically designed to handle such a fire. Equip fuel sources that are supplied under pressure with safety shutoff valves and controls. Use flammable, combustible, liquefied petroleum, and natural gases as fuels as long as the training simulator is designed to handle that type of fuel fire. Provide fuel separators to remove unburned fuel from the water runoff at the end of the evolution.

Training Safety

In general, all of the safety considerations listed in the structural fire section apply here as well. The only difference is in the water supply area. While separate sources are still required for both attack and backup lines, no recommended minimum amounts of water are given according to NFPA 1403. The minimum water supply and delivery for live-fire training evolutions cited in this standard are based on the criteria found in NFPA 1142. NFPA 1403 establishes the requirements for the minimum size of attack and says that all hoselines must be capable of flowing at least 95 gpm (360 L/min).

Training Foam Concentrates

Foam concentrates used in training may be either regular fire-fighting foam concentrates or training foam concentrates. The choice between the two is based on the following factors: availability, cost, and desire for realism. Many jurisdictions are prohibited from conducting foam concentrate use training exercises for one of the following two major reasons: (1) high cost of the fire-fighting foam concentrates or (2) difficulty in separating the foam concentrate from the water at the end of the evolution and the long period of time it takes for the separation to occur naturally. If necessary, foam training on the operation of foam equipment and the application of foam streams can be successfully performed without the use of any foam concentrate at all, but these evolutions cannot involve live fires. Also, some alternatives to purchasing foam concentrates are available. Fire-fighting foam concentrates provide the most realistic results, however, and should be used periodically for training. Large-scale fires or fires that require a thick foam blanket, CAFS training, and other specialist training situations should also be considered as appropriate for the use of fire-fighting foam concentrates.

Fire and emergency services organizations that cannot justify the cost of using fire-fighting foam concentrates for entry-level training should consider the purchase of training foam concentrates. Several manufacturers have foam concentrates available that are designed especially for training (**Figure 9.38**). Training concentrates are specifically formulated for hydrocarbon fuel fire training. Training concentrates generally reproduce the white color, appearance, expansion ratio, and drain time of aqueous film forming foam (AFFF) concentrate. It can be mixed with water at 3- or 6-percent proportions by any eductor or proportioning system or it can be premixed for use in a training exercise.

Figure 9.38 Most manufacturers of foam concentrate produce training foams such as the one shown in this photograph.

It has no film-forming properties or burnback resistance. Some of the other benefits of training concentrates include the following:

- Defoaming agents are not necessary to separate the foam concentrate from the water.
- Training concentrates may be used at temperatures as low as 35°F (2°C).
- Training concentrates are available in 5-gallon (19 L) pails, 55-gallon (208 L) drums, and 275-gallon (1 041 L) bulk drums. The containers are usually a different color from fire-fighting foam concentrate containers and labeled to avoid using the wrong ones. Never store training foam concentrates with fire-fighting foam concentrates.

Training Foam Characteristics

Training finished foam breaks down much faster than AFFF finished foam, which minimizes delays in training evolutions. In many cases, the training foam blanket breaks down in less than 3 minutes. This characteristic also eliminates the need to add more fuel to burn finished foam residue. The training foam concentrate has the initial knockdown characteristics of AFFF concentrate but less burnback resistance and no real vapor-sealing capabilities.

Training foam concentrates have an indefinite storage life if kept in their original shipping containers or airtight tanks that are designed for such storage. Training foam concentrates are subject to natural biodegradation when they are exposed to air. They may be stored in stainless steel, polyethylene, or fiberglass storage containers. If frozen, thawing renders them serviceable again. However, it is recommended that training foam concentrates be stored at temperatures between 35 and 120°F (2°C and 49°C).

WARNING

Never use training foam concentrates for actual fire combat. Avoid hazardous locations when training with foam concentrate surrogates. Closely supervise training evolutions involving these concentrates.

Never mix, store, or use training foam concentrates with any other type of foam concentrate. Proportioning and application equipment (including nozzles and hoses) must be flushed clean after use and before using a different type of foam concentrate. Training foam concentrates are nontoxic and biodegradable. However, prolonged exposure to the skin causes dryness; avoid contact with the eyes. Refer to the product's material safety data sheet (MSDS) for exposure remedies. Do not discharge training foam concentrates or solutions directly into waterways or storm sewers without diluting them first. When addressing health and environmental concerns, treat training foam concentrates the same way that fire-fighting foam concentrates are treated. Dispose of training foam concentrates according to the manufacturer's recommendation and the applicable hazardous waste regulations. Even the empty 5-gallon (19 L) foam concentrate pails are considered hazardous waste in some states/provinces. The most logical procedure is to retain the pails and refill them from a bulk supply of foam concentrate.

Cost-Saving Alternatives

Obviously, the high cost of fire-fighting foam concentrates does not permit its use for training on a regular basis. However, foam concentrates that have become contaminated, remained in an apparatus tank for an excessive period of time, been premixed and then not used, or been drained from an apparatus or other storage tank during maintenance can be used for training.

In place of training foam concentrate, Class A foam concentrate can be diluted and used to simulate Class B foam on simulated flammable and combustible training fires. Because the Class A foam concentrates manufactured by different companies may act differently when diluted, conduct tests to determine the best dilution to create a 3-percent foam solution that resembles Class B foam. Conduct tests at 1:1, 2:1, 4:1, and so on until the finished foam blanket has the correct appearance. Consult the MSDS for the Class A foam concentrate to determine the environmental impact that it will have on the waste water and surrounding soil.

Training Without Concentrates

As mentioned earlier, training in the operation of foam equipment and the application of foam streams can be successfully performed without the use of any foam concentrate at all; however, these evolutions must not involve live fires (**Figure 9.39**). In these situations, mark an area of a parking lot or other similar surface and designate it as the "spill" area. Add a significant amount of food coloring to water to use in place of a foam concentrate.

When the foam distribution system operates, the dyed water is proportioned into the fire stream, and the final water that discharges is tinted slightly toward the color of the "concentrate." Foam proportioning equipment is designed to proportion foam concentrate that typically is more viscous than water. Therefore, the proportioning rate for the dyed water may vary slightly from the actual setting. Usually, the manufacturer of the proportioning equipment can provide conversion factors that ensure the proportioning equipment is operating within the design limitations. Apply the final "agent" to the "spill" area with the same techniques that would be used for real finished foam.

Figure 9.39 Water fog may be used to simulate the application of foam during basic nonfire exercises such as blanketing a simulated fuel spill area.

Summary

Training is the basis for all fire and emergency services organizations. A thorough and detailed training program helps to ensure that organizations provide their services in a safe, efficient, and effective manner. When new protocols or systems are incorporated into an organization's structure, training is essential to ensure the acceptance of the new methods and equipment. Therefore, all fire and emergency services organizations that are responsible for structural fire fighting, flammable and combustible liquids spill containment, hazardous materials leak confinement, or wildland fire fighting must have training programs that include the selection and use of foam concentrates and the selection, care, and use of foam proportioning and delivery systems. Training must begin with entry-level personnel and include both advanced and specialized training as appropriate. It must be remembered that no matter how realistic the training evolutions, they will never duplicate the actual conditions that are encountered at emergency incident scenes. In all cases, safety takes precedence over realism in training. Training facilities must be designed to support foam concentrate use training programs by providing both safe and realistic training evolutions.

Appendices

Appendix A
Standards Related to Foam

A variety of standards, tests, and certifications relate to foam and come from various groups: National Fire Protection Association (NFPA), Underwriters Laboratories Inc. (UL), Underwriters' Laboratories of Canada (UL Canada), American Society for Testing and Materials (ASTM), American National Standards Institute (ANSI), Factory Mutual Research Corporation (FMRC), National Wildfire Coordinating Group (NWCG), federal government military specifications (milspecs), International Civil Aviation Organization (ICAO), and federal and state/provincial departments. Standards are basically divided into two groups: (1) those that apply to foam concentrates, equipment, and systems and (2) those that apply to foam information required of fire-fighting personnel.

Foam Concentrates, Equipment, and Systems

NFPA standards relating to fire-fighting foams serve as industry guides to selecting foam fire-fighting agents and their delivery systems. NFPA does not approve or recommend specific products for fire protection and suppression, but rather it sets guidelines for engineered systems, foam fire-fighting appliances, and mobile fire apparatus whose intended applications are primarily one of fire protection and suppression. The following standards and guides apply to foam concentrates, equipment, and systems:

- NFPA 10, *Standard for Portable Fire Extinguishers* (2002)
- NFPA 11, *Standard for Low-, Medium-, and High-Expansion Foam* (2002)
- NFPA 11A, *Standard for Medium- and High-Expansion Foam Systems* (1999)
- NFPA 16, *Standard for the Installation of Foam-Water Sprinkler and Foam-Water Spray Systems* (2002)
- NFPA 18, *Standard on Wetting Agents* (1995)
- NFPA 295, *Standard for Wildfire Control* (1998)
- NFPA 403, *Standard for Aircraft Rescue and Fire Fighting Services at Airports* (1998)
- NFPA 409, *Standard on Aircraft Hangars* (2001)
- NFPA 412, *Standard for Evaluating Aircraft Rescue and Fire Fighting Foam Equipment* (1998)
- NFPA 414, *Standard for Aircraft Rescue and Fire Fighting Vehicles* (2001)
- NFPA 1150, *Standard on Fire-Fighting Foam Chemicals for Class A Fuels in Rural, Suburban, and Vegetated Areas* (1999)
- NFPA 1145, *Guide for the Use of Class A Foams in Manual Structural Fire Fighting* (2000)
- NFPA 1405, *Guide for Land-Based Fire Fighters Who Respond to Marine Vessel Fires,* (2001)
- NFPA 1901, *Standard for Automotive Fire Apparatus* (1999)
- NFPA 1906, *Standard for Wildland Fire Apparatus* (2001)
- NFPA 1914, *Standard for Testing Fire Department Aerial Devices* (2002)
- NFPA 1925, *Standard on Marine Fire Fighting Vessels* (1998)

Professional Qualifications

Consult any of the listed standards from NFPA and NWCG for specific requirements related to foam and its use.

National Fire Protection Association

The following professional qualifications standards have requirements related to foam knowledge:

- NFPA 1001, *Standard for Fire Fighter Professional Qualifications* (2002)
- NFPA 1002, *Standard for Fire Apparatus Driver/Operator Professional Qualifications* (1998)
- NFPA 1003, *Standard for Airport Fire Fighter Professional Qualifications* (2000)

- NFPA 1021, *Standard for Fire Officer Professional Qualifications* (1997)
- NFPA 1051, *Standard for Wildland Fire Fighter Professional Qualifications* (2002)
- NFPA 1081, *Standard for Industrial fire Brigade Member Professional Qualifications* (2001)

National Wildfire Coordinating Group

The NWCG produces material relating to procedures and equipment used in fighting wildland fires. The following documents concern the use of Class A foams:

- PMS 445-1 *Introduction to Class A Foam*
- PMS 445-2 *The Properties of Foam*
- PMS 446-1 *Foam vs. Fire, Class A Foam for Wildland Fires*
- PMS 446-2 *Foam vs. Fire Primer*

Appendix B
Foam Concentrate Certification/Regulartory Organizations and Testing Methods

Foam concentrates are available that are both *rated* and *nonrated*. Rated foam concentrates have been manufactured to meet specific hazards, have been tested by a third-party testing agency, and are certified by that agency to meet the claims of the manufacturer. Nonrated foam concentrates are also available, but they have no certification or rating that they will perform as claimed by the manufacturer. A list of testing and certification organizations is provided in the Foam Certification/Regulatory Organizations section.

Rated foam concentrates meet and pass the testing criteria established by National Fire Protection Association (NFPA), Underwriters Laboratories Inc. (UL), Underwriters' Laboratories of Canada (UL Canada), American Society for Testing and Materials (ASTM), and other organizations. A sample of the test procedures is included in the Foam Concentrate/Solution Quality Assurance and Testing section. When specifying foam concentrate for purchase, fire and emergency services organizations should establish a procedure for evaluating competing products. A general guideline for this process is included in the Foam Concentrate Product Evaluation section.

In addition to the quality-control tests that foam concentrates must meet, other tests that in-service foam must meet annually are required to guarantee that the foam has not deteriorated. These requirements are established in NFPA 11, *Standard for Low-, Medium-, and High-Expansion Foam* (2002), and are provided later in the Annual Foam Concentrate Quality Assurance Testing section.

Foam Certification/Regulatory Organizations

Organizations that affect the use of foam fire-fighting concentrates and materials include Underwriters Laboratories Inc., Underwriters' Laboratories of Canada (UL Canada), American Society for Testing and Materials, American National Standards Institute (ANSI) member organizations, and Factory Mutual Research Corporation (FMRC). Although some of these organizations do not create standards (like NFPA does), all provide testing and certification criteria and third-party certification of foam concentrates and delivery systems. Brief descriptions of these organizations and agencies are in the sections that follow.

Underwriters Laboratories Inc.

Underwriters Laboratories Inc. (UL) in the United States and Underwriters' Laboratories of Canada (UL Canada) do testing and certification of foam concentrates and equipment. The tests are based on standard UL 162, *Foam Equipment and Liquid Concentrates, 7th edition* (1994). In addition, standard UL 8, *Foam Fire Extinguishers, 5th edition* (1995), provides certification for this type of equipment. Equipment and concentrates that meet these requirements are said to be *UL listed.*

These tests provide consumers with independent documentation on product-performance characteristics. These independent tests provide ongoing, nonbiased assurance that fire-fighting products perform as advertised. Information contained in the UL *Fire Protection Equipment Directory* is often used by progressive fire departments for prepurchase comparisons and can also be used to compose purchasing specifications that ensure minimum performance and suitability for a product's intended use.

American Society for Testing and Materials

The ASTM is a standards-developing organization similar to UL. Currently the only standard that the organization has that relates to foam is ASTM F1994-99, *Standard Test Method for Shipboard Fixed Foam Fire Fighting Systems*.

American National Standards Institute

Although ANSI does not write American National Standards (ANSs), it does facilitate the development of standards by qualified member organizations. ANSI-accredited developers create consensus standards that are used in the production, testing, and certification of foam concentrates and delivery systems.

Factory Mutual Research Corporation

FMRC, like UL, is a third-party testing organization. Foam concentrates and delivery systems are tested to ensure that they meet NFPA, ASTM, or ANSI-accredited standards requirements.

Government Agencies

Government agencies and departments in the United States, Canada, and the states and provinces regulate the use of fire-fighting foams. Agencies that regulate foam use are the U.S. Department of Transportation, U.S. Department of Agriculture, U.S. Coast Guard, U.S. Federal Aviation Agency, U.S. Environmental Protection Agency, Canadian Standards Association, state water resources boards, and state forestry agencies. Federal specifications are included in O-F-555C, *Foam Liquid, Fire Fighting Mechanical*, 1990. Products purchased for use by the military establishments of the governments must meet military specifications (milspecs). Class A foam is regulated by U.S. milspec MIL-F-24385F. The International Civil Aviation Organization, which is responsible for coordinating aviation safety worldwide, also regulates the use of fire-fighting foam on airport property in member nations.

Foam Concentrate/Solution Quality Assurance and Testing

NFPA requires that all foam concentrates and delivery systems be tested and certified for the specific hazard that they are intended to protect against. Testing and certification is performed by nationally listed testing organizations such as Underwriters Laboratories Inc., Underwriters' Laboratories of Canada, or Factory Mutual Research Corporation. Manufacturers submit samples of their products to the laboratories where they are tested against the requirements of the standards. Once the products have passed the tests, they are certified or approved. A certified or approved product may then be produced and sold for the purpose specified. It should be kept in mind that a product that is certified for use with Class B fires is not necessarily certified for other uses.

Foam concentrates are tested for performance before their manufacturer makes them available on the market. Random tests for quality assurance may also be performed on a regular basis during the manufacturing process. Seldom, if ever, will fire departments perform foam concentrate quality testing. NFPA 1150, *Standard on Fire-Fighting Foam Chemicals for Class A Fuels in Rural, Suburban, and Vegetated Areas* (1999) defines the testing and certification criteria for foam concentrates used for Class A fires. NFPA 1145, *Guide for the Use of Class A Foams in Manual Structural Fire Fighting* (2000) provides guidelines for the use of Class A foams in structural applications.

The foam that fire departments purchase can usually be classified into one of three general categories. Listed in order of increasing product quality, these classifications are as follows:

- Meets only the manufacturer's specifications
- Meets a third-party testing agency's standards (such as UL)
- Meets military specifications (commonly called milspec foams) or other government specifications (such as those of the General Services Administration [GSA] or Bureau of Land Management [BLM])

Concentrates that are not rated by a testing agency or do not meet milspecs may be suitable for most fire and emergency services incidents. However, they may require a longer application time or a slightly greater flow rate to achieve fire extinguishment than do higher quality foams. They will also require more frequent application to maintain a foam blanket. Rated foam concentrates provide a level of assurance that should be taken into consideration when purchasing or specifying a particular type of foam.

There are a number of tests involving foam concentrates and foam distribution systems that fire and emergency services personnel should be aware of. The sections that follow discuss these tests.

Foam Concentrate Displacement Test

This test measures the accuracy of foam proportioning equipment by checking the volume of foam concentrate that is drawn through the system while it is in operation. The foam system is operated at a predetermined flow using water as a substitute for foam concentrate. The water is drawn from a calibrated tank instead of the normal foam concentrate tank or foam concentrate pails. The volume of water drawn from the calibrated tank over a measured period of time is then correlated to the actual percentage of foam concentrate that the system would be proportioning at the test flow rate. This is the same test method that is used for testing the proportioning equipment mentioned in Chapter 3, Foam Proportioning and Delivery Equipment.

NOTE: Water has a different viscosity than foam concentrate and may be drawn into the system at a different rate. The manufacturer of the proportioning system and/or the foam concentrate normally used in the system will be able to provide a correction factor to assure that the results of testing with water are accurate.

Foam Concentrate Pump Discharge Volume Test

This procedure tests the volume of foam concentrate that is proportioned into the fire stream in some direct injection type foam proportioning systems. As with the previous test, water may be used in place of foam concentrate for testing purposes. With the foam system operating at a predetermined flow, the discharge from the foam concentrate pump is collected in a calibrated container for a specified time period. This volume can then be correlated to the actual percentage of foam concentrate that the system would be proportioning at the test flow rate.

Foam Solution Refractivity Testing

Foam solution refractivity testing tests the quality of a foam solution after it has been created by a foam proportioning system. This test method is recommended for protein- and fluoroprotein-based foam solutions. It will not be accurate for synthetic-based foams because they typically have a very low refractive index reading. The conductivity test methods described in the Foam Solution Conductivity Testing section are also used for synthetic-based foams.

The amount of foam concentrate in the solution is measured using a device called a *refractometer*. This device operates on the principle of measuring the velocity of light that travels through a medium. In this case the medium is the foam solution. The refractometer compares samples of foam solution drawn from the foam system being tested to carefully prepared base reading solutions. Deviations in the foam concentrate content of the foam solution results in different bending of the light beams through the refractometer. Because the scale on the refractometer does not reflect the actual foam concentrate proportioning percentage, the results of the tests must be plotted on a graph to be interpreted.

The first step in the testing process is to develop a base calibration curve. This calibration curve is based on the recommended proportioning rate of the foam concentrate being used (usually either 1, 3, or 6 percent). Foam concentrate and water are taken from the system being tested to make the base curve solutions. Three standard solutions are made for each concentration being tested. One solution contains the exact recommended concentration, one solution contains 0.3 percent less concentrate than recommended, and the third contains 0.3 percent more concentrate than recommended. For example, if testing 3-percent foam, one solution would have exactly 3-percent concentrate in solution, one would have 2.7 percent, and the third would have 3.3 percent.

The following procedure details the preparation of samples for testing 3-percent foam concentrate.

The numbers are adjusted if 1- or 6-percent concentrates are tested. The solutions can be mixed as follows:

Step 1: Gently add foam concentrate to each of three labeled, plastic 100 milliliter (ml) or larger graduated bottles. Place 2.7 ml of concentrate into one, 3 ml into the second, and 3.3 ml into the third. A pipette or syringe may be used.

Step 2: Fill each bottle with water to the 100 ml mark.

Step 3: Add a plastic stirring bar to each bottle, and cap them tightly.

Step 4: Shake each bottle thoroughly to mix the water and concentrate.

When testing proportioning equipment that can be operated at more than one setting (such as an in-line eductor that has 1-, 3-, and 6-percent settings), take three samples for each and prepare separate charts.

After the samples are prepared, a refractive index is taken of each sample. Place a few drops of the sample on the refractometer prism, close the cover plate, and observe the reading. It may take 10 to 20 seconds to get an accurate reading because the refractometer must adjust for temperature fluctuations. Each of these readings is then plotted on graph paper. One axis contains the refractometer reading and the other contains the proportioning percentage. A line is drawn between the three points to establish a baseline curve.

Once this procedure is complete, samples of the actual foam solution produced by the system being tested are taken. These samples are tested on the refractometer and then plotted on the graph. The results must fall within the parameters previously discussed for the various NFPA standards.

Foam Solution Conductivity Testing

Foam solution conductivity testing is used to check the quality of synthetic-based foams that are produced by foam proportioning equipment and systems. Because synthetic-based foam concentrates are very light in color, the refractivity tests listed above are not very accurate for them. Conductivity testing does not rely on the colors of the foam, but rather on their ability to conduct electricity to verify their actual composition.

Conductivity is the ability of a substance to conduct an electrical current. Water and foam concentrate both conduct electricity. When proportioning foam concentrate into water, the conductivity of the resulting foam solution is somewhere in between the figures for plain water and foam concentrate. These figures can be used to measure the amount of foam concentrate in the solution.

Three methods of performing conductivity testing on foam solutions are available: direct reading, conductivity comparison, and conductivity calibration curve. Each of these methods is detailed in the sections that follow.

Direct Reading Conductivity Testing

This method is used when a direct reading conductivity meter is available. The readout on this meter may or may not indicate the actual percentage of foam concentrate in the solution. If the meter does not have an actual percentage reading, a calibration curve will need to be developed. The procedure for developing this curve would be the same as that described for refractivity testing.

To perform this test, the meter must first be calibrated using plain water. This procedure can be done in either one of the following two ways:

- Collect a sample of the water that will be used in the test in a container and immerse the sensor head to calibrate the meter.
- Mount the sensor directly into the pump discharge line and calibrate the meter while flowing water.

Once the meter has been calibrated, a sample of foam solution from the proportioning system is obtained. As with the water, a reading may be obtained in either a container or the meter may be mounted directly to the pump discharge. The sensor is immersed in the solution, and a reading is taken from the meter. If the meter does not give readings in percentage of foam concentrate in the solution, the reading must be plotted on the calibration curve to get the final results.

Conductivity Comparison Testing

This method is used when a conductivity meter that reads in units of microSiemens per centimeter (microS/cm) is available. The procedure is fairly simple. First, take a reading from the plain water that is to be used for the test. Then take a reading from the foam solution produced by the system being tested. The percentage of foam concentrate in the solution can then be determined using the following formula:

$$\%\ of\ concentrate\ in\ solution = \frac{(conductivity\ of\ solution) - (conductivity\ of\ water)}{500}$$

The constant divisor of 500 is used only if the meter is incremented in units of microS/cm. If a different type of meter is used, the divisor has to be adjusted. The manufacturer of the meter should provide that information.

Conductivity Calibration Curve Testing

Conductivity calibration curve testing is performed by using a handheld, temperature-compensated conductivity meter. The procedure for this method is very similar to that described earlier for foam solution refractivity testing. A calibration curve is developed following the same guidelines as those for refractivity testing. The only difference is that in this case the readings will be taken using the conductivity meter. Once the calibration curve has been developed, samples of foam solution from the proportioning equipment being tested may be taken and analyzed. The readings from these tests are then plotted on the calibration curve to determine the percent of concentrate in the solution.

Foam Concentrate Product Evaluation

Purchasing foam concentrate follows the same procedures established by the authority having jurisdiction to ensure that the purchase process meets all federal, state/provincial, and local laws or ordinances. Generally, the steps are fairly simple.

First, establish the performance criteria for the foam desired. Determine if the foam is to be effective on Class A fires, Class B, polar solvents, etc., or is it to be a multipurpose concentrate. Determine the quantity required, the size of the storage units, required delivery time, technical support requirements, and training requirements in the use of the foam. Determine whether the product is rated or nonrated for a specific hazard or use.

The performance criteria are included in a Request for Proposal (RFP) that is sent by the purchasing authority to the manufacturers or their designated vendors. Part of the RFP can include a performance test that includes a simulated fire test for each of the products. Companies responding to the RFP agree to provide foam and technical support for this test although the fire and emergency organization services personnel will perform the actual test.

The evaluation test occurs with all vendors present. All fire tests should be consistent with each other. A performance evaluation sheet is completed for each product. The fire and emergency services representatives determine which product best meets the organization's needs and submits its recommendation to its purchasing department. If more than one product meets the criteria, then the criteria are submitted as part of the purchase request and all approved vendors may submit bids.

It is important that the evaluation meet the requirements of a fair and open test, which means that each product is tested to the same standards or operational criteria and that each test is conducted in front of impartial witnesses. Subjective results can cause difficulties in the purchasing process and may result in the purchase of materials that do not fully meet the need of the organization.

Annual Foam Concentrate Quality Assurance Testing

NFPA 11 recommends that all foam systems be thoroughly inspected and checked annually for proper operation. Fixed, mobile, and portable systems are inspected and maintained by manufacturer-trained and certified technicians. Even though foam concentrates are designed to have a long shelf life, certain conditions can cause them to deteriorate. Annual sampling and testing of the foam concentrate ensures that it has not deteriorated during the preceding year. In addition to detecting deterioration, annual sampling and testing also helps locate and prevent the accumulation of sediment in tanks, containers, and systems that can cause the proportioner to clog and fail.

Foam Concentrate Deterioration Causes

Deterioration can be avoided if care is taken in the storage, use, and handling of the foam concentrate. Some of the causes of foam concentrate deterioration include the following:

- *Evaporation* — Occurs when the foam tanks are left partially full, exposing the surface to air contact
- *Dilution* — Occurs when water is allowed to enter the foam tank or container
- *Contamination with inferior or incompatible concentrates* — Occurs when different types of foam are mixed together in the apparatus or delivery systems tank
- *Excessively high or low storage temperatures* — Occurs when storage containers are left in the open, in unheated structures, or in climates with temperature extremes
- *Unsuitable storage conditions* — Occurs when foam concentrate is stored in containers other than those supplied by the manufacturer, which causes a reaction with the concentrate; also when concentrate is repeatedly agitated in the apparatus tank, it mixes with air in the tank or tank opening, creating an aerated solution that can cause a congealed mass on the exposed surface

Sampling Procedures

To ensure an accurate evaluation of the stored foam concentrate, samples need to be representative of the entire concentrate quantity. Circulate the concentrate through the system and back into the storage tank, thoroughly mixing it. If it is not possible to circulate the concentrate, then take test samples from the top of the tank, the middle of the tank, and the bottom of the tank. Mix these samples together to form a single concentrate and submit it for analysis. If the construction of the tank does not permit the taking of three samples, then take a sample from the top and a sample from the bottom and mix them together. If the sample from the bottom of the tank is obtained from a drain-off point or valve, allow the concentrate to flow long enough to ensure that accumulated sludge, residue, or other impurities are removed and pure foam concentrate is obtained for the analysis. Although this method provides a test sample for analysis, it does not provide an accurate representation of the entire contents of the tank.

Testing laboratories will specify the quantity of concentrate required for the test and the manner for labeling and shipping. Generally, a minimum of 1 pint (500 ml) is required for testing. Samples are placed in clean, tightly sealed containers make of polyethylene plastic. The label indicates the source of the concentrate, the name of the organization, and the type of concentrate.

Laboratory Analyses

Most laboratories that test for foam concentrate condition perform the following tests:

- Specific gravity
- pH value
- Sediment ratio
- Expansion ratio
- Quarter drainage time

Some testing laboratories have the ability to test foam concentrate based on the UL 162 standard fire test. This test provides information on how well the concentrate will function under fire conditions.

The laboratory provides a complete written evaluation of the quality of the stored concentrate. Foam concentrate that does not meet the test criteria should be removed from service and only used for training purposes.

Appendix C
Foam Properties

Table C.1
Foam Properties

Type	Characteristics	Storage Range	Application Rate	Application Techniques	Primary Uses
Protein Foam (3% and 6%)	• Protein based • Low expansion • Good reignition (burnback) resistance • Excellent water retention • High heat resistance and stability • Performance can be affected by freezing and thawing • Can freeze protect with antifreeze • Not as mobile or fluid on fuel surface as other low-expansion foams	35–120°F (2°C to 49°C)	0.16 gpm/ft^2 (6.5 L/min/m^2)	• Indirect foam stream; do not mix fuel with foam • Avoid agitating fuel during application; static spark ignition of volatile hydrocarbons can result from plunging and turbulence • Use alcohol-resistant type within seconds of proportioning • Not compatible with dry chemical extinguishing agents	• Class B fires involving hydrocarbons • Protecting flammable and combustible liquids where they are stored, transported, and processed
Fluoroprotein Foam (3% and 6%)	• Protein and synthetic based; derived from protein foam • Fuel shedding • Long-term vapor suppression • Good water retention • Excellent, long-lasting heat resistance • Performance not affected by freezing and thawing • Maintains low viscosity at low temperatures • Can freeze protect with antifreeze • Use either freshwater or saltwater • Nontoxic and biodegradable after dilution • Good mobility and fluidity on fuel surface • Premixable for short periods of time	35–120°F (2°C to 49°C)	0.16 gpm/ft^2 (6.5 L/min/m^2)	• Direct plunge technique • Subsurface injection • Compatible with simultaneous application of dry chemical extinguishing agents • Deliver through air-aspirating equipment	• Hydrocarbon vapor suppression • Subsurface application to hydrocarbon fuel storage tanks • Extinguishing in-depth crude petroleum or other hydrocarbon fuel fires

Continued on next page.

Table C.1 (continued)

Type	Characteristics	Storage Range	Application Rate	Application Techniques	Primary Uses
Film Forming Fluoroprotein Foam (FFFP) (3% and 6%)	• Protein based; fortified with additional surfactants that reduce the burnback characteristics of other protein-based foams • Fuel shedding • Develops a fast-healing, continuous-floating film on hydrocarbon fuel surfaces • Excellent, long-lasting heat resistance • Good low-temperature viscosity • Fast fire knockdown • Affected by freezing and thawing • Use either freshwater or saltwater • Can store premixed • Can freeze protect with antifreeze • Use alcohol-resistant type on polar solvents at 6% solution and on hydrocarbon fuels at 3% solution • Nontoxic and biodegradable after dilution	35–120°F (2°C to 49°C)	***Ignited Hydrocarbon Fuel:*** 0.10 gpm/ft^2 (4.1 L/min/m^2) ***Polar Solvent Fuel:*** 0.24 gpm/ft^2 (9.8 L/min/m^2)	• Cover entire fuel surface • May apply with dry chemical agents • May apply with spray nozzles • Subsurface injection • Can plunge into fuel during application	• Suppressing vapors in unignited spills of hazardous liquids • Extinguishing fires in hydrocarbon fuels
Aqueous Film Forming Foam (AFFF) (1%, 3%, and 6%)	• Synthetic based • Good penetrating capabilities • Spreads vapor-sealing film over and floats on hydrocarbon fuels • Can use nonaerating nozzles • Performance may be adversely affected by freezing and storing • Has good low-temperature viscosity • Can freeze protect with antifreeze • Use either freshwater or saltwater • Can premix	25–120°F (-4°F to 49°C)	0.10 gpm/ft^2 (4.1 L/min/m^2)	• May apply directly onto fuel surface • May apply indirectly by bouncing it off a wall and allowing it to float onto fuel surface • Subsurface injection • May apply with dry chemical agents	• Controlling and extinguishing Class B fires • Handling land or sea crash rescues involving spills • Extinguishing most transportation-related fires • Wetting and penetrating Class A fuels • Securing unignited hydrocarbon spills

Continued on next page.

Table C.1 (continued)

Type	Characteristics	Storage Range	Application Rate	Application Techniques	Primary Uses
Alcohol-Resistant AFFF (3% and 6%)	• Polymer has been added to AFFF concentrate • Multipurpose: Use on both polar solvents and hydrocarbon fuels (use on polar solvents at 6% solution and on hydrocarbon fuels at 3% solution) • Forms a membrane on polar solvent fuels that prevents destruction of the foam blanket • Forms same aqueous film on hydrocarbon fuels as AFFF • Fast flame knockdown • Good burnback resistance on both fuels • Not easily premixed	25–120°F (-4°C to 49°C) (May become viscous at temperatures under 50°F [10°C])	***Ignited Hydrocarbon Fuel:*** 0.10 gpm/ft^2 (4.1 L/min/m^2) ***Polar Solvent Fuel:*** 0.24 gpm/ft^2 (9.8 L/min/m^2)	• Apply directly but gently onto fuel surface • May apply indirectly by bouncing it off a wall and allowing it to float onto fuel surface • Subsurface injection	Fires or spills of both hydrocarbon and polar solvent fuels
High-Expansion Foam	• Synthetic detergent based • Special-purpose, low water content • High air-to-solution ratios: 200:1 to 1,000:1 • Performance not affected by freezing and thawing • Poor heat resistance • Prolonged contact with galvanized or raw steel may attack these surfaces	27–110°F (-3°C to 43°C)	Sufficient to quickly cover the fuel or fill the space	• Gentle application; do not mix foam with fuel • Cover entire fuel surface • Usually fills entire space in confined space incidents	• Extinguishing Class A and some Class B fires • Flooding confined spaces • Volumetrically displacing vapor, heat, and smoke • Reducing vaporization from liquefied natural gas spills • Extinguishing pesticide fires • Suppressing fuming acid vapors • Suppressing vapors in coal mines and other subterranean spaces and concealed spaces in basements • Extinguishing agent in fixed extinguishing systems • Not recommended for outdoor use

Continued on next page.

Table C.1 (continued)

Type	Characteristics	Storage Range	Application Rate	Application Techniques	Primary Uses
Class A Foam	• Synthetic • Wetting agent that reduces surface tension of water and allows it to soak into combustible materials • Rapid extinguishment with less water use than other foams • Use regular water stream equipment • Can premix with water • Mildly corrosive • Requires lower percent-age of concentration (0.2 to 1.0) than other foams • Outstanding insulating qualities • Good penetrating capabilities	25–120°F (-4°C to 49°C) (Concentrate is subject to freezing but can be thawed and used if freezing occurs)	Same as the minimum critical flow rate for plain water on similar Class A Fuels; flow rates are not reduced when using Class A foam	• Can propel with compressed-air systems • Can apply with conventional nozzles	Extinguishing Class A combustibles only

Appendix D
Hydraulic Calculations Chart

Foam Eductor Performance Calculations (English Units)

Foam Eductor (gpm/psi)	Spill Fire Type and Control Size (ft²)	Concentrate Flow (gpm)	Nozzle or Hose (psi)	Distance 1.5-inch Hose ID (ft)	Distance 1.75-inch Hose ID (ft)	Distance 2.0-inch Hose ID (ft)	Distance 2.5-inch Hose ID (ft)	Distance 3.0-inch Hose ID (ft)	Distance 4.0-inch Hose ID (ft)	Distance 5.0-inch Hose ID (ft)
60 gpm	Hydrocarbon: 600	1% = 0.6	100	330	705	857				
200 psi	Polar Solvent: 300	3% = 1.8	75	611	1,294	1,571				
		6% = 3.6	50	888	1,882	2,285				
			10	1,330	2,823	3,428				
95 gpm	Hydrocarbon: 950	1% = 1.0	100	136	200	375	1,200			
200 psi	Polar Solvent: 425	3% = 3.0	75	250	350	687	2,200			
		6% = 6.0	50	363	500	1,000	3,200			
			10	545	800	1,500	4,800			
125 gpm	Hydrocarbon: 950	1% = 1.0	100	83	176	214	750			
200 psi	Polar Solvent: 425	3% = 3.0	75	152	323	392	1,375			
		6% = 6.0	50	220	407	572	2,000			
			10	330	705	857	3,000			
250 gpm	Hydrocarbon:	1% = 2.5	100				200	480	3,000	
200 psi	2,500	3% = 7.5	75				360	880	5,500	
	Polar Solvent:	6% = 15.0	50				530	1,280	8,000	
	1,250		10				800	1,920	12,000	
300 gpm	Hydrocarbon:	1% = 3.0	100				150	330	1,666	
200 psi	3,000	3% = 9.0	75				275	611	3,055	
	Polar Solvent:	6% = 18.0	50				400	888	4,444	
	1,500		10				600	1,333	6,666	
500 gpm	Hydrocarbon:	1% = 5.0	100				50	120	600	1,200
200 psi	5,000	3% = 15.0	75				91	220	1,100	2,200
	Polar Solvent:	6% = 30.0	50				133	320	1,600	3,200
	2,500		10				200	480	2,400	4,800

ID = Internal Diameter

Modified from *Foam Eductor Performance* by Jim Cottrell, Cottrell Associates, Inc. © 2001, Combat Support Services.

Calculations based on 35% pressure drop across the eductor and common fire hose using the Hazen-Williams calculation for friction loss.

Foam Eductor Performance Calculations (Metric Units)

Foam Eductor (L/min and kPa)	Spill Fire Type and Control Size (m^2)	Concentrate Flow (L/min)	Nozzle or Hose (kPa)	Distance 38-mm Hose ID (m)	Distance 45-mm Hose ID (m)	Distance 50-mm Hose ID (m)	Distance 65-mm Hose ID (m)	Distance 77-mm Hose ID (m)	Distance 100-mm Hose ID (m)	Distance 125-mm Hose ID (m)
227.12 L/min 1 378.95 kPa	Hydrocarbon: 55.74 Polar Solvent: 27.87	1% = 2.27 3% = 6.81 6% = 13.62	689.47 517.10 344.73 68.94	100.58 186.23 270.66 405.38	214.88 394.41 578.63 860.45	261.21 478.84 696.46 1 044.85				
359.61 L/min 1 378.95 kPa	Hydrocarbon: 88.25 Polar Solvent: 39.48	1% = 3.78 3% = 11.35 6% = 22.71	689.47 517.10 344.73 68.94	41.45 76.20 110.64 166.11	60.96 106.68 152.40 243.84	114.30 209.39 304.80 457.20	365.76 670.56 975.36 1 463.04			
473.17 L/min 1 378.95 kPa	Hydrocarbon: 88.25 Polar Solvent: 39.48	1% = 3.78 3% = 11.35 6% = 22.71	689.47 517.10 344.73 68.94	25.29 46.32 67.05 100.58	53.64 98.45 124.05 214.88	65.22 119.48 174.34 261.21	228.60 419.10 609.60 914.40			
946.35 L/min 1 378.95 kPa	Hydrocarbon: 232.25 Polar Solvent: 116.12	1% = 9.46 3% = 28.39 6% = 56.78	689.47 517.10 344.73 68.94				60.96 109.72 161.54 243.84	146.30 268.22 390.14 585.21	914.40 1 676.40 2 438.40 3 657.60	
1 135.62 L/min 1 378.95 kPa	Hydrocarbon: 278.704 Polar Solvent: 139.35	1% = 11.35 3% = 34.06 6% = 68.13	689.47 517.10 344.73 68.94				45.72 83.82 121.92 182.88	100.58 186.23 270.66 406.29	507.79 931.16 1 354.53 2 031.79	
1 892.70 L/min 1 378.95 kPa	Hydrocarbon: 464.50 Polar Solvent: 232.25	1% = 18.92 3% = 56.78 6% = 113.56	689.47 517.10 344.73 68.94				15.24 27.73 40.53 60.96	36.57 67.05 97.53 146.30	182.88 335.28 487.68 731.52	365.76 670.56 975.36 1 463.04

ID = Internal Diameter

Modified from *Foam Eductor Performance* by Jim Cottrell, Cottrell Associates, Inc. © 2001, Combat Support Services.

Calculations based on 35% pressure drop across the eductor and common fire hose using the Hazen-Williams calculation for friction loss.

Appendix E

Sample Class A Foam Concentrate Material Safety Data Sheet

National Foam

MATERIAL SAFETY DATA SHEET #NMS700

KNOCKDOWN

CLASS A FIRE FIGHTING FOAM Concentrate

Section 1. CHEMICAL PRODUCT/COMPANY IDENTIFICATION

Material Identification

Product: KnockDown Class A Foam
Synonyms: Synthetic Detergent, Wetting Agent
CAS No: Mixture - No single CAS # applicable

Company Identification
Manufacturer:
National Foam, Inc.
150 Gordon Drive
P.O. Box 695
Exton, PA 19341-0695
Emergency Phone Number (Red Alert): (610) 363-1400 (U.S.A.)
Fax (610) 524-9073
www.nationalfoam.com
www.Kidde-Fire.com

Section 2. COMPOSITION / INFORMATION ON INGREDIENTS

Components	CAS Number
Water	7732-18-5
Proprietary mixture of synthetic detergents	No single CAS # applicable
1, 2 Propanediol	57-55-6
(2-Methoxymethylethoxy) Propanol	34590-94-8
Proprietary mixture of corrosion inhibitors	No single CAS # applicable

NMS# 700 Page 1 of 8 08/10/00

Section 3. HAZARDS IDENTIFICATION

Potential Health Effects

Inhalation
Vapors are minimal at room temperature. If product is heated or sprayed as an aerosol, airborne material may cause respiratory irritation.

Skin Contact
Contact with liquid may cause moderate irritation or dermatitis due to removal of oils from the skin.

Eye Contact
Product is an eye irritant.

Ingestion
Not a hazard in normal industrial use. Small amounts swallowed during normal handling operations are not likely to cause injury; swallowing large amounts may cause injury or irritation.

Additional Health Effects
Existing eye or skin sensitivity may be aggravated by exposure.

Carcinogenicity Information
No data available.

Section 4. FIRST AID MEASURES

Inhalation
No specific treatment is necessary since this material is not likely to be hazardous by inhalation. If exposed to excessive levels of airborne aerosol mists, remove to fresh air. Seek medical attention if effects occur.

Skin Contact
In case of skin contact, wash off in flowing water or shower. Launder clothing before reuse.

Eye Contact
In case of eye contact, flush eyes promptly with water for 15 minutes. Retract eyelids often to ensure thorough rinsing. Consult a physician if irritation persists.

Ingestion
Swallowing less than an ounce is not expected to cause significant harm. For larger amounts, do not induce vomiting. Give milk or water. Never give anything by mouth to an unconscious person. Seek medical attention.

Section 5. FIRE FIGHTING MEASURES

Flammable Properties
Flash Point – Not applicable

Fire and Explosion Hazards
Avoid contact with water reactive materials, burning metals and electrically energized equipment.

Extinguishing Media
Product is an extinguishing media. Use media appropriate for surrounding materials.

Special Fire Fighting Instructions
This product will produce foam when mixed with water.

Section 6. ACCIDENTAL RELEASE MEASURES

Safeguards (Personnel)
NOTE: Review FIRE FIGHTING MEASURES and HANDLING (Personnel) sections before proceeding with clean-up. Use appropriate Personal Protective Equipment during clean-up.

Accidental Release Measures

Concentrate
Stop flow if possible. Use appropriate protective equipment during clean up. For small volume releases, collect spilled concentrate with absorbent material; place in approved container. For large volume releases, contain and collect for use where possible. Flush area with water until it no longer foams. Exercise caution, surfaces may be slippery. Prevent discharge of concentrate to waterways. Disposal should be made in accordance with federal, state and local regulations.

Foam/Foam Solution
See above. Flush with water. Prevent discharge of foam/foam solution to waterways. Do not discharge into biological sewer treatment systems without prior approval. Disposal should be made in accordance with federal, state and local regulations.

Section 7. HANDLING AND STORAGE

Handling (Personnel)
Avoid contact with eyes, skin or clothing. Avoid ingestion or inhalation. Rinse skin and eyes thoroughly in case of contact. Review HAZARDS and FIRST AID sections.

Storage
Recommended storage environment is between 20°F (-7°C) and 120°F (49°C). Store product in original shipping container or tanks designed for product storage.

Section 8. EXPOSURE CONTROLS/PERSONAL PROTECTION

Engineering Controls
Special ventilation is not required.

Personal Protective Equipment

Respiratory
Recommended exposure limits (OSHA-PEL and ACGIH-TLV) have not been determined for this material. The need for respiratory protection should be evaluated by a qualified health specialist.

Protective Clothing
Rubber or PVC gloves recommended.

Eye Protection
Safety glasses, face shield or chemical splash goggles must be worn when possibility exists for eye contact. Contact lenses should not be worn. Eye wash facilities are recommended.

Other Hygienic Practices
Use good personal hygiene practices. Wash hands before eating, drinking, smoking, or using toilet facilities. Promptly remove soiled clothing and wash thoroughly before re-use.

Exposure Guidelines

Exposure Limits
(2-Methoxymethylethoxy) Propanol (34590-94-8)

PEL(OSHA)
100 ppm, 8 hr. TWA Skin
150 ppm, 15 min. STEL Skin

TLV (ACGIH)
100 ppm, 8 hr. TWA Skin
150 ppm, 15 min. STEL Skin

Section 9. PHYSICAL AND CHEMICAL PROPERTIES

Physical Data

Boiling Point:	Not applicable
Vapor Pressure:	Not applicable
Vapor Density:	Not applicable
Melting Point:	Not applicable
Evaporation Rate:	<1 (Butyl Acetate = 1.0)

Solubility in Water:	100%
pH:	9.0
Specific Gravity:	1.05 @ 25°C
Odor:	Bland
Form:	Liquid
Color:	Light Green

Section 10. STABILITY AND REACTIVITY

Chemical Stability
Stable.

Incompatibility, Materials to Avoid
Avoid use of product on burning metals, electrically-energized equipment and contact with water reactive materials.

Polymerization
Will not occur.

Section 11. TOXICOLOGICAL INFORMATION

Mammalian Toxicity

	Concentrate	Mixed Fire Chemical
Acute Oral Toxicity – Sprague-Dawley Rats	LD_{50}>5000mg/kg	LD_{50}>5000mg/kg
Acute Dermal Toxicity – New Zealand White Rabbits	LD_{50}>2000mg/kg	LD_{50}>2000mg/kg
Primary Dermal Irritation – New Zealand White Rabbits	Slightly Irritating (Toxicity Category IV)	Non-Irritating (Toxicity Category IV)
Primary Eye Irritation – Unwashed Eyes New Zealand White Rabbits	Moderately Irritating (Toxicity Category I)	Minimally Irritating (Toxicity Category IV)
Primary Eye Irritation – Washed Eyes New Zealand White Rabbits	Mildly Irritating (Toxicity Category III)	Practically Non-Irritating (Toxicity Category IV)

Section 12. ECOLOGICAL INFORMATION

Ecotoxicological Information Aquatic Toxicity
No data available at this time.

Environmental Fate

	Concentrate	0.5% Solution	1% Solution
BOD_5	389,000 mg/kg	2,140 mg/kg	4,220 mg/kg
COD	782,000 mg/kg	3,900 mg/kg	7,960 mg/kg

Section 13. DISPOSAL CONSIDERATIONS

"KnockDown, as sold, is not a RCRA-listed waste or hazardous waste as characterized by 40 CFR 261. However, State and local requirements for waste disposal may be more restrictive or otherwise different from Federal regulations. Therefore, applicable local and state regulatory agencies should be contacted regarding disposal of waste foam concentrate or foam/foam solution.

Concentrate
Do not discharge into biological sewer treatment systems without prior approval. Specific concerns are high BOD load and foaming tendency. Low dosage flow rate or antifoaming agents acceptable to the treatment plant may be helpful. Do not flush to waterways. Disposal should be made in accordance with federal, state and local regulations.

Foam/Foam Solution
KnockDown Class A Foam solution can be treated by waste water treatment facilities. Discharge into biological sewer treatment facilities may be done with prior approval. Specific concerns are high BOD load. Dilution will reduce BOD and COD factors proportionately. Low dosage flow rate or antifoaming agents acceptable to the treatment plant may be helpful. Do not flush to waterways. Disposal should be made in accordance with federal, state and local regulations.

NOTE: As a service to our customers, National Foam has approvals in place with disposal facilities throughout the U.S. for waste water treatment and solidification and landfill of our foam liquid concentrates and foam solutions. If required, National Foam, Inc. can also provide information on the disposal of drums used for shipping our concentrates. Please contact National Foam's Risk Management Administrator at (610) 363-1400 for additional information.

Section 14. TRANSPORTATION INFORMATION

Shipping Information
Proper Shipping Name: Fire Extinguisher Charges or Compounds N.O.I., Class 60
National Motor Freight Code: 69160 Sub 0
Hazard Class: None
UN Number: None

Section 15. REGULATORY INFORMATION

U.S. Federal Regulations

Toxic Substances Control Act (TSCA)
All components of this product are listed in the TSCA inventory.

Superfund Amendments and Reauthorization Act of 1986 (SARA), Title III

Section 302/304
There are no components of this material with known CAS numbers which are on the Extremely Hazardous Substances (EHS) list.

Section 311 & 312
Based on available information, this material contains the following components which are classified as the following health and/or physical hazards according to Section 311 & 312:
(2-Methoxymethylethoxy) Propanol 34590-94-8 (Flammability)

Section 313
This material does not contain any chemical components subject to Section 313 reporting requirements.

COMPREHENSIVE ENVIRONMENTAL RESPONSE, COMPENSATION, AND LIABILITY ACT (CERCLA)
This material does not contain any components subject to the reporting requirements of CERCLA.

OTHER REGULATORY INFORMATION
None.

STATE REGULATIONS

PENNSYLVANIA RIGHT-TO-KNOW HAZARDOUS SUBSTANCES LIST

PA Hazardous Substances present at levels greater than 1%:
1, 2 Propanediol 57-55-6
(2-Methoxymethylethoxy) Propanol 34590-94-8

Section 16. OTHER INFORMATION

NFPA Rating

Health 0
Flammability 0
Reactivity 0

ADDITIONAL INFORMATION

Revision Summary

1/20/00	Added Mammalian Toxicity data to Section 11
2/5/99	Added MSDS number, updated Sections 1,3,4,6,7,9,11,12,13,15
9/18/97	Temporary issue
08/10/00	Revised Section 1

For further information, see National Foam Product Data Sheet for KnockDown Class A Foam.

The information contained herein is furnished without warranty either expressed or implied. This data sheet is not a part of any contract of sale. The information contained herein is believed to be correct or is obtained from sources believed to be generally reliable. However, it is the responsibility of the user of these materials to investigate, understand and comply with federal, state and local guidelines and procedures for safe handling and use of these materials. National Foam, Inc. shall not be liable for any loss or damage arising directly or indirectly from the use of this product and National Foam, Inc. assumes no obligation or liabilities for reliance on the information contained herein or omissions herefrom.

August 10, 2000

Glossary and Index

Glossary

A

Acid — Corrosive chemical that reacts with water to produce hydrogen ions. *Also see* Base.

Aerial Attack (Wildland) — Use of aircraft to apply extinguishing agents to wildland fires. *Also see* Attack Methods (2).

AFFF — Abbreviation for aqueous film forming foam.

Aircraft Attitude — Angle of the aircraft front to rear while in flight. *Also see* Aircraft Velocity.

Aircraft Hangar, Group I — Classification of aircraft hangar that has a single fire area in excess of 40,000 feet (3 716 m), has an access door height in excess of 28 feet (8.5 m), houses aircraft with a tail height in excess of 28 feet (8.5 m), and/or houses strategically important military aircraft.

Aircraft Hangar, Group II — Classification of aircraft hangar that has a single fire area that is less than 40,000 feet (3 716 m) and an access door height that is less than 28 feet (8.5 m). Construction type and fixed fire suppression systems also are used to determine Group II qualifications.

Aircraft Hangar, Group III — Classification of aircraft hangar that has aircraft access doors that are less than 28 feet (8.5 m) in height and a single fire area that is less than those given for the various types of building construction found in NFPA 409, *Standard on Aircraft Hangars* (2001).

Aircraft Rescue and Fire Fighting (ARFF) — Operations that involve aircraft incidents.

Aircraft Rescue and Fire Fighting (ARFF) Apparatus — Fire-fighting vehicles specifically designed for incidents involving aircraft.

Aircraft Velocity — Speed of an aircraft relative to its surrounding air mass; also called *airspeed. Also see* Aircraft Attitude.

Alcohol-Resistant AFFF Concentrate (AR-AFFF) — Aqueous film forming foam that is designed for use with polar solvent fuels. *Also see* Aqueous Film Forming Foam and Foam Concentrate

Application Rate — Minimum amount of foam solution that must be applied to an unignited spill, spill fire, or fire to control vapor emission or extinguish the fire.

Aqueous Film Forming Foam (AFFF) — Synthetic foam concentrate that can be used to form a complete vapor barrier over fuel spills; also called *Light Water. Also see* Alcohol-Resistant AFFF Concentrate and Foam Concentrate.

AR-AFFF — Abbreviation for Alcohol-Resistant Aqueous Film Forming Foam.

ARFF — Acronym for Aircraft Rescue and Fire Fighting.

Around-the-Pump Proportioner — Apparatus-mounted foam proportioner; small quantity of water is diverted from the apparatus pump through an inline proportioner where it picks up the foam concentrate and carries it to the intake side of the pump; most common apparatus-mounted foam proportioner in service. *Also see* Foam Proportioner and Proportioning.

Aspect — Compass direction that a slope faces.

Aspirating Foam Nozzle — Nozzle designed for the application of all types of foam concentrates; may be air-aspirating (air is introduced into the foam stream at the nozzle and draws in foam solution by the Venturi process) or water-aspirating (draws air into the back of the nozzle and forces the foam solution through a screen to create finished foam). *Also see* Nonaspirating Foam Nozzle and Venturi Principle.

Aspiration — Adding air to a foam solution as the solution is discharged from a nozzle; also called *aeration.*

Atmospheric Temperature — Measure of the warmth or coldness of the air.

Attack Methods — (1) Tactics for interior fire-suppression operations; includes direct, indirect, and combination attacks. (2) Tactics for wildland fire-suppression operations; includes aerial, direct, flank, frontal, indirect, mobile, pincer, and tandem attacks.

Automatic Oscillating Foam Monitor — Large-capacity foam system that is designed to operate automatically when a fire-detection system activates; may be found in aircraft hangars, tank farms, and loading racks. *Also see* Foam Monitor, Manual Foam Monitor, and Remote-Controlled Foam Monitor.

B

Backdraft — Instantaneous explosion or rapid burning of superheated gases that occur when oxygen is introduced into an oxygen-depleted confined space. It may occur because of inadequate or improper ventilation procedures. *Also see* Flashover, Rollover, and Explosion.

Backpressure — Pressure loss or gain created by changes in elevation between the nozzle and pump.

Bank-Down Application Method (Deflect) — Method of foam application that may be employed on an unignited or ignited Class B fuel spill. The foam stream is directed at a vertical surface or object that is next to or within the spill area. The foam deflects off the surface or object and flows down onto the surface of the spill to form a foam blanket. *Also see* Roll-On Application Method and Rain-Down Application Method.

Base — Corrosive water-soluble compound that reacts with an acid to form a salt. *Also see* Acid.

Batch Mixing — Production of foam solution by adding an appropriate amount of foam concentrate to a water tank before application. The resulting solution must be used or discarded following the incident. *Also see* Premixing.

Bladder-Tank Balanced-Pressure Proportioner — Type of mobile or fixed foam system that uses water to displace foam concentrate in a storage tank and force it into a proportioning system. *Also see* Foam Proportioner and Proportioning.

Blended Gasoline — Gasoline that has oxygen added to it to increase the efficiency of the combustion of the fuel; also known as *reformulated gasoline.*

BLEVE — Acronym for Boiling Liquid Expanding Vapor Explosion.

Boiling Liquid Expanding Vapor Explosion (BLEVE) — Superheating of a flammable liquid within a container that generates sufficient vapor pressure to rupture the container. The resulting explosion and fire can have a devastating effect on the immediate area. *Also see* Explosion.

British Thermal Unit (Btu) — Amount of heat needed to raise the temperature of 1 pound of water 1°F.

Btu — Abbreviation for British Thermal Unit.

Bypass-Type Balanced-Pressure Proportioner — Foam proportioning system that discharges foam through a pump separate from the water supply pump. It is one of the most accurate types of foam proportioning systems in use; most commonly found in airport crash vehicles and in fixed-site facilities. *Also see* Foam Proportioner and Proportioning.

C

CAFS — Abbreviation for Compressed-Air Foam System.

Calorie — Amount of heat required to raise the temperature of 1 gram of water 1°C. *Also see* Joule.

Canadian Transportation Emergency Centre (CANUTEC) — Centre that provides fire and emergency responders with 24-hour information for incidents involving hazardous materials; similar to the U.S. CHEMTREC® organization except that it is a part of Transport Canada, a department of the Canadian government.

CANUTEC — Acronym for the Canadian Transportation Emergency Centre.

Check Valve — Inline valve that prevents water from flowing into a foam concentrate container when the nozzle is turned off or there is a kink in the hoseline.

Chemical Foam — Foam produced as a result of a reaction between two chemicals. *Also see* Mechanical Foam.

Chemical Heat Energy — Creation of heat through a chemical reaction; includes heat of combustion, spontaneous heating, heat of decomposition, and heat of solution. *Also see* Electrical Heat Energy, Mechanical Heat Energy, Nuclear Heat Energy, and Solar Heat Energy.

Chemical Reaction — Combining of compounds that may result in the creation of a water-reactive material. *Also see* Reactivity.

Chemical Transportation Emergency Center (CHEMTREC®) — U.S. Chemical industry resource for fire and emergency services that supplies 24-hour information for incidents involving hazardous materials. *Also see* Canadian Transportation Emergency Centre.

CHEMTREC® — Acronym for the Chemical Transportation Emergency Center.

Class A Foam Concentrate — Foam fire-suppression agent that is designed for use on Class A combustible fires; essentially a wetting agent that reduces the surface tension of water and allows it to soak into combustible materials more readily than plain water. *Also see* Foam Concentrate.

Class B Foam Concentrate — Foam fire-suppression agent that is designed for use on unignited or ignited Class B flammable or combustible liquids. *Also see* Foam Concentrate.

Closed Sprinkler Head — Sprinkler head that is equipped with a heat-sensitive element such as a fusible link or frangible bulb rated at a fixed temperature that causes the head to open when the link melts or the bulb bursts due to heat rise past the preset temperature; may be used on wet-pipe, dry-pipe, or preaction sprinkler systems. *Also see* Open Sprinkler Head and Foam-Water Sprinkler Head.

Combination Attack (Structural) — Attack method that involves moving the water or foam stream around a compartment in an *O, T,* or *Z* pattern. This movement allows the extinguishing agent to be applied to the fire and to the surrounding uninvolved fuel. *Also see* Attack Methods (1).

Combined-Agent Vehicle — Type of rapid intervention vehicle that is designed to apply multiple types of fire-extinguishing agents on aircraft crash incidents.

Combustible Liquid — Liquid having a flash point at or above 100°F (38°C) and below 200°F (93°C). *Also see* Flammable Liquid.

Combustion — Self-sustaining process of rapid oxidation (chemical reaction) of a fuel that produces heat and light.

Compressed-Air Foam System (CAFS) — High-energy foam generating system that depends on an air compressor at the pump to inject air into the foam solution before it enters a hoseline.

Conduction — Transfer of heat from one body to another by direct contact of the two bodies or an intervening heat-conducting medium. *Also see* Convection, Radiation, and Law of Heat Flow.

Cone Roof Tank — Fixed-site vertical storage tank used to store flammable, combustible, and corrosive liquids; also called *dome roof tank. Also see* External Floating Roof Tank and Internal Floating Roof Tank.

Confinement — (1) Process of controlling the flow of a spill and capturing it at some specified location. (2) Fire-fighting operation to prevent fire from extending to uninvolved areas or structures.

Consistency — Quality of finished foam that has small bubbles of equal size — an important quality for all types of foam. *Also see* Finished Foam.

Containment — Act of stopping the further release of a material from its container.

Convection — Transfer of heat by the movement of heated fluids or gases, usually in an upward direction. *Also see* Conduction, Radiation, and Law of Heat Flow.

Crosswind — Wind that is blowing in a direction from the side of an aircraft or foam stream; can affect a foam distribution pattern. *Also see* Wind, Headwind, and Downwind.

D

Dead-Man Valve — Spring-loaded valve that controls the flow of fuel from a loading rack into a tank vehicle or rail tank car; designed to immediately turn off the flow when the operator releases the handle.

Decomposition — Result of oxygen acting on a material that results in a change in the material's composition. Oxidation occurs slowly, sometimes resulting in the rusting of metals. *Also see* Oxidation and Pyrolysis Process.

Deluge Sprinkler System — Fire-suppression system that consists of piping and open sprinkler heads. A fire detection system is used to activate the water or foam control valve. When the system activates, the extinguishing agent expels from all sprinkler heads in the designated area. *Also see* Dry-Pipe Sprinkler System, Preaction Sprinkler System, and Wet-Pipe Sprinkler System.

Density — Weight per unit of volume of a substance. The density of any substance is obtained by dividing the weight by the volume.

Diked Area — Area surrounding storage tanks or loading racks that is designed to retain spilled fuel and fire-extinguishing agents such as water and foam. *Also see* Nondiked Area.

Direct Attack (Structural) — Attack method that involves the discharge of water or a foam stream directly onto the burning fuel. *Also see* Attack Methods (1).

Direct Attack (Wildland) — Operation where action is taken directly on burning fuels by applying an extinguishing agent to the edge of the fire or close to it. *Also see* Attack Methods (2).

Direct Injection — Application method where foam concentrate is injected directly into the water stream at the pump before it enters the hoseline. *Also see* Subsurface Injection and Semisubsurface Injection.

Directional Foam Spray Nozzle — Foam delivery device that consists of a small foam nozzle used to protect areas such as loading racks by applying the extinguishing agent onto the surface beneath tanker trucks.

Discharge Outlet, Type I — Foam delivery device that conducts and delivers finished foam onto the burning surface of a liquid without submerging it or agitating the surface; no longer manufactured and considered obsolete but may still be found in some fixed-site applications.

Discharge Outlet, Type II — Foam delivery device that delivers finished foam onto the surface of a burning liquid, partially submerges the foam into the surface, and produces limited agitation on the surface of the burning liquid.

Discharge Outlet, Type III — Foam delivery device that delivers finished foam in a manner that causes it to fall directly onto the surface of the burning liquid and does so in a way that causes general agitation; includes master streams and handlines.

Downwind — Wind that is blowing in a direction from behind a person or aircraft. *Also see* Wind, Headwind, and Crosswind.

Drainage Time — Amount of time that it takes for a foam blanket to break down or dissolve; also called *drainage rate, drainage dropout rate,* or *drainage. Also see* Quarter-Life.

Drop Height — Most effective and safest altitude of an aircraft when fire-extinguishing agents are dropped on wildland fires.

Dry-Pipe Sprinkler System — Fire-suppression system that consists of closed sprinkler heads attached to a piping system that contains air under pressure. When a sprinkler head activates, air is released that activates the water or foam control valve and fills the piping with extinguishing agent. *Also see* Deluge Sprinkler System, Preaction Sprinkler System, and Wet-Pipe Sprinkler System.

E

Eduction — Process used to mix foam concentrate with water in a nozzle or proportioner; concentrate is drawn into the water stream by the Venturi method; also called *induction. Also see* Venturi Principle.

Eductor (Inductor) — Portable foam proportioning device that injects foam concentrate into the water flowing through a hoseline. *Also see* Proportioning and Foam Proportioners.

Electrical Heat Energy — Energy in the form of heat that is created when an electric current is passed through a conductor such as a wire; includes resistance heating, dielectric heating, leakage current heating, heat from arching, and static electricity. *Also see* Chemical Heat Energy, Mechanical Heat Energy, Nuclear Heat Energy, and Solar Heat Energy

Emulsifier — Foam concentrate that is designed to mix with the fuel that it is covering, break the fuel into small droplets, and encapsulate it. The resulting emulsion is rendered nonflammable. *Also see* Foam Concentrate.

Environmental Protection Agency (EPA) — U.S. government agency that regulates potential hazards to the environment such as water or air pollution.

EPA — Abbreviation for Environmental Protection Agency.

Expansion Ratio — Ratio of the finished foam volume to the volume of the original foam solution. *Also see* Foam Expansion.

Explosion — Instantaneous oxidation of a material; heat, light, and energy are released with great force. *Also see* Backdraft and Boiling Liquid Expanding Vapor Explosion.

External Floating Roof Tank — Fixed-site vertical storage tank that has no fixed roof but relies on a floating roof to protect the contents and prevent evaporation; also called *open-top floating roof tank.* *Also see* Cone Roof Tank and Internal Floating Roof Tank.

F

FFFP — Abbreviation for Film Forming Fluoroprotein.

Film Forming Fluoroprotein (FFFP) Foam — Foam concentrate that combines the qualities of fluoroprotein foam with those of aqueous film forming foam. *Also see* Foam Concentrate.

Filter Basket — Fine mesh screen that is attached to the end of the foam concentrate pickup hose to prevent sediment from entering the system and clogging the proportioner.

Finished Foam — Extinguishing agent formed by mixing foam concentrate with water and aerating the solution for expansion; also known as *foam. Also see* Foam Solution and Foam Concentrate.

Fire Point (Burning Point) — Temperature at which a liquid fuel produces sufficient vapors to support combustion once the fuel is ignited. The fire point is usually a few degrees above the flash point. *Also see* Flash Point.

Fire Tetrahedron — Model of the four elements required to have a fire. The four sides represent fuel, heat, oxygen, and chemical chain reaction.

Fixed-Flow Direct-Injection Proportioner — Fixed-site foam system that provides foam at a preset rate. Foam concentrate and water are provided to the proportioner by two separate pumps that operate at preset rates. *Also see* Foam Proportioner and Proportioning.

Fixed Foam Extinguishing System — Complete installation of piping, foam concentrate storage, water supply, pumps, and delivery systems used to protect a specific hazard such as a petroleum storage facility. *Also see* Portable Foam Extinguishing System and Semifixed Foam Extinguishing System.

Flame — Visible, luminous body of a burning gas.

Flammable/Explosive Range — Percentage of a gas vapor concentration in the air that will burn if ignited.

Flammable Liquid — Any liquid having a flash point below 100°F (38°C) and having a vapor pressure not exceeding 40 psi (276 kPa). *Also see* Combustible Liquid.

Flank Attack (Wildland) — Attack on the flanks or sides of a wildland fire and working along the fire edge toward the head of the fire. *Also see* Attack Methods (2).

Flashover — Stage of a fire where all surfaces and objects within a space have been heated to their ignition temperature and flame erupts almost at once over the surface of all objects in the space. *Also see* Backdraft and Rollover.

Flash Point — Minimum temperature at which a liquid gives off enough vapors to form an ignitable mixture with air near the liquid's surface. *Also see* Fire Point.

Fluoroprotein Foam — Protein foam concentrate with synthetic fluorinated surfactants added. *Also see* Foam Concentrate.

Foam — See *Finished Foam.*

Foam Chamber — Foam delivery device that is mounted on storage tanks; applies foam onto the surface of the fuel in the tank.

Foam Concentrate — Raw foam liquid as it rests in its storage container before the introduction of water and air. *Also see* Foam Solution, Finished Foam, Alcohol-Resistant AFFF Concentrate, Aqueous Film Forming Foam, Film Forming Fluoroprotein Foam, Class A Foam Concentrate, Class B Foam Concentrate, Synthetic Foam Concentrate, Fluoroprotein Foam, Protein Foam Concentrate, and Emulsifier.

Foam Expansion — Result of adding air to a foam solution consisting of water and foam concentrate. Expansion creates the foam bubbles that result in finished foam or foam blanket. *Also see* Expansion Ratio.

Foam Monitor — Master stream appliance used for the application of foam solution. *Also see* Manual Foam Monitor, Remote-Controlled Foam Monitor, and Automatic Oscillating Foam Monitor.

Foam Proportioner — Device that introduces foam concentrate in the proper ratio into a water stream to create a foam solution. *Also see* Proportioning, Around-the-Pump Proportioner, Bladder-Tank Balanced-Pressure Proportioner, Bypass-Type Balanced-Pressure Proportioner, In-line Proportioner, Fixed-Flow Direct-Injection Proportioner, Eductor, Variable-Flow Demand-Type Balanced-Pressure Proportioner, and Variable-Flow Variable-Rate Direct-Injection System.

Foam Solution — Result of mixing the appropriate amount of foam concentrate with water. Foam solution exists between the proportioner and the nozzle or aerating device that adds air to create finished foam. *Also see* Foam Concentrate, and Finished Foam.

Foam Tanker (Tender/Mobile Water Supply Apparatus) — See *Foam Tender.*

Foam Tender — Apparatus that is specially designed to transport large quantities of foam concentrate to the scene of an incident; vehicle may be equipped with transfer pumps and have a tank ranging from 1,500 gallons (5 678 L) to 8,000 gallons (30 283 L). *Also see* Foam Trailer and Mobile Foam Extinguishing System.

Foam Trailer — Foam concentrate tank mounted on a trailer; can be easily towed to the desired location; usually found at fixed-site facilities and used to supply master stream and subsurface injection systems. *Also see* Foam Tender and Mobile Foam Extinguishing System.

Foam-Water Sprinkler Head — Deluge-type open sprinkler head that mixes water with the foam solution as it passes through the head. *Also see* Open Sprinkler Head and Closed Sprinkler Head.

Fog Nozzle — Nozzle that can provide either a fixed or variable spray pattern. The nozzle breaks the foam solution into small droplets that mix with air to form finished foam. *Also see* Multiagent Nozzle.

Frontal Attack (Wildland) — Attack directed at the head of a wildland fire. *Also see* Attack Methods (2).

Fuel Spill — Unintentional release of a flammable or combustible liquid from its container.

H

Headwind — Wind that is blowing in a direction toward the face of a person or the front of an aircraft. *Also see* Wind, Crosswind, and Downwind.

Heat — Form of energy associated with the motion of atoms or molecules in solids or liquids.

Heat Stratification — See *Thermal Layering of Gases.*

Heat Wave — Movement of radiated heat through space until it reaches an opaque object.

High-Expansion Foam — Foam concentrate that is mixed with air in the range of 200 parts air to 1 part foam solution (200:1) to 1,000 parts air to 1 part foam solution (1,000:1). *Also see* Low-Expansion Foam, Medium-Expansion Foam, and Mechanical Blower.

Hydrocarbon Fuel — Petroleum-based organic compound that contain only hydrogen and carbon. *Also see* Polar Solvent Fuel and Liquefied Petroleum Gas.

I

IBC — Abbreviation for Industrial Bulk Container.

Ignition Stage (Incipient Phase) — First phase of the burning process where the substance being oxidized is producing some heat, but it has not spread to other substances nearby. During this stage, the oxygen content of the air has not been significantly reduced. *Also see* Smoldering Stage.

Ignition Temperature — Point at which flammable/combustible vapors are released from a solid when it is heated.

Indirect Attack (Structural) — Attack where the fire stream is directed at the ceiling level of a compartment in order to generate a large amount of steam. The steam helps to cool the area enough so that firefighters may safely enter and make a direct attack to extinguish a fire. *Also see* Attack Methods (1).

Indirect Attack (Wildland) — Method of controlling a wildland fire where a control line is constructed or located some distance from the edge of the main fire and the fuel between the two points is burned. *Also see* Attack Methods (2).

Induction — See *Eduction.*

Industrial Bulk Container (IBC) — Large-capacity bulk storage container used for foam concentrate, usually in quantities of 250 to 450 gallons (950 L to 1 710 L).

Injection — Method of proportioning foam that uses an external pump or head pressure to force foam concentrate into the fire stream at the correct ratio for the flow desired. *Also see* Proportioning.

In-Line Proportioner — Type of foam delivery device that is located in the water supply line near the nozzle. The foam concentrate is drawn into the water line using the Venturi method. *Also see* Foam Proportioning, Foam Proportioner, and Venturi Principle.

Internal Floating Roof Tank — Fixed-site vertical storage tank that combines both the floating roof and the closed roof design. Also called *covered floating roof tank. Also see* Cone Roof Tank and External Floating Roof Tank.

International Shore Connection (ISC) — Pipe flange with a standard size and bolt pattern allowing land-based fire department personnel to charge and supply a vessel's fire main.

ISC — Abbreviation for International Shore Connection.

J

Jet Ratio Controller — Type of foam eductor that is used to supply self-educting master stream nozzles; may be located at distances up to 3,000 feet (914 m) from the nozzle.

Joule (J) — Unit of work or energy in the International System of Units; the energy (or work) when unit force (1 newton) moves a body through a unit distance (1 meter); takes the place of calorie for heat measurement (1 calorie = 4.19 J). *Also see* Calorie.

L

Law of Conservation of Mass — Theory that states that mass is neither created nor destroyed in any ordinary chemical reaction; mass that is lost is converted into energy in the form of heat and light.

Law of Heat Flow — Theory that states that heat will transfer from a hot surface to a cold surface. *Also see* Convection, Conduction, and Radiation.

LFL — Abbreviation for Lower Flammable Limit.

Liquefied Petroleum Gas (LPG) — Gas containing certain specific hydrocarbons that are gaseous under normal atmospheric conditions but can liquefy under moderate pressure at normal temperatures; propane and butane are the principal examples. *Also see* Hydrocarbon Fuel.

Loading Rack — Fixed facility where flammable and combustible liquids are bulk loaded into tank vehicles or rail tank cars.

Local Application System — Fixed-site fire-suppression system that is required to cover a protected area with 2 feet (0.6 m) of foam depth within 2 minutes of system activation; foam supply must support the continuous operation of the system for at least 12 minutes. *Also see* Total Flooding System.

Lower Flammable Limit (LFL) — Lower limit at which a flammable gas or vapor will ignite; below this limit the gas or vapor is too lean or thin to burn (too much oxygen and not enough gas). *Also see* Upper Flammable Limit

Low-Expansion Foam — Foam concentrate that is mixed with air in the range of less than 20 parts air to 1 part foam solution (20:1). *Also see* High-Expansion Foam and Medium-Expansion Foam.

LPG — Abbreviation for Liquefied Petroleum Gas.

M

Manual Foam Monitor — Foam monitor that is operated by hand; may be found mounted on apparatus, in fixed locations to protect target hazards, or as a portable unit. *Also see* Foam Monitor, Automatic Oscillating Foam Monitor, and Remote-Controlled Foam Monitor.

Material Safety Data Sheet (MSDS) — Form provided by the manufacturer and blender of chemicals that contains information about chemical composition, physical and chemical properties, health and safety hazards, emergency response procedures, and waste disposal procedures of the specified material.

Mechanical Blower — High-expansion foam generator that uses a fan to inject the air into the foam solution as it passes through the unit. *Also see* High-Expansion Foam.

Mechanical Foam — Foam produced by a physical agitation of a mixture of foam concentrate, water, and air. *Also see* Chemical Foam.

Mechanical Heat Energy — Heat that is generated by friction or compression. *Also see* Chemical Heat Energy, Electrical Heat Energy, Nuclear Heat Energy, and Solar Heat Energy.

Medium-Expansion Foam — Foam concentrate that is mixed with air in the range of 20 parts air to 1 part foam solution (20:1) to 200 parts air to 1 part foam solution (200:1). *Also see* Low-Expansion Foam and High-Expansion Foam.

Military Specifications (milspecs) — Specifications developed by the U.S. Department of Defense for the purchase of materials and equipment.

Milspecs — Acronym for Military Specifications.

Mobile Attack (Wildland) — Using mobile fire apparatus in a direct attack along the fire edge; also referred to as *pump and roll*. *Also see* Attack Methods (2).

Mobile Foam Apparatus — See *Mobile Foam Extinguishing System*.

Mobile Foam Extinguishing System — Foam delivery system that is mounted on a fire apparatus or trailer. *Also see* Foam Tender and Foam Trailer.

Moisture Content — Amount of moisture that is available in the environment.

MSDS — Abbreviation for Material Safety Data Sheet.

Multiagent Nozzle — Device that is capable of applying foam, halon substitute, or dry chemical extinguishing agents simultaneously. *Also see* Fog Nozzle.

N

National Fire Danger Rating System (NFDRS) — System used to classify wildland fuels based on similar burning characteristics.

Negative Heat Balance — Condition that occurs in a fire when heat is dissipated faster than it is generated and, therefore, will not sustain combustion. *Also see* Positive Heat Balance.

NFDRS — Abbreviation for National Fire Danger Rating System.

Nonaspirating Foam Nozzle — Nozzle that does not draw air into the foam solution stream. The foam solution is agitated by the nozzle design causing air to mix with the solution after it has exited the nozzle. *Also see* Aspirating Foam Nozzle.

Nondiked Area — Any location where flammable or combustible liquids might be spilled but not contained within a system of predesigned barriers. *Also see* Diked Area.

Nonintervention Strategy — Strategy for handling fires involving hazardous materials where the fire is allowed to burn until all of the fuel is consumed.

Nonrated Concentrate — Foam concentrate that has not been tested or certified by Underwriters Laboratories Inc. *Also see* Rated Concentrate and Foam Concentrate.

Nuclear Heat Energy — Creation of heat through the splitting apart or combining of atoms. *Also see* Chemical Heat Energy, Electrical Heat Energy, Mechanical Heat Energy, and Solar Heat Energy.

O

Open Sprinkler Head — Sprinkler head that lacks a heat-sensitive element and is open at all times; used on the deluge-type sprinkler system. *Also see* Closed Sprinkler Head and Foam-Water Sprinkler Head.

Oxidation — Chemical process that occurs when a substance combines with oxygen; a common example is the formation of rust on metal. *Also see* Decomposition and Pyrolysis Process.

Oxidizer — Substance that yields oxygen readily and may stimulate the combustion of organic and inorganic matter.

P

pH — Measure of acidity of an acid or the level of alkaline in a base.

Pickup Tube — Solid or flexible tube used to transfer foam concentrate from a storage container to the in-line eductor or proportioner.

Pincer Attack (Wildland) — Simultaneous attack on two sides of a wildland fireline; fire is enveloped from multiple sides; similar to a wildland flank attack. *Also see* Attack Methods (2).

Plug — (1) Safety device that grounds a tank vehicle or rail tank car during the loading and unloading process and prevents static electricity buildup. (2) Patch to seal a small leak in a container.

Polar Solvent Fuel — Flammable liquid that has an attraction for water, much like a positive magnetic pole attracts a negative pole; examples include alcohols, ketones, and lacquers. *Also see* Hydrocarbon Fuel.

Portable Foam Extinguishing System — Foam extinguishing system that can be carried by hand such as a handheld foam fire extinguisher. *Also see* Fixed Foam Extinguishing System and Semifixed Foam Extinguishing System.

Positive Heat Balance — Situation that occurs when heat is fed back to a fuel. A positive heat balance is required to maintain combustion. *Also see* Negative Heat Balance.

Preaction Sprinkler System — Fire-suppression system that consists of closed sprinkler heads attached to a piping system that contains air under pressure and a secondary detection system; both must operate before the extinguishing agent is released into the system; similar to the dry-pipe sprinkler system. *Also see* Dry-Pipe Sprinkler System, Deluge Sprinkler System, and Wet-Pipe Sprinkler System.

Premixing — Mixing premeasured portions of water and foam concentrate in a container. Typically, the premix method is used with portable extinguishers, wheeled extinguishers, skid-mounted multiagent units, and vehicle-mounted tank systems. *Also see* Batch Mixing.

Pretreating — Exposure protection tactic that involves the application of foam onto unburned materials near the fire. The foam soaks the material making it less likely to ignite.

Proportioning — Mixing of water with an appropriate amount of foam concentrate to form a foam solution. *Also see* Foam Proportioner.

Protein Foam Concentrate — Foam concentrate that consists of a protein hydrolysate plus additives to prevent the concentrate from freezing, prevent corrosions on equipment and containers, control viscosity, and prevent bacterial decomposition of the concentrate during storage. *Also see* Foam Concentrate.

Pyrolysis Process (Sublimation) — Chemical decomposition of a substance through the action of heat. *Also see* Decomposition and Oxidation.

Q

Quarter-Life — Time required in minutes for one-fourth of the total liquid solution to drain from a foam blanket. Also referred to as *25 percent drain time* and *quarter drain time*. *Also see* Drainage Time.

R

Radiation — Transfer of heat energy through light by electromagnetic waves without an intervening medium. Also referred to as *radiated heat*. *Also see* Convection, Conduction, and Law of Heat Flow.

Radiative Feedback — Radiant heat providing energy for continued vaporization. *Also see* Vaporization.

Rail Tank Car — Railroad car that is designed to carry liquids in pressurized or unpressurized cylinders; may be constructed of steel, stainless steel, or aluminum.

Rain-Down Application Method (Raindrop) — Method of foam application in which the foam stream is directed into the air above the unignited or ignited spill and allowed to gently float down onto the surface of the liquid. *Also see* Bank-Down Application Method and Roll-On Application Method.

Rapid Intervention Vehicle (RIV) — Type of mobile foam apparatus that is used to provide a rapid attack on an aircraft crash incident for quick extinguishment and rescue.

Rated Concentrate — Foam concentrate that has been tested and certified by Underwriters Laboratories Inc. *Also see* Nonrated Concentrate and Foam Concentrate.

Rate of Vaporization — Speed at which a liquid evaporates or vaporizes. *Also see* Vaporization.

Reactivity — Ability of a substance to react with water creating an explosion potential. *Also see* Chemical Reaction.

RECEO Model — One of many models for prioritizing activities at an emergency incident: Rescue, Exposures, Confine, Extinguish, and Overhaul.

Reformulated Gasoline — See *Blended Gasoline.*

Relative Humidity — Measure of the moisture content (water quantity expressed in a percentage) in both the air and solid fuels.

Remote-Controlled Foam Monitor — Large-capacity foam system that is operated by a remote control located away from the monitor; usually found on aircraft rescue and fire fighting apparatus and fireboats. *Also see* Foam Monitor, Automatic Oscillating Foam Monitor, and Manual Foam Monitor.

Retention — Characteristic of Class A foam and foam solution to remain on and in the fuel, reduce the fuel temperature, and increase the fuel moisture content.

Roll-On Application Method (Bounce) — Method of foam application in which the foam stream is directed at the ground at the front edge of the unignited or ignited liquid fuel spill; foam then spreads across the surface of the liquid. *Also see* Bank-Down Application Method and Rain-Down Application Method.

Rollover (Flameover) — Stage of a fire where superheated, unburned combustible gases accumulating at the upper level of a compartment are pushed into uninvolved areas, mix with fresh oxygen, and erupt into flames. *Also see* Backdraft and Flashover.

S

Self-Educting Master Stream Foam Nozzle — Large-capacity nozzle with built-in foam eductor.

Self-Educting Nozzle — Handline nozzle that has the foam eductor built into it.

Semifixed Foam Extinguishing System — Foam system that is designed to provide fire-extinguishing capabilities to an area but is not automatic in operation and depends on human intervention to place it into operation. *Also see* Fixed Foam Extinguishing System and Portable Foam Extinguishing System.

Semisubsurface Injection — Application method that discharges foam through a flexible hose that rises from the bottom of a storage tank, up through the fuel, and to the surface of the fuel; foam then blankets the surface of the fuel. *Also see* Direct Injection and Subsurface Injection.

Slope — Natural or artificial topographic incline; may be slight or very steep.

Smoldering Stage — Phase of combustion when the level of oxygen in a confined space is below that needed for flaming combustion; characterized by glowing embers, high heat at all levels of a room, and heavy smoke and fire gas production. *Also see* Ignition Stage.

Solar Heat Energy — Creation of heat from the sun in the form of electromagnetic radiation. *Also see* Chemical Heat Energy, Electrical Heat Energy, Mechanical Heat Energy, and Nuclear Heat Energy.

Solubility — Ability of a substance to dissolve in a liquid.

Specific Gravity — Weight of a substance compared to the weight of an equal volume of water at a given temperature. A specific gravity less than 1 indicates a substance lighter than water; a specific gravity greater than 1 indicates a substance heavier than water. *Also see* Vapor Density.

Structure Fire — Fire that involves a building, enclosed structure, vehicle, vessel, aircraft, or like property. *Also see* Wildland Fire.

Subsurface Injection — Application method where foam is pumped into the bottom of a burning fuel storage tank and allowed to float to the top to form a foam blanket on the surface of the fuel. *Also see* Direct Injection and Semisubsurface Injection.

Surface Application — Application method where finished foam is applied directly onto the surface of the burning fuel or unignited fuel spill.

Surface Tension — Force minimizing a liquid surface's area.

Surface-to-Mass Ratio — Ratio of the surface area of the fuel to the mass of the fuel.

Surfactant — Chemical that lowers the surface tension of a liquid; allows water to spread more rapidly over the surface of Class A fuels and penetrate organic fuels. *Also see* Surface Tension.

Synthetic Foam Concentrate — Foam concentrate that is composed of a synthetically produced material (such as a fluorochemical or hydrocarbon surfactant) that forms a foam blanket across a liquid; performs similarly to a protein-based foam concentrate; examples are aqueous film-forming foam and alcohol-resistant aqueous film-forming foam. *Also see* Foam Concentrate.

T

Tandem Attack (Wildland) — Attack on a wildland fire using multiple resources including apparatus, hand crews, or aircraft. *Also see* Attack Methods (2).

Tank (Full-Trailer) — Any vehicle with or without auxiliary motive power, equipped with a cargo tank mounted thereon or built as an integral part thereof, used for the transportation of flammable and combustible liquids or asphalt; constructed so that practically all of its weight and load rests on its own wheels; a trailer with axles at the front and rear of the frame capable of supporting the entire weight of the tank.

Tank (Semitrailer) — Any vehicle with or without auxiliary motive power, equipped with a cargo tank mounted thereon or build as an integral part thereof, used for the transportation of flammable and combustible liquid or asphalt; constructed so that when drawn by a tractor by means of a fifth-wheel connection, some part of its load and weight rests upon the towing vehicle.

Tank Truck — Any single self-propelled motor vehicle equipped with a cargo tank mounted thereon and used for the transportation of flammable and combustible liquids or asphalt; a tank truck with two or three axles on which the tank is permanently affixed.

Tank Vehicle — Any tank truck, tank full-trailer, or tractor and tank semitrailer combination.

Thermal Blocking — Phenomenon that occurs when a concealed hot spot contains enough heat to turn small amounts of penetrating water into steam, thereby preventing the water from cooling the material thoroughly.

Thermal Layering of Gases — Tendency of gases to form layers according to temperature; also called *thermal balance* or *heat stratification.*

Thermal Updraft — Convection column of hot gases, smoke, and flames rising above a high-intensity fire; can create unsafe operating conditions for aircraft and reduce the effectiveness of finished foam.

Topography — Physical configuration of the land or terrain.

Total Flooding System — Fire-suppression system designed to protect hazards within enclosed structures. Foam is released into a compartment or area and fills it completely to extinguish the fire. *Also see* Local Application System.

Transit Time — Time that it takes a foam solution to pass from the proportioner to the nozzle.

U

UFL — Abbreviation for Upper Flammable Limit.

UL — Abbreviation for Underwriters Laboratories, Inc.

UL Canada — Abbreviation for Underwriters' Laboratories of Canada.

Underwriters Laboratories Inc. (UL) — Independent research and testing laboratory that certifies equipment and materials. Equipment and materials are approved only for the specific use for which it is tested.

Underwriters' Laboratories of Canada (UL Canada) — Independent, nonprofit third-party product safety testing and certification organization that is accredited by the Standards Council of Canada.

Upper Flammable Limit (UFL) — Upper limit at which a flammable gas or vapor will ignite. Above this limit the gas of vapor is too rich to burn (lacks the proper quantity of oxygen). *Also see* Lower Flammable Limit.

Urban/Wildland Interface — Term used to describe the area where an undeveloped wildland area meets a human development area. *Also see* Wildland Fire.

V

Vapor Density — Weight of a given volume of pure vapor or gas compared to the weight of an equal volume of dry air at the same temperature and pressure. A vapor density less than 1 indicates a vapor lighter than air; a vapor density greater than 1 indicates a vapor density heavier than air. *Also see* Specific Gravity.

Vaporization — Process of evolution that changes a liquid into a gaseous state; rate of vaporization depends on the substance involved, heat, and pressure. *Also see* Radiative Feedback.

Vapor Mitigating Foam — Foam concentrate that is designed solely for use on unignited spills of hazardous liquids and is not effective for fire-suppression operations. *Also see* Foam Concentrate.

Vapor Suppression — Action taken to reduce the emission of vapors at a fuel spill.

Variable-Flow Demand-Type Balanced-Pressure Proportioner — Foam proportioning system that is used in both fixed and mobile applications. A variable speed mechanism drives the foam pump and automatically monitors the flow of foam to produce an effective foam solution. *Also see* Foam Proportioner and Proportioning.

Variable-Flow Variable-Rate Direct-Injection System — Apparatus-mounted foam system that injects the correct amount of foam into the pump piping, thereby supplying all discharges with foam; automatically monitors the operation of the hoselines and maintains a consistent quality of foam solution. *Also see* Foam Proportioner and Proportioning.

Venturi Principle — Liquid or gas will increase in speed when forced through a tube that decreases in diameter. When applied to a proportioner, water that is passed through the restricted orifice increases velocity and creates a vacuum, pulling foam concentrate up the foam tube into the water stream. *Also see* In-Line Proportioner and Aspirating Foam Nozzle.

Viscosity — Liquid's thickness or ability to flow.

Volatility — Ability of a substance to vaporize easily at a relatively low temperature.

W

Wet-Pipe Sprinkler System — Fire-suppression system that is built into a structure or site; piping contains either water or foam solution continuously; activation of a sprinkler head causes the extinguishing agent to flow from the open head. *Also see* Deluge Sprinkler System, Preaction Sprinkler System, and Dry-Pipe Sprinkler System.

Wildland Fire — Unplanned and unwanted fire requiring suppression action or an uncontrolled fire spreading through vegetative fuels. *Also see* Structure Fire and Urban/Wildland Interface.

Wind — Air in motion. *Also see* Downwind, Crosswind, and Headwind.

Index

I

J

Indexed by Kari Kells.